Fundamentals of Engineering Design

Barry Hyman
Department of Mechanical Engineering
University of Washington

Prentice Hall
Upper Saddle River, New Jersey 07458

Library of Congress Cataloging-in-Publication Data

Hyman, Barry I.,
    Fundamentals of engineering design / Barry Hyman.
      p. cm.
    Includes bibliographical references and index.
    ISBN 0-13-531385-6
    1. Engineering design. I. Title
TA174.H96   1998
620'.0042—dc21   97-39989
                                    CIP

Acquisitions editor: *Bill Stenquist*
Editorial/Production services: *Pine Tree Composition, Inc.*
Editor-in-Chief: *Marcia Horton*
Assistant Vice President of Production and Manufacturing: *David W. Riccardi*
Managing Editor: *Bayani Mendoza de Leon*
Full Service/Manufacturing Coordinator: *Donna Sullivan*
Manufacturing Manager: *Trudy Pisciotti*
Creative director: *Jayne Conte*
Cover designer: *Patricia Woscyk*
Editorial Assistant: *Meg Weist*

© 1998 by Prentice-Hall, Inc.
Simon & Schuster/A Viacom Company
Upper Saddle River, New Jersey 07458

Printed in the United States of America
10 9 8 7 6 5 4 3 2 1

ISBN: 0-13-531385-6

Prentice-Hall International (UK) Limited, *London*
Prentice-Hall of Australia Pty. Limited, *Sydney*
Prentice-Hall Canada Inc., *Toronto*
Prentice-Hall Hispanoamericana, S.A., *Mexico*
Prentice-Hall of India Private Limited, *New Delhi*
Prentice-Hall of Japan, Inc., *Tokyo*
Simon & Schuster Asia Pte. Ltd., *Singapore*
Editora Prentice-Hall do Brasil, Ltda., *Rio de Janeiro*

To
Joyce, Celia, Rachel, and Cassandra

Contents

Preface

This is a comprehensive textbook for teaching design to undergraduate engineering students. It devotes several chapters to design process and methodology, with a particular emphasis on problem formulation and concept generation. In addition, the book includes chapters on engineering economics, project planning, professional and social context of design, information acquisition and communication skills, probabilistic considerations, decision making, and optimization. I believe that these topics are fundamental to engineering design regardless of the specific engineering discipline.

Virtually every undergraduate engineering curriculum has a design component, although the extent and structure of that component varies widely. From the mid-1950s to about 1990, traditional engineering curricula have been front-loaded with mathematics and engineering science. Design was often left to upper division courses on the assumption that meaningful engineering design cannot occur without a solid analytical foundation. While the more recent trend is to start teaching design earlier in the curriculum, there is no clear indication of how wide-spread and long-lasting that trend will be. Regardless of how and when a given curriculum addresses design, this book will meet a substantial part of its needs. There is sufficient material in the textbook for a year-long treatment of design, especially if students have the opportunity to apply the topics covered in the book in a design project context. Shorter offerings can be based on selective chapters or selective sections within chapters. Some schools may adopt the book as a text in a lower division course, others in an upper division course. In other cases, the topics covered in this book may be spread out over several different courses.

My approach to this book is based on six premises:

1. Engineering design concepts and principles are as fundamental to undergraduate engineering education as traditional fundamentals such as mathematics and engineering sciences.

2. Engineering design is not just doing engineering analysis in reverse order. Design contains its own body of knowledge that is independent of the science-based engineering analysis tools it is usually coupled with. In this sense, design is a field of study similar to calculus. It has its own internal generic methods and rules. Like calculus, design is more interesting to engineering students when those techniques are applied within a particular disciplinary context.

3. Engineering design fundamentals are common to all engineering disciplines—mechanical, chemical, electrical, civil. However, several of the topics covered in this book (engineering economics, project planning, optimization) are already integral to typical industrial engineering curricula, so the coverage herein may be too superficial for that discipline.

4. Many of these engineering design fundamentals can be taught early in the engineering curriculum, even at the freshman level. However, in many curricula, these topics may not be introduced until the junior or senior year.

5. Engineering students can be motivated to learn many of the theoretical and abstract concepts of engineering science and analysis by embedding those concepts in practical, real world, design problems. This careful melding of design and analysis can help engineering students strengthen their synthesis skills and appreciate the practical value of much of their analysis-oriented courses.

6. The best way to learn design is to practice. I've included a large number of design exercises and projects throughout this book.

I approach all topics covered in the book in the least abstract way possible. Whenever feasible, I introduce each new concept by posing a design problem. I then develop the method as a tool to solve the design problem, rather than using the more conventional approach of developing an abstract theory, then illustrating the theory with an example. In this way, the reader can see the usefulness of the technique as a design tool from the very beginning. Once having developed the tool in the context of a particular design problem, I discuss its generalization and applicability to other situations. Some rigorous proofs I omit entirely, and others I relegate to appendices. I also rely heavily on geometrical interpretations to make concepts less abstract. See, for example, the discussion of the Simplex method in Section 10.4 and the treatment of convexity in Section 10.6.

This book is organized so that a given topic can be taught very early in the curriculum, without sacrificing the rigor and sophistication needed to make the same topic appropriate for teaching in an upper division course. To a large extent, the rigor and sophistication needed to use the book for an upper division design course is provided by the mathematics and engineering analytical tools required to solve specific exercises and design projects. Most of the material in every chapter will be suitable for use at any level, including the freshman year. The exceptions are Chapter 5, Section 7.5, and Section 9.6, which require understanding of basic probability theory; Section 9.3 which requires knowledge of matrix algebra; and Section 10.8 and the latter part of Section 10.6, where a basic knowledge of calculus is needed.

Many engineering professors (including myself) find it more difficult to teach design than to teach traditional engineering science-based analytical courses. That is because a design course does not have the structured, logical, linear character of a typical analysis-based engineering course. Design topics can be taught in almost any order. Some of them can be omitted entirely from the course without affecting the viability of the treatment of other topics. Thus, decisions regarding what topics to include, and in what sequence, and in how much detail, are more difficult than for a traditional course. This book is organized to accommodate a wide variety of topic selection, depth of coverage, and sequence. Each chapter is essentially self-contained. While I use cross-references to other chapters as a way to provide continuity, any chapter other than Chapter 1 can be omitted without affecting the remaining chapters. The sequence in which the topics are taught is independent of the order in which the chapters are arranged. I have experimented with the order in which I cover the topics in my course. Sometimes I'll cover optimization (Chapter 10) midway through the quarter and teach probabilistic considerations (Chapter 5) as the last topic.

I believe it is important to place fundamental design concepts in a disciplinary focus and to integrate them with the appropriate level of engineering science knowledge. I use two methods in this book to accomplish this objective.

First, I present mini case studies throughout the book that are drawn from diverse engineering disciplines. For example, the discussion of the design process in Chapter 1 is keyed to the design of an automobile bumper (mechanical engineering). In Chapter 3, I discuss the design of a band-pass filter circuit (electrical). The project planning discussion in Chapter 7 includes an example of a major project within a chemical plant. In Chapter 8, I provide details of the economic analysis of a construction project that has civil, electrical, and mechanical engineering aspects. The Chapter 10 discussion of optimal design methods is in the context of locating a transmission line (electrical, civil), designing a water purification system (chemical, civil), designing a metal alloy (materials), and a street repaving project (civil).

I have also prepared several hundred exercises and projects, many of which apply the design fundamentals to specific engineering disciplines. There are a set of generic exercises associated with each section in the text that give students an opportunity to practice applying that specific technique. In addition, there are variations of those generic exercises. These variations, or specialty exercises and projects, have two characteristics. The first characteristic is a disciplinary context (electrical, civil, mechanical.). The second characteristic is the level of analytical sophistication required. For example, consider the topic of incorporating discount rate into a life-cycle cost comparison of design options. The book includes a suggested design project that applies the concept of discount rate to a life-cycle cost comparison of an insulated container for shipping perishables, in which analytical capabilities familiar to third-year mechanical engineering students are needed to develop the design of the container.

Both the body of the text and the exercises and projects are an outgrowth of my own teaching experiences. I distributed early versions of much of the material as lecture notes, and since June 1995 I have used the then current version of the manuscript as the text and encouraged students to provide feedback to me. I've made many changes in response to that input from students.

The entire draft manuscript was reviewed in detail by Professor Michael Wells of Montana State University, Professor Gary Bernstein of Notre Dame University, Professor Peter Frise of the University of Windsor, and an anonymous reviewer. I greatly appreciate their feedback and the final book incorporates many of their suggestions. In particular, Figure 1–11 was included at Professor Frise's suggestion. Several others contributed to specific portions of the book. Harold Federow, LLD, of the University of Washington's Program in Engineering and Manufacturing Management provided a thorough critique of Chapter 4. I am indebted to Ms. Patricia Cardinal of the United States Corps of Engineers for the documentation she provided for the Chapter 8 discussion of the Chittenden Locks rehabilitation project. Mr. Raymond Baculi was of enormous assistance in organizing the exercises for each chapter and the design projects, as well as helping with library searches. Graduate students at the University of Washington who served as teaching assistants for me over the years have contributed to formulating some of the exercises and projects. While my memory is fuzzy on who did precisely what, and since many revisions and modifications have been made since the students were involved, I do want to thank Messrs. Michael Safoutin, Scot Freeman, Dan Billingsley, Steve Ellison, and Ms. Sandra Pailey for their contributions.

An important impetus for this project was my involvement in the early stages of the NSF-sponsored engineering ECSEL Coalition project involving the University of Washington and five other engineering schools in the early 1990s. The opportunity to interact with colleagues at the ECSEL participating schools helped shaped my views on engineering design education. A major theme of the book and a key organizing principle are an outgrowth of an ECSEL instructional model I developed and used for a new freshman-level course in engineering design. Steve Ellison was an important contributor to the first draft of that module.

Ms. Linda Ratts-Engleman of Prentice Hall provided crucial guidance to this project during its formative stages. She brought the project to the attention of William Stenquist, Executive Editor at Prentice Hall who saw the potential for this book and provided the nurturing needed to transform an idea into reality. I thank him for his confidence, insights, advice, and moral support. It was also a pleasure to work with Ms. Marianne Hutchinson, who masterfully supervised the transformation of the draft manuscript into the finished book.

Barry Hyman

CHAPTER 1

Introduction to Engineering Design

1.1. OVERVIEW

This chapter is intended to familiarize engineering students with the nature of engineering design and how it is done. We start by drawing the distinction between classical mathematics and science problem solving on the one hand and engineering design on the other. We then examine a definition of engineering design for use in this book and present a nine-step model of the engineering design process. We introduce the notion of engineering systems and address design issues at the system level and explore the evolution of design ideas from their inception to implementation. There we emphasize the critical importance of investing sufficient intellectual resources to the early phases of design in order to reduce the need to make costly changes at later stages. We close this introductory chapter with a philosophical reflection on the intricacies, and indeed, the beauty of the engineering design enterprise.

1.2. SCIENCE, MATHEMATICS, AND ENGINEERING

Many students cite their interest in science and mathematics as one of their reasons for wanting to study engineering. They base this perception of engineering on their high-school experience in solving abstract mathematics problems such as:

$$\text{Find } x \text{ if } x^2 - 8x + 15 = 0$$

or more realistic word problems such as:

The force of the wind on a flat surface varies directly with the area of the surface and the square of the velocity of the wind. When the wind is blowing 16 miles per hour, the force on an area of 4 square feet is 5 pounds. What is the force on a square yard when the wind is blowing 12 miles per hour?

They also recall their experience with science based problems that require combining mathematical manipulations with applications of scientific principles. For example, correct application of the gas laws is needed to solve the following problem:

> A balloon is filled with 1200 ml of H_2 at a pressure of 740 mm of Hg and a temperature of 30°C. The balloon is allowed to ascend 1 mile, where the pressure is 640 mm of Hg and the temperature 7°C. Calculate the volume of the balloon at a height of 1 mile.

This perception that engineering is only "applied science and mathematics" is reinforced by traditional engineering curricula, which emphasize science and mathematics courses during the first two years and specialized applications of science and mathematics in what are typically called "engineering science" courses.[1] This emphasis continues in many upper division engineering courses. Whatever the particular field of mathematics or science used in these types of problems, they tend to have four features in common.

First, the problems are well-posed in a very compact form. By well-posed we mean that the statement of the problem is complete, unambiguous, and free from internal contradictions. If it didn't have these features, the students would complain vigorously and the teacher would apologize profusely for presenting a poorly stated problem.

Second, the solutions to each problem are unique and compact. There generally is a single correct answer available, that is, a number, a set of numbers, or symbols. In fact, many textbooks publish the answers to the odd-numbered problems in the back of the book.

Third, these problems have a readily identifiable closure. It is easy to recognize when the answer has been obtained (not necessarily the correct one).

Fourth, these problems require application of very specialized areas of knowledge and there is little doubt what the subject is for each problem. Clearly a problem at the end of Chapter 4 in the calculus book is going to require application of the concepts addressed in that chapter. Some end-of-chapter problem sets are even coded so that the student knows which section of the chapter to focus on. A problem in Chapter 4 is not going to require you to apply the material covered in Chapter 7. And you can bet that a problem in your calculus book is not going to require knowledge of physics to get the solution.

Solving problems that have some or all of these four characteristics is an important part of engineering education. It develops and strengthens specific analytical skills that are essential in most engineering design situations. However, most real-world engineering design problems do not share these characteristics. In particular, many real engineering design problems are poorly posed, do not have a unique solution or a readily identifiable closure, and almost always will require integration of

---

[1]Engineering science courses deal with applications of scientific principles and mathematical concepts for analyzing a wide variety of engineering problems such as: motion of objects, current in electric circuits, deflection of beams, temperature in fluids, and efficiency of engines.

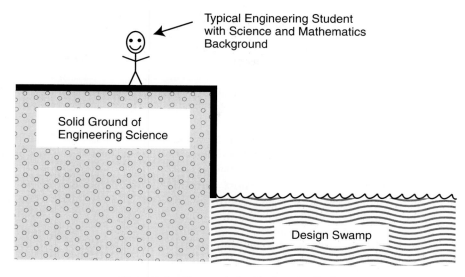

Figure 1–1.   Contemplating Engineering Design

knowledge from several subject areas. Students who have mastered the skills of solving traditional mathematics and science problems but have not had prior exposure to design may find it difficult to adjust to this less precisely defined world of real engineering design problems. Much of our emphasis in the remainder of this chapter and in several subsequent chapters of this book is on appreciating the less-precise nature of engineering design and on developing and utilizing new skills for being a successful design engineer.

1.3. MOVING FROM THE CLIFF TO THE SWAMP

Professor Donald A. Schön from MIT uses a graphic notion to highlight the difference between solving traditional science and mathematics problems and what is required to address real engineering design problems.[2] He compares the world of traditional mathematics and science problem solving to standing on a rock-solid surface at the top of a cliff. The firm foundation provided by the unambiguous and never changing laws of science and rules of mathematics is a comfortable place for most students about to embark on an engineering curriculum. In contrast, the world of engineering design involves many uncertainties, ambiguities, and inconsistencies. Professor Schön compares this world to a swamp at the base of the cliff (see Fig. 1–1). It is very difficult to get a firm footing in the swamp, and a completely different set of survival skills are needed.

[2]Schön, p. 42.

Because subjective considerations tend to be much more prevalent down in the swamp of engineering design as compared to the objective nature of analytical life up on the cliff, the relationship of the engineering design instructor to engineering students is fundamentally different.

The mathematics, science, or engineering science instructor is an expert in their field, and education consists of a one-way transfer of some of that expertise to the student. While there are many modes for facilitating that transfer (lectures, interactive problem solving, discussion sessions, textbooks) the dominant direction is that the instructor transmits, and the student receives, objective information. The instructor presumably knows the answers, and with luck by the end of the course the student will have learned enough of the right answers.

But since design is much more subjective, there rarely is a single "correct" answer. Judgments as to whether one design alternative is superior to another may be highly dependent on the values and preferences of the evaluator. The design instructor is not so much a transmitter of facts, but a facilitator of the design process and a partner with the students in searching for successful solutions of design problems (see Fig. 1–2).

The design instructor is less like a basketball referee who determines whether the actions are consistent with the rules and more like a fishing guide whose experience can make a fishing trip more enjoyable and productive. The guide can point out the logs and boulders that are scattered throughout the swamp, and provide you with a pair of hip boots to make your journey more pleasant. Studying design will help ease entry into the swamp and make your experience not only survivable but enjoyable (see Fig. 1–3). It won't remove the subjective considerations and uncertainties associ-

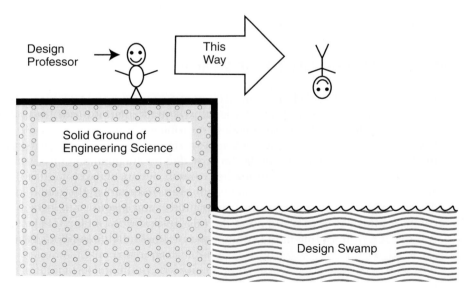

Figure 1–2.   Guidance Provided by Engineering Design Instructor

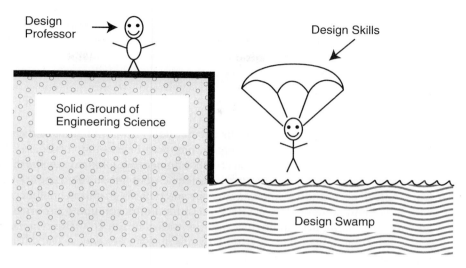

Figure 1–3. Benefits of Understanding Engineering Design

ated with design problems, but it will help you adapt to this new environment and function effectively in it.

1.4. DEFINITION OF ENGINEERING DESIGN

Solving Real Everyday Problems

Engineering design is a more advanced version of a problem solving technique that many people use routinely. The general procedure for solving real everyday problems is straightforward: A problem is encountered, information about the problem is obtained, alternate solutions are formulated, and the best alternative is adopted. Some problems are so straightforward and solutions so obvious that people solve those problems without being consciously aware of the specific steps in the process. For example, a serious problem confronts a child whose pants get caught in his/her bicycle chain. Three ways in which the child may resolve this dilemma are: rolling up the pants cuff, installing a chain guard, or securing the pants with a rubber band. The approach a child uses depends on many factors, including his or her familiarity with the bike, available materials, experience with the problem, and creativity. What works best for Johnny may not work best for Susie. What works best today may not work best six months from now. Whatever the solution, the child progresses through a design process without hesitation. When the problem is more complex (as most engineering problems are), an organized and methodical approach is needed.

Engineering design is a methodical approach to solving a particular class of large and complex problems. How can we distinguish engineering design from other kinds of problem solving activities and from other kinds of design? A few moments of reflection on how to answer this question should make you aware that you are already in the swamp! There is no single correct answer to this question—there is no universally agreed upon definition of engineering design. However, we should not let this situation paralyze us into inaction. Let us select one reasonable definition in order to get moving (if you remain stationary in the swamp, you will surely drown).

ABET Definition of Design

For the subsequent discussion, let us use the definition of engineering design adopted by the Accreditation Board for Engineering and Technology (ABET). ABET is the organization that evaluates and accredits engineering curricula in the United States. Since many U.S. firms and government agencies will only hire engineers who graduate from ABET-accredited schools, and since many U.S. graduate engineering programs will only admit students from ABET-accredited undergraduate programs, ABET has an enormous influence on engineering education in the U.S. Equivalent certification and quality control entities play similar roles in other countries. Therefore, the ABET definition of engineering design is an appropriate starting place for our discussion. ABET defines engineering design as follows:[3]

> Engineering design is the process of devising a system, component, or process to meet desired needs.[4] It is a decision-making process (often iterative), in which the basic sciences, mathematics, and engineering sciences are applied to convert resources optimally to meet a stated objective. Among the fundamental elements of the design process are the establishment of objectives and criteria, synthesis, analysis, construction, testing, and evaluation.

> The engineering design component of a curriculum must include at least some of the following features: development of student creativity, use of open-ended problems, development and use of design methodology, formulation of design problem statements and specifications, consideration of alternative solutions, feasibility considerations, and detailed system descriptions. Further, it is essential to include a variety of realistic constraints such as economic factors, safety, reliability, aesthetics, ethics, and social impact.

Note that engineering design is not a single isolated action, but a "process" (we will examine the nature of this process in detail in Sec. 1.5). ABET identifies the goal of engineering design as, ". . . devising a system, component, or process to meet desired needs." Note also that the result of design might not be a physical piece of hardware; it can be a process. This latter kind of design is of particular interest to chemical engineers, materials engineers, industrial engineers, and computer software engineers.[5] The ABET

---

[3]ABET, p. 7.

[4]ABET uses the word 'process' in two different ways in this sentence: first, to describe the ongoing design activity; and second, to describe a particular type of outcome or product of the design effort. The intended meaning should be clear from the sentence context.

[5]For example, a chemical engineer might be faced with the design problem of selecting the temperature and pressure at which two chemicals are to be mixed in order for the chemical reaction to proceed in the desired manner. As another example, an industrial engineer may be asked to design the order in which the components of a system should be assembled so that the assembly costs are minimized.

definition also hints at some of the analytical tools engineers use in their design activities; "basic sciences, mathematics, and engineering sciences." Finally, the ABET definition identifies some elements of the design process; "the establishment of objectives and criteria, synthesis, analysis, construction, testing, and evaluation." As a whole, the ABET definition serves as a guide from which to start. The ABET definition gives the design engineer the freedom and responsibility to determine what is appropriate and necessary to create a design and solve a problem. There are no absolute rules for what to do, when to do it, or how to do it; only experience and blurred, soft, marshy "rules of thumb." Engineering design is a swamp.

The Centrality of Design

In spite of the difficulties we encounter when trying to reduce the complexities of engineering design to simple, universally agreed upon definitions and models, the centrality of design to the engineering profession is unchallenged. Design is the culmination of all engineering activities, embodying engineering analysis and other engineering activities as tools to achieve design objectives.

1.5. A MODEL OF THE ENGINEERING DESIGN PROCESS

Just as the ABET statement is only one of many definitions of engineering design, there are many approaches to describing how design is done. Some of these descriptions have been formalized into simplified step-by-step "models" of the design process. While no one model is universally accepted by the engineering community, it is helpful to organize our discussion using one model. In doing so, we recognize that there are many other approaches that are just as useful.

In this section we briefly outline a nine-step model of the engineering design process. Before discussing each of the nine steps in this model, a few general comments are in order. First, it is important to recognize that any model is a simplified description of a more complicated reality. The value of a model lies in its ability to help us organize our thoughts and gain insight into important aspects of reality. So keep in mind while we discuss these nine steps that actual designs do not necessarily evolve in a linear, orderly progression from step one through step nine. Not every step will be used to the same extent in every design, and some steps may be performed out of order.

We defer more detailed discussion of each step to later sections in the book. In fact, some of the steps are the topics of entire chapters. The nine steps and the parts of the book in which they are further elaborated are summarized in Table 1–1. The fact that the sequence of topics presented in the book doesn't match up exactly with the nine steps in our design process model reinforces the non-sequential nature of the model.

Many engineering designs are performed by teams of engineers and not every team member participates in every step of the process. Some team members may be specialists in one or more of the nine steps. In many situations, design engineers unconsciously blend some of these steps together. Also, each step may be revisited sev-

TABLE 1–1.   RELATION BETWEEN DESIGN PROCESS MODEL AND ORGANIZATION
OF THIS BOOK

Steps in Design Process Model	Location of Detailed Discussion
1. Recognizing the need	Chapter 2: Problem Formulation
2. Defining the problem	Chapter 2: Problem Formulation
3. Planning the project	Chapter 7: Project Planning
4. Gathering information	Chapter 3: Information and Communication
5. Conceptualizing alternative approaches	Chapter 6: Concept Generation
6. Evaluating the alternatives	Chapter 8: Engineering Economics
7. Selecting the preferred alternative	Chapter 9: Decision Making
8. Communicating the design	Chapter 3: Information and Communication
9. Implementing the preferred design	Section 1.7: Life Cycle of Engineering Designs

eral times during the evolution of a design. However, even experienced engineers will
regularly step back from their immersion in design details and rely on such a model to
assure themselves that they haven't overlooked key elements in their search for a design solution. The map of the design swamp shown in Figure 1–4 depicts the relationships among the nine steps.

Note the absence of a "STOP" activity in the Design Swamp. Does the design
process continue without end? Possibly. It continues as long as the need continues,
and it ends only when the cost of continuing the design process exceeds the value of
an improved design. The decision to stop the design process is difficult and requires
careful thought. It may be made by the engineer or the client, or it may be a result of a
schedule constraint. This is what we mean when we say design is an open-ended
process; there frequently is no readily identifiable closure point.

The automobile, for example, is a solution to the need for transportation, and
automobile design has evolved continuously since its invention. From a longer range
perspective, automobiles evolved from horse drawn wagons or from ancient push
carts. In any case, automobile design continues to evolve because no automobile is
perfect and because the needs themselves change. Even the best-selling automobiles
are redesigned regularly because a need exists for new or different features such as
pollution control equipment, airbags, and anti-lock brakes.

Step 1: Recognizing the Need

The first step in the design process establishes the ultimate purpose of the project via
a general statement of the client's dissatisfaction with a current situation. Consider
the hypothetical conversation between Jane, a design engineer for an automotive engineering company, and Sandra, her immediate supervisor.

Sandra: "Jane, we need you to design a stronger bumper for our new passenger car."

Jane: "Why do we need a stronger bumper?"

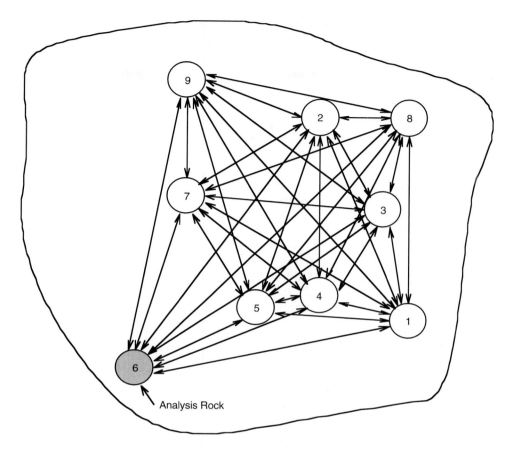

Figure 1–4.   A Map of the Design Swamp

Sandra: "Well, our current bumper gets easily damaged in low-speed collisions, such as those that occur in parking lots."

Jane: "Well, a stronger bumper may be the way to go, but there may be better approaches. For example, what about a more flexible bumper that absorbs the impact but then returns to its original shape?"

Sandra: "I never thought of that. I guess I was jumping to conclusions. Let's restate the need as 'there is too much damage to bumpers in low-speed collisions.' That should give you more flexibility in exploring alternative design approaches."

Notice that Sandra's revised needs statement is more general than her initial one, focuses on what is unsatisfactory with the present situation, and is silent in terms of the design approach to use. See Chapter 2 for a more detailed discussion of this topic.

Step 2: Defining the Problem

Once the need has been established, the next step is to translate the needs statement into one that addresses how we propose to satisfy the need. A problem statement generally consists of three components:

- Goals
- Objectives
- Constraints

Each of these components of a problem statement are examined in more detail in Chapter 2.

Step 3: Planning the Project

After the problem has been defined, an overall plan should be developed. A good project plan will help you identify what tasks need to be accomplished and in what order. Jane might develop the plan shown in Table 1–2 for the bumper design project. Note that none of the tasks listed in Table 1–2 correspond to any of the nine steps in the design process model. This is because each task is results oriented. It corresponds to a stage in the evolution of the design (see Sec. 1.7) rather than on the mental processes or activities which contribute to that result. Each task may incorporate some or all of the nine steps in the design process. Tools for conducting project planning are presented in Chapter 7.

Step 4: Gathering Information

Each design problem requires a unique combination of information sources. When designing in an area which is well developed, there is most likely a wealth of information on similar designs that have already been completed, and on what constraints are imposed on design possibilities by codes and standards. Additionally, there may be a solution to a similar problem which may offer insight and save time. Certainly, it would be prudent for Jane to collect information on the current approach to automobile bumper design.

On the other hand, there may be very little information available if the design task is in an area which has never been considered before. Even if there is some information on the topic, it may be theoretical research and not directly applicable to de-

TABLE 1–2. PLAN FOR BUMPER DESIGN PROJECT

Task	Starting Date	Completion Date	Cost ($)
Preliminary Design	July 1	July 10	1,000
Build Prototype	July 8	July 15	2,000
Test Prototype	July 13	July 17	1,500
Final Design	July 20	July 31	3,000

sign situations. The engineer may need to perform experiments or develop new models in order to obtain the necessary information, but we defer discussion of that until Step 6–Evaluating the Alternatives. See Chapter 3 for a thorough examination of the information gathering activity.

Step 5: Conceptualizing Alternative Approaches

This step is where possible design solutions are first envisioned. The principal effort is to generate a wide range of design options. This is where you call upon your creativity and imagination to develop design approaches that have the potential to satisfy the objectives and constraints defined in Step 2. This activity involves a mode of thinking that is quite different from the analytical mindset so widely used by engineers. Chapter 6 explores this topic in more detail, including a discussion of specific techniques to enhance your creative design skills.

Step 6: Evaluating the Alternatives

Once alternatives have been conceived, they must be evaluated to determine the extent to which they satisfy the design objectives and constraints. This is done by using mathematics and the engineering sciences to evaluate the performance characteristics of each design alternative. We label Step 6 as Analysis Rock; the only comfortable, firm ground within the design engineering swamp. Much of your engineering education is devoted to learning a wide variety of analytical tools for use at Analysis Rock. These range from formulas and experiments for determining the stresses in a beam to finding the voltage drop across an electrical device.

Other analytical techniques that are important parts of the engineering design process, but which are not included in many engineering curricula, include estimating the cost of production and determining the probability of the design suffering a catastrophic failure. For this reason, we devote two chapters in this book to these topics. Economic analysis techniques are covered in Chapter 8, and probabilistic considerations are examined in Chapter 5.

Since Jane's project plan (Table 1–2) contains a budget and projected completion date, she may not have the luxury of conducting the most thorough and rigorous analysis. She may have to make a lot more simplifying assumptions than she is initially comfortable with. This may cause some frustration, because she has the natural desire of most young engineers to demonstrate their skill at using the analytical techniques learned in school. But with experience, she will appreciate that the most valuable design aids are the analytical techniques that can be adapted for the design problem at hand. The difficult part of Step 6 is selecting which analyses to conduct and at what level of detail and sophistication to conduct them.

Step 7: Selecting the Best Alternative

Now that you've had a chance to rest and refresh yourself on Analysis Rock near the edge of the swamp, it's time to dive back into the murky subjective soup and continue with the remainder of our journey through the design swamp. In Step 6 you've carefully and thoroughly analyzed each of the alternative designs. Now it is time to select the best alternative. At first glance this might seem to be trivial; after all, won't the results of Step 6 tell us which alternative performs the best? Well, maybe and maybe not!

Suppose from Step 2 that the four characteristics that Jane (and by proxy, Sandra) are most interested in are initial cost, damage control, drivability, and recyclability. Jane may now have to decide which performance criteria are more important, and how much more important. But remember, we've slipped off Analysis Rock and are back in the swamp. There is no $F = MA$ or other objective way to establish the relative importance of the various performance criteria. It is an inherently subjective decision that can only reflect the values, preferences, and priorities of the decision maker (see Fig. 1–5). Actually, the choice that Jane makes should reflect the client's values, preferences, and priorities as articulated in this case by Sandra serving as the client's surrogate. There are systematic ways for Jane to ascertain the relative importance to Sandra of the various design objectives (synonym for performance criteria) and Sandra's willingness to make trade-offs between them. The techniques for doing this are examined in Chapter 9.

Step 8: Communicating the Design

Communication is essential in all aspects of design in order to inform interested parties of project progress and status. Communication also provides a record for the design engineer to reconstruct and justify what he/she did and why they did it. This is especially important after a prolonged period of inactivity on a project (a common occurrence for engineers with several simultaneous projects). Good communication clears the fog and allows other engineers to readily follow the route you forged through the swamp. Communication also serves as a bridge between the design engineers and the rest of the world; it helps clients understand what comes out of the design swamp without having to get muddy themselves (see Fig. 1–6). Without communication, everyone would be stuck in the same swamp; each reinventing the wheel in their own way.

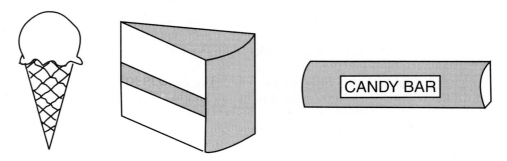

Figure 1–5. The Dilemma of Selecting the Best Alternative

Figure 1–6.   The Communications Bridge

Developing effective technical communications skills requires hard work, attention to details, and lots of practice. We can only scratch the surface in this section, but most engineering students will take at least one required course in technical communications as part of their engineering education.

Formal communication has two main forms: written records and oral presentations. These are discussed in depth in Chapter 3.

Step 9: Implementing the Preferred Design

Some people consider design and implementation to be separate activities. But design activities are intended to solve a problem and to do so, action must occur. Implementation refers to this translation of design concepts into actions; it is a natural step in the design process. The ABET definition of engineering design in Section 1.4 refers to implementation using the words construction and testing.

Traditionally, designers tossed their designs "over the wall" to manufacturing or construction engineers who were responsible for implementation, but had no influence on the design. As a result, costly designs were implemented when less costly approaches might have been available. Modern approaches that integrate implementation considerations into the design process include "design for manufacturing" and "design for assembly." These and other approaches are discussed further in Section 1.7.

1.6. ENGINEERING SYSTEMS DESIGN

Our discussion in Section 1.5 regarding the design of an automobile bumper treated the bumper as a single entity. Since the bumper probably consists of a steel core with a chromium plated surface, it may be useful to think of the steel bumper core and the plating material as separate entities that are combined in a plating operation. In addition, the bumper may have brackets welded to it for mounting the bumper onto the vehicle chassis. The brackets and the bumper can also be considered separate entities until they are welded together.

To avoid confusion, we can refer to the combined bumper and bracket as the bumper/bracket system. It has two components: the bumper and the bracket. Upon reflection, we should also include the weld material as a third component of the bumper/bracket system. In turn the bumper can be thought of as a subsystem that has two components; the bumper core and the plating. We can illustrate the relationships by using the hierarchical tree diagram shown in Figure 1–7.

While this description of an automobile bumper/bracket system meets the needs for our introduction of system design concepts, it is a gross simplification of many real contemporary automobile bumper designs. For example, Figure 1–8 shows a shock-absorbing bumper system utilizing plastic honeycomb.

To prevent things from getting too complicated, we'll stick with our simplified model but continue our line of thinking by recognizing that the bumper/bracket system is itself part of a larger system (the automobile). In turn, we can think of an automobile as just one part of an integrated transportation system. A graphical hierarchical arrangement including these broader systems might look something like Figure 1–9. The exercise of constructing Figures 1–7 and 1–9 should indicate that there are many ways in which a given object can be disaggregated into smaller parts, or combined with other objects to form a more complex object. Any hierarchical diagram such as Figure 1–9 can, in principle, be expanded to include the entire universe. That is, everything is connected to everything else. This illustrates that the engineering design swamp has a very complex ecology.

System Concepts

For the purpose of generalizing the preceding discussion, let us introduce some definitions. We start by defining a system as a collection of elements that interact with each other to fulfill a function. As mentioned, the object of any engineering design effort can be envisioned as part of a very large complicated system. However, for practical

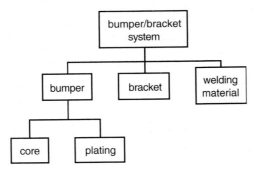

Figure 1–7. Three-level Hierarchical Diagram of Automobile Bumper/Bracket System

Fascia
Retainer
Center honeycomb
Corner honeycomb
Armature
Stay

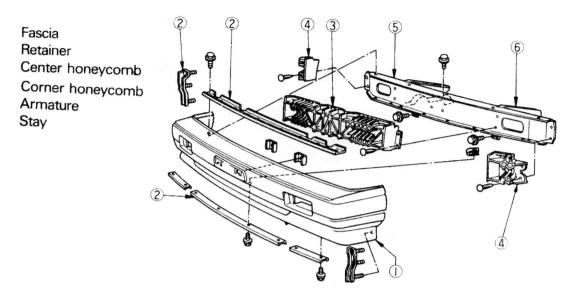

Figure 1–8. Bumper System *(Reprinted from Nosho, et. al., p. 16, with permission from the Society of Automotive Engineers, Inc.)*

reasons we must recognize that we cannot take on the whole universe; we must limit our attention to those system elements that are the most important factors for the specific design activity. We do this by establishing a system boundary or set of boundaries that separates the universe into two parts: the system under consideration and the external environment.

These boundaries either isolate the system from the environment or provide the mechanism by which the system interacts with the environment. Engineers and their clients define systems and their boundaries as the object of their design efforts during the problem formulation stage of design. If, for example, our immediate design efforts are focused on the automobile bumper, we might decide to limit our attention to the elements depicted in Figure 1–7.

The smallest identifiable element of a system is called a component. Collections of components are called subsystems when not all elements of the system are included in the collection. It is frequently helpful to display a system and its subsystems and components in the form of a hierarchical tree such as shown in Figure 1–8. Sometimes it may be helpful to use terminology such as assembly, subassembly, or member to distinguish between subsystems and components at different levels of a hierarchy.

The manner in which any system is subdivided depends on the perspectives of the engineer and the customer and the nature of the specific design activities that are being facilitated by the subdivision. The number of levels in a hierarchy and the number of elements in each level are selected primarily for convenience and as an organiz-

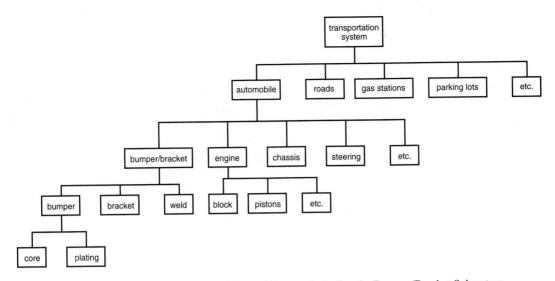

Figure 1–9. Five-level Hierarchical System Diagram Including the Bumper/Bracket Subsystem

ing tool for planning the design activities (see Chap. 7 for a more detailed discussion of project planning). A given system may be subdivided in many different ways simultaneously. Consider the five-level hierarchical depiction of an automobile system in Figure 1–8 that includes the immediate object of our design efforts—the bumper/bracket subsystem—as one of its subsystems. Another system depiction that also includes the bumper/bracket subsystem might look like Figure 1–10.

It is important to be able to identify and document the function of each system element. Some elements exist solely to serve as an interface between other elements. For example, the weld element in Figure 1–7 exists only to connect the bracket to the bumper, and the function of the bracket is to connect the bumper/bracket subsystem to the chassis. Design of these connecting elements requires information about each of the elements to be connected. One engineer or team of engineers may be designing the bumper, another designing the bracket, and a third designing the chassis. They must communicate with each other regularly so that the elements can properly interact with each other (see Chap. 3 for a further discussion of design communications).

Another aspect of system design does not directly involve the design of any particular element; that is the design of the spatial relationships of the elements to each other. Take a look under the hood of your automobile. How did they ever decide where to put the battery, the windshield washer, or the alternator? And how did they ever get everything to fit? We refer to this aspect of system design as configuration design. While the immediate focus of configuration design is the arrangement of system elements, clearly decisions about the location, size, and shape of the space available for an element could affect the design of that element.

The process of addressing requirements on the system level (the top tier of a hierarchical diagram such as Fig. 1–9) creates requirements at the first subsystem level.

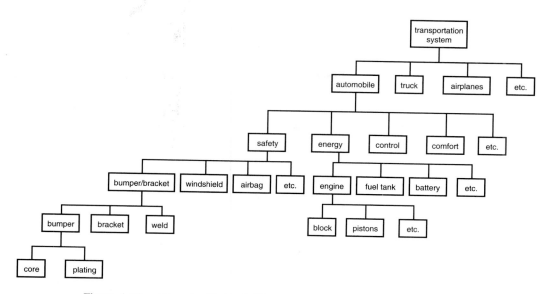

Figure 1–10. Alternate Six-level Hierarchical System Diagram Including Bumper/Bracket Subsystem

Turn the design crank for each requirement on subsystem level 1, and objectives and constraints are created for subsystems at level 2. Depending on the complexity of the system, this progression can happen hundreds or thousands of times. Sometimes the design activity reverses direction as decisions made at a lower level in the system hierarchy require that prior decisions made at higher levels be reconsidered. Design of complex engineering systems requires a great deal of patience to wait for the overall solution to emerge from solutions to many individual pieces. Another important ingredient of system design is detailed documentation of ongoing modifications to system components and continuous communication among team members as these modifications are being considered and implemented. Much of this project structure is developed and maintained as part of the project planning aspect of design, discussed in more detail in Chapter 7.

1.7. LIFE CYCLE OF ENGINEERING DESIGNS

Our nine-step model of engineering design addresses the nature of the activities that design engineers engage in over the course of a design project. We made a special point to note that the nine-step process is not sequential; many steps are encountered out of order and iteratively.

This does not imply that the object or system being designed doesn't evolve in an orderly fashion. Evolution of a design is the focus of our attention in this section.

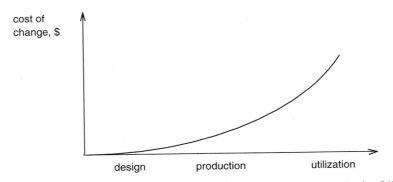

Figure 1–11.   Cost of Making Changes During Different Phases of the Design Life Cycle

We'll organize our discussion using Simon's[6] eight-step description of the life cycle of engineering designs:

1. Needs Analysis
2. Feasibility Study
3. Preliminary Design
4. Detailed Design
5. Production
6. Distribution
7. Consumption
8. Retirement

This model identifies a natural progression of a design through stages, although we may need to retrace steps at any stage to modify decisions made at earlier stages. The sequential perspective emphasizes the advantages of systematically studying, understanding, and focusing our energies on the early stages of design since the cost of retracing steps increases dramatically in the latter stages (Fig. 1–11). The proverb that "an ounce of prevention is worth a pound of cure" applies as much to engineering design as it does to anything.

We recognize the first stage of Simon's model, Needs Analysis, as the counterpart to the first step in our design process model—Recognizing the Need. We will postpone further discussion of it until the next chapter and concentrate here on key aspects of the other stages of Simon's model.

[6]Simon, p. 4.

Feasibility Study

In this phase, the engineer addresses the fundamental questions of whether to proceed with the project. Just because someone has articulated a need doesn't mean that it is reasonable, desirable, or possible to fulfill that need. There may be technological and economic limitations: It just might not be feasible to design an automobile bumper that costs less than $10 and can withstand a 50 mph head-on collision with a concrete wall without causing damage to the vehicle. There may be no point in spending any more time attempting to achieve these unrealistic objectives.

Other feasibility considerations include assessing the market. If the object to be designed is envisioned as a mass-produced product, a market study of the sales potential for the product may be warranted before justifying the cost of establishing a manufacturing capability. On the other hand, the feasibility of a large public works project, such as a highway bridge or a sewage treatment plant involves assessing the reaction of the community in which the facility is to be located.

Clearly, other professionals besides engineers are likely to be involved in feasibility studies. Some of the important skills needed in this phase of design are financial analysis, marketing, and community relations. Design engineers involved in feasibility studies have to interact with these other professionals, so good people-to-people skills and communications skills are extremely important.

Many times the question to ask is not whether a project is feasible, but whether it is feasible for the firm to take on the project at a specific time. It may involve an unfamiliar technical area, or other projects to which the firm is currently committed may preclude allocating sufficient resources to the proposed project. The company may decide that it doesn't want to deal with the specific uncertainties associated with the project. All these considerations should be dealt with during the feasibility study before a commitment is made to pursue extensive design activities.

Preliminary Design

Design alternatives begin to take shape in the minds of engineers, or as part of conversations among engineers exploring ways to satisfy the stated needs. A key step in the early phases of design is capturing these ideas in the form of free-hand sketches. These sketches, which at first, may be not much more than doodles, may contain suggestions of key features, such as overall form and ways to decompose the system into components. They may also indicate aspects of the loading and other operational conditions. The design engineer's thoughts about possible materials, size, and other features may begin to crystallize and be jotted down along with the pictorial representation. Generally these sketches and supporting notations precede any formal analysis of the design.

As described in more detail in Chapter 6, we strive to generate a large number of preliminary designs. If we are successful, more design concepts will emerge than we have time to formally analyze or to begin detailed design. An important part of the preliminary design phase is to winnow down the design alternatives to a relatively small number. In this process of narrowing down the options, new options are bound to arise as we refine the concepts. This temporary enlargement of the number of options is followed by another contraction phase until we converge on a single approach or small number of approaches. This process of alternating between concept generation and elimination is called controlled convergence and is depicted schematically in Figure 1–12.

Detailed Design

Once a single design concept or a small number of design options are selected, we move to the phase of refining it (them) in greater detail. Formal analysis helps to define sizes, choose materials, estimate costs, and plan procurement and fabrication requirements. These decisions are made at the individual component level and overall system level. Generally this phase results in a set of formal engineering drawings (either hard copy or computer files) for each component. These drawings provide complete unambiguous instructions for making that component. Also, assembly drawings are produced that display how the components fit together to form subsystems and how the subsystems fit together to form the system.

Models and prototypes.   As the main features of the design are defined, it will be helpful to construct either a scale model or a full-scale prototype. Models and prototypes provide a three-dimensional treatment to designs so that they can be examined from any perspective to inspect for conformity to expectations, and so the design can be viewed in the context of its surroundings. Since these models and proto-

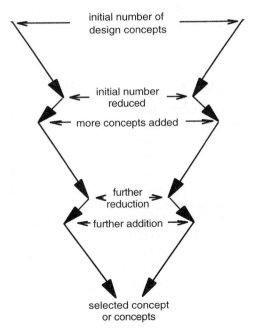

Figure 1–12. Controlled Convergence During Preliminary Design *(Adapted from Pugh, p. 75, with permission from Addison-Wesley Longman Ltd.)*

types are intended primarily to display shape, they usually are made of an inexpensive material that is easy to work with.

Sophisticated computer-based three-dimensional modeling techniques now allow engineers to create images of designs that can be maneuvered and rotated so that the design can be viewed from any angle and against any backdrop. These effects include shading from light sources located virtually anywhere relative to the object and animated "walk-throughs" to examine designs from the inside. These latter effects are particularly useful for designs of buildings and other large structures. The availability of these techniques substantially reduce, but do not totally eliminate, the need for prototypes. In particular, clients may need a physical object that they can inspect at their convenience, and they may not be content with images on a computer screen.

Rapid prototyping.   Several new technologies for rapidly producing prototypes gained popularity within the engineering design community in the early 1990s. These techniques produce three-dimensional scale models in hours instead of days or weeks. Most rapid prototyping methods automatically produce models directly from computer-aided design (CAD) files of the object.

One of the earliest approaches to rapid prototyping is stereolithography (STL). The first step in the process converts CAD files into STL files which describe the object in terms of thin layers. Next, these files are downloaded into an apparatus that literally constructs the model of the object one layer at a time. The original STL technology consists of a movable elevator table submerged in a vat of liquid polymer. The polymer cures and solidifies when exposed to ultraviolet light. At the start of the prototype construction process, the table is just below the liquid surface. Guided by the data in the STL file, an ultraviolet light source directs a fine laser beam onto the surface of the polymer and solidifies a thin layer of the polymer between the surface and the elevator table as it traces out the shape of the first layer of the prototype. After this first layer is formed, the table is lowered and the second layer is formed on top of the first layer using the same process.

Other forms of rapid prototyping include spray deposition of plastic layers, and hand assembly of paper or foam prototypes whose individual layers are produced by a sign-making machine from the STL files. One of the advantages of these rapid prototyping technologies is that the prototypes are constructed by automated machinery that doesn't require continuous human attention. Depending on the particular technique and materials used in rapid prototyping, surface finish or tolerances may be of low quality.

Production prototypes.   Once design details have been established, another type of prototype may be built. This may be a full-scale prototype fabricated from the same material as the anticipated production version. Its role may be to verify fabrication requirements or to be tested for strength or other performance characteristics.

Testing.   Part of the detailed design phase may also involve a testing program. Many different kinds of tests can be carried out as part of this phase of design. Testing scale models of airplanes in wind tunnels help airplane designers decide on wing shape and placement of engines. Tests on production prototypes may be used to verify the performance analyses. In some cases, applicable codes and standards may require that production prototypes be tested to failure.

Production

Once the design has been finalized, it is time to implement it by producing the product. Very large one-of-a-kind systems, such as buildings, tunnels, and petroleum refineries, are constructed on-site, although some components and subassemblies may be fabricated off-site and assembled on-site. Very small products intended for mass production are produced in a manufacturing facility and shipped to the customer. In between these two extremes are large, mass-produced products such as automobiles and airplanes that involve many components and subassemblies produced in dispersed locations and then shipped to a central assembly facility. For example, the engines, landing gear, and major segments of the fuselage and wings of airplanes may be built by subcontractors and assembled by the prime manufacturer.

Design for manufacturing and assembly.   We can think of the latter phases of detailed design in which the design starts to "move off the paper (or computer screen)" as the beginning of the implementation step (Step 9) of the design process model discussed in Section 1.5. Consider a design process that does not include implementation. We saw earlier that our friend Jane worked with her boss to identify a need and develop a problem definition. Suppose that without concern for implementing the solution, Jane gathers information, conceptualizes possible solutions, evaluates the alternatives, and selects a preferred approach. She then proceeds to a detailed design, produces all the supporting documentation, and passes the completed design to the manufacturing department. The manufacturing people review the design and question Jane's sanity for proposing a design that is impractical to build: "Doesn't she know that drilling a standard size hole of 0.50″ diameter is much less expensive than drilling the 0.517″ diameter hole she specified in her design?" They sit down to catch their breath and stop laughing at this and other ludicrous features of Jane's design. They then contact Jane to suggest how her design can be modified to reduce manufacturing costs. Jane, now depressed because her revolutionary design cannot be built, returns to the drawing board to incorporate the suggestions of the manufacturing people; and much of her first design is discarded.

This is a common scenario for engineers who fail to design for implementation. If Jane had asked, "How will they build this?" early in the project, and had discussed her design ideas with the manufacturing people (Step 8!), her first design might have incorporated features that simplified implementation. This approach would have saved both Jane and the manufacturing engineers frustration, kept the project on schedule, and saved money. If the design has to be substantially modified at the later stages of development, it will be a bandaged design and it will perform like a leaky boat in a swamp.

How does one plan for implementation? It may be helpful to ask, "How would I implement this?" and, "Is there an easier way to do this?" Another approach is to seriously consider implementation during Step 2 of the design process, and explicitly incorporate these requirements into the Problem Statement. It also might be helpful to assume that you were responsible for monitoring and supporting your design solution over its entire life. Then before committing to a solution, consider its implications. Foresight will help you identify the benefits and shortcomings of alternative designs. You will find more elegant solutions to problems, and you will be able to avoid implementation quagmires before they arise.

Of course, no matter how thoroughly implementation is planned, minor adjustments and refinements will still be necessary at Step 9. But if implementation requirements were anticipated, this will be a time of pride and accomplishment.

Techniques for incorporating these production considerations into early phases of design are labeled "design for manufacturing and assembly." Design and production can be integrated even more tightly so that certain production activities can begin even before all design details have been completed. This approach is known as "concurrent design."

Taguchi method.   When size and other features of designs are specified, we recognize that it is virtually impossible to make the product so that these specifications are exactly satisfied. The traditional approach to dealing with these deviations is to incorporate tolerances into the detailed design. Thus, the diameter of the hole to be drilled in the bumper mounting bracket might be specified as $0.500'' \pm 0.005''$. Product quality is maintained by an inspection procedure during production which rejects brackets in which the hole diameters fall outside of this acceptable range.

A more sophisticated approach using statistical concepts to incorporate product quality into the early stages of design is known as the Taguchi method. This technique recognizes that a bumper bracket with an actual hole diameter of $0.505''$ is not as high a quality product as one whose hole diameter is $0.501''$. By quantifying the differences in performance of these two brackets, and estimating the cost of producing the bracket to a tighter tolerance, the appropriate amount of quality can be designed into the product. More generally, the Taguchi method allows engineers to control design parameters so that the design is less sensitive to variations that can degrade product quality.[7]

Distribution

Many mass-produced products are delivered to customers through complicated distribution channels. The distribution process may include shipment by several transportation modes and temporary storage in warehouses or other distribution facilities. The products themselves and/or their packaging must be designed to withstand the sometimes very rough handling and extreme environments they experience during distribution.

Consumption

Many complex engineered systems require regular sophisticated maintenance programs after they have been delivered to the customer. Examples include commercial airplanes, bridges, electric power generating plants, and petroleum refineries.

[7]Pugh, pp. 214–218.

For such systems, ease of maintenance may be an important design objective. Design techniques and approaches that focus on this issue are labeled "design for maintainability."

During their operational phase, many engineering systems consume resources and generate waste products. Many state and federal environmental regulations constrain the nature and amount of adverse environmental impact associated with operation of these systems. Keeping environmental performance as a major focal point during design comes under the heading of "design for the environment."

Retirement

As an engineered system approaches the end of its useful lifetime, consideration must be given to its ultimate disposition. The importance of reusing and recycling parts of the system has increased as we have become more sensitive to problems associated with solid waste disposal. This leads to design philosophies and procedures that focus on the ease of dismantling the system and recycling its components. The terms "design for disassembly" and "design for sustainibility" have been adopted to refer to this emphasis.

1.8. CLOSURE

In this chapter we introduced the concept of engineering design. We saw that many aspects of design, including definitions of what it is and how it is done, are amorphous. This lack of solid physical laws and rigid rules regarding engineering design activities may lead many engineering students to adjust their conceptions of this most fundamental of engineering activities.

Now that you've had a thorough guided tour of the engineering design swamp, we hope that you appreciate the rich diversity of activity that occurs in that territory. This new-found insight into the complexities and intricacies of engineering design is similar to the recent awareness in society-at-large of the ecological significance of wetlands (formerly called swamps!). What at first glance is perceived as unproductive and hostile territory reveals, after closer examination, a fascinating and mysterious array of interconnected activities that are essential to the success of the enterprise. We hope you will revisit the engineering design wetlands often and that each visit will be personally enjoyable, productive, and professionally fulfilling.

1.9. REFERENCES

ABET, 1990. *Criteria for Accrediting Programs in Engineering in the United States.* New York: Accreditation Board for Engineering and Technology, Inc.

DIETER, G. E. 1991. *Engineering Design: A Materials and Processing Approach, 2nd Edition.* New York: McGraw-Hill, Inc.

EDER, W. E. 1988. Education for Engineering Design—Applications of Design Science. *Int. J. Appl. Engng. Ed.* Vol. 4, No. 3.

Meredith, Dale D., et. al. 1985. *Design and Planning of Engineering Systems, 2nd Edition.* Englewood Cliffs, NJ: Prentice-Hall, Inc.

Middendorf, William H. 1990. *Design of Devices and Systems, 2nd Edition.* New York: Marcel Dekker, Inc.

Nosho, H., et. al. Application of Plastics Technology to Automotive Components. In Geaman, Gregory N., et. al., eds. 1989. *Plastics in Automotive Applications: An Overview.* PT-32. Warrendale, PA: Society of Automotive Engineers, Inc.

Pugh, S. 1991. *Total Design.* Reading, MA: Addison-Wesley Publishing Co.

Schön, Donald A. 1983. *The Reflective Practitioner: How Professionals Think in Action.* New York: Basic Books Inc.

Simon, H. A. 1975. *A Student's Introduction to Engineering Design.* New York: Pergamon Press, Inc.

von Oech, R. 1990. *A Whack on the Side of the Head, 2nd Edition.* New York: Warner Books.

Walton, Joseph W. 1991. *Engineering Design: From Art to Practice.* St. Paul, MN: West Publishing Company.

1.10. EXERCISES

1. Compare the following ten definitions of engineering design with each other and with the ABET definition of design. Which of these are you most comfortable with?
    a. Engineering design is a purposeful activity directed towards the goal of fulfilling human needs, particularly those which can be met by the technological factors of our culture.
    b. Mechanical engineering design is the use of scientific principles, technical information and imagination in the definition of a mechanical structure, machine, or system to perform prespecified functions with maximum economy and efficiency. The designer's responsibility covers the whole process from conception to the issue of detailed instructions for production and his interest continues throughout the designed life of the product in service.
    c. Engineering design is the process of applying various techniques and scientific principles for the purpose of defining a device, a process, or a system in sufficient detail to permit its physical realization.
    d. Engineering design is simulating what we want to make (or do) before we make (or do) it as many times as may be necessary to feel confident in the final result.
    e. Engineering design is the optimum solution to the sum of the true needs of a particular set of circumstances.
    f. Engineering design is the imaginative jump from present facts to future possibilities.
    g. Engineering design is a creative activity—it involves bringing into being something new and useful that has not existed previously.
    h. Engineering design is the performing of a very complicated act of faith.
    i. Engineering design is a goal-directed problem solving activity.
    j. Engineering design is a process performed by humans aided by technical means through which information in the form of requirements is converted into information in the form of descriptions of technical systems, such that this technical system meets the needs of mankind.
   ◆ *Source:* Eder, pp. 167–184.

2. List at least six professions other than engineering, in which design is a major activity.

3. Read the attached newspaper column and prepare a letter to the editor in response. Your letter should be brief and be aimed at the general readership of the newspaper (avoid technical jargon!). In your letter you should address the issue of whether Don Lambert does "back-shop engineering" by explaining the nature of engineering design and its relationship to Don Lambert's activities; and why engineers at, say Boeing and Washington State University, approach the design of new energy technologies from a different perspective than Don Lambert.

How to Steam Toward a Revolution

Jon Hahn

Don Lambert built the little engine that could. And it can.

Less than $60 of scrap steel from Boeing Surplus Sales and a lifetime of know-how by the retired Boeing machinist came together in a homemade steam engine that can light a house, pump water, grind grain, run various tools, and even cut its own boiler wood.

This country's biggest untapped energy reserve isn't pooled underground. It's so close to the surface we can't see it: the hands-on skills and mental juices of the so-called little people, like Lambert.

Retired Boeing workers probably represent one of the largest pools of on-the-hoof American know-how. But no one's thought to tap it, or put it to work on the thinking we need to get out of our foreign oil straitjacket.

We put a whole lot of hope in big dreams like atomic energy when the reality of something like steam power is as plain as the nose on our national face. And just as easy to reach out and touch.

"I've been tinkering with steam engines ever since I was a little boy on the farm," said Lambert, 66, of Bonney Lake. He was one of seven children in a Wisconsin farm family that didn't even have electricity until the late 1930s.

Building his own steam engine came naturally to Lambert, who says: "I've always been concerned about things going to waste. Every time I see logging slash being burned, it makes me mad . . . all that potential power going to waste."

All the know-how from 25 years as a machinist wasn't about to be wasted when Lambert retired after heart surgery. He'd already been tinkering out back in his workshops, and he got down to

some pretty serious tinkering. "I wanted to make something that anyone with a workshop could make," he said.

Well, maybe not anyone. His steam engine—like his other projects—has grown from rough drawings to cardboard and paper scale models to actual parts. "I've built it and rebuilt it a number of times," he said. "Each time I discovered how I might've done something just a bit better."

There aren't any cast-off castings in his engine. "All from stock iron, except for the valves. I didn't want to be dependent on anyone else," he said. That might sound a little stubborn. It also sounds self-reliant.

Lambert taught himself how to case-harden stock steel after reading an old book about blacksmithing. He made his own steel-hardening furnace for parts like pistons and cylinders. Of course, he has the tools and skills, but it's more than that.

He has to machine special rigs and jigs so that he can make parts, so that he can make a machine. But after several years, he has a little marvel of back-shop engineering that puts the Lambert family comfortably close to energy independence. He really doesn't have to depend on the engine, so most of its extended runs have been at shows such as the annual July old-engine show at the Roy, WA fairgrounds.

Lambert's engine, which weighs about 250 pounds with a full head of steam, can run 10 hours on about two grocery bags full of wood scraps. "I ran it for a half-day once just on pine cones," he said. At about 80 pounds per square inch of working steam pressure, it generates

(continued)

◆ *Source: Seattle Post-Intelligencer,* January 29, 1991. Reprinted with permission from the *Seattle Post-Intelligencer.*

"about the same amount of power as a half-horsepower electric motor," he said.

The engine has built-in efficiencies. Even the welded pipe frame helps condense the used steam and feed it back into the system. Stoking the boiler every 40 minutes or so keeps the little engine humming at about 1,800 power strokes per minute.

It was even more efficient when it has an insulating jacket, but Lambert removed it so the little engine would also heat his workshop. It was a trade-off in efficiencies.

"If someone asked Boeing to make one of these, they'd need at least $100,000. They'd have their engineers draw up plans and then turn it over to someone like me to actually build it." Lambert said. "I built it for less than $100, if you don't count my labor time.

"But when the small guy wants to develop something like this, no one will give him the time of day. I tried getting some people from Washington State University to look at this engine, and they couldn't have been less interested. I figured that with all the slash timber being burned there was a terrible waste of energy that someone could turn into something good."

Lambert has already proved that.

4. Many large engineering design projects require the engineer to interact with many different kinds of people. Consider the project of designing an underground mass transit system in a large city. List ten different types of personnel required to work on this project and identify the specific skills they would bring to the project.

5. Consider a feasibility study for a new sports arena in a downtown metropolitan area. List the types of personnel required to conduct this study and the specific skills they would bring to the project.

6. Generally there is no single "correct" answer to engineering design problems. Each of the alternatives may be "best" from a different perspective. Consider the following set of figures and the problem of selecting the one that is different from all the others. Show that there are five possible "correct answers," depending on your perspective.

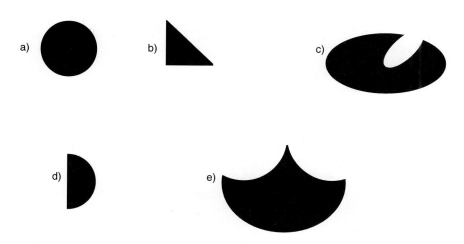

♦ *Source:* von Oech, pp. 22–24. Reprinted with permission from Warner Books.

7. Compare the nine-step model of the design process to the ABET definition of design. Can you identify portions of the ABET definition that correspond to each one of the nine steps?

8. What is the relationship between each of the steps in the following model of the design process and the nine steps presented in this chapter?
   1. Define the Problem
   2. Consider Alternative Solutions
   3. Refine Ideas
   4. Analyze Ideas
   5. Decision
   6. Implementation
◆ *Source:* Walton, pp. 80–83.

9. What is the relationship between each of the steps in the following model of the design process and the nine step model discussed in this chapter?
   1. Problem Definition
   2. Plan Approach
   3. Allocate Resources
   4. Model and Analyze
   5. Design and Evaluate Alternatives
   6. Select Preferred Alternative
◆ *Source:* Meredith, pp. 13–15.

10. What is the relationship between each of the steps in the following model of the design process and the nine steps discussed in this chapter?
    1. Problem Definition
    2. Problem Evaluation
    3. Synthesis
    4. Analysis
    5. Communication for Manufacture
◆ *Source:* Middendorf, pp. 6–10.

11. What is the relationship between each of the steps in the following model of the design process and the nine steps discussed in this chapter?
    1. Problem Statement and Needs Formulation
    2. Information Collection
    3. Modeling
    4. Value Statement
    5. Synthesis of Alternatives
    6. Analysis and Testing
    7. Evaluation and Decision
    8. Optimization
    9. Iteration
    10. Communication and Implementation
◆ *Source:* Simon, pp. 4–5.

12. Construct a plausible, seven-level hierarchical system diagram which includes the operating system for personal computers at some intermediate level. Each level (except the highest) of your diagram must have at least two entries.

13. Construct a plausible, seven-level hierarchical system diagram which includes the Mona Lisa at some intermediate level. Each level (except the highest) of your diagram must have at least two entries.

14. Construct a plausible, seven-level hierarchical system diagram which includes the cargo door for a commercial airliner at some intermediate level. Each level (except the highest) of your diagram must have at least two entries.

15. Discuss the aspect of engineering design that is described by the term "controlled convergence."

CHAPTER 2

Problem Formulation

2.1. OVERVIEW

The importance of devoting adequate time to properly formulating design problems cannot be over-emphasized. More design project failures can be traced to inadequate problem formulation than to any other cause. These failures, when they come later in the life cycle, such as during testing or production, can be very expensive. We can significantly reduce the chances of such problems occurring by devoting sufficient time early in the project to ensuring that the problem is properly formulated. For this reason, we dedicate this chapter to closely examining Steps 1 and 2 of the design process. Our focus is to learn how to clarify what we want our design efforts to achieve, without prematurely committing ourselves to particular approaches to achieve those objectives.

The best way to enhance our survival chances in the design swamp is to provide formal, structured frameworks for problem formulation. Think of these frameworks as rafts constructed from tree branches that are initially randomly arranged as they float in the swamp. Arranging these branches in a disciplined fashion provides structure to our thinking. This structure includes clearly separating the need recognition and problem definition aspects of problem formulation.

We also provide several other organizing principles which further assist in structuring our consideration of problem formulation. These frameworks provide tools for clarifying various elements of design projects, and help us navigate one of the most challenging regions of the design swamp. We also sort out the tangled swamp grass of conflicting terminology associated with design problem formulation.

2.2. RECOGNIZING THE NEED

The ABET definition of engineering design in Section 1.4 states that any engineering design must satisfy "desired needs." Therefore, a key step in the design process is identifying a client's unfulfilled need. This is by no means a trivial task. It deserves the special attention we've given it by devoting one of the nine steps in the design process model to identifying the need.

The identification of a client's need implies that (a) we know who the client is; and (b) the client knows what their needs are. Neither of these assumptions are necessarily true when we first tackle a design problem. We first turn our attention to the issue of identifying the client.

Identifying the Client

Your client is the individual or group whose needs you are responsible for satisfying. This may be your immediate supervisor, your company's sales department, an individual customer, the local community that will use or be affected by your design, or society at large. More frequently than not, the engineer on a given project has several clients, each of whom has needs that may differ from those of other clients. This creates a tension since it may not be possible for a design to satisfy everyone to the same extent; and you may be confronted with conflicting loyalties and obligations. Some of these conflicts are addressed in detail in Chapter 4. In the following discussion, we will assume that, in the case of multiple clients, we have struck an appropriate balance among them.

Interacting with the Client

The client's need defines the purpose for the design. When design engineers focus too narrowly on specific design tasks and lose sight of the client's overall need, they can waste much time, money, and effort on design activities that do not satisfy, or worse, which conflict with the need.

Keeping focused on the need is not necessarily an easy task for the design engineer. Frequently clients may not clearly articulate the need; they might even be unaware of the real need. They may pose the problem too narrowly to the design engineer, describing what they perceive as a solution to their need, rather than the need itself. This can unnecessarily discourage the engineer from conceiving alternative approaches that may be superior to the solution that the client envisions. Further, since clients are real people, they are not necessarily consistent; their perception of the need may change with time. In some cases, insight into the need comes only after other design activities have occurred. Sometimes there are circumstances beyond the client's control, like budget cuts, that can force the need to be reexamined.

Engineers should work closely with their clients to clearly identify the actual need as early as possible in the design process. If a client states, "I need *XYZ*," the engineer should ask "why?" until the client identifies the *reason* they need XYZ. This exchange will help the client and engineer mutually understand the need for the project; then the engineer can design a solution.

With respect to the conversation between Sandra and Jane that we overheard in Section 1.5, it's also possible that the response to Jane's question of 'why?' could proceed along these lines: "Jane, here you go again with your 'why' questions! I'm getting tired of it. I just came out of a three-hour meeting with the chief engineer where we explored all the alternatives, and we concluded that what we really need is a stronger bumper. I don't have time to justify that decision to you. Are you willing and ready to take on this assignment or not?"

As this little drama illustrates, the nature of the dialogue with which the engineer engages his/her client has to be sensitive to the relationship between the individuals. Don't get your client mad at you because your questions are annoying or offensive. Great skill may be required by the engineer to establish the atmosphere where an appropriate exploration of needs can be pursued. Yes, interpersonal relationships are an important element of design engineering (see Fig. 2–1). The stereotype that engineering is for inarticulate nerds is way off base!

Format for Needs Statement

The term "need" comes from the ABET definition of engineering design, so it has a meaning in the context of problem formulation that is different from its meaning in ordinary conversation. In particular, recall that when she started the dialogue with, "... we *need* ... a stronger bumper ... ," Sandra was focusing on a particular solution. The fact that the client uses the word 'need' in conversation does not mean that the client has properly articulated the *need* in the sense of the formal problem formulation framework discussed in this chapter. As design engineers, we have to be alert and avoid falling into the semantic trap of confusing the meaning of terms used in lay conversation with their specialized meanings in the context of formulating design problems.

Perhaps the most useful device for distinguishing between a need and a goal (see Sec. 2.3 for a detailed discussion of goals) is to articulate the need as an expression of dissatisfaction with the current situation. This negative emphasis and focus on the present situation contrasts with the positive, future orientation of a goal.

DILBERT® by Scott Adams

Figure 2–1.   The Importance of People Skills *(Reprinted with permission from United Media.)*

Market Assessment

We don't want to leave this discussion with the impression that all design problems originate with a client. It is not unusual for the client to play a very passive role and not be aware of the need until it is brought to their attention by the design engineer or by a marketing campaign. Many creative individual engineers conceive innovative technologies without a clearly defined market.

An example of this phenomenon is the development of the personal computer in the mid-1970s. At that time, all serious computing was conducted on mainframe computers. There was no active or foreseen need for a stand-alone desktop computer. A group of young hobbyists developed the first personal computers, not because they saw the enormous business potential, but strictly for the fun of it! The concept was so revolutionary that it took some time for both the developers and users to realize the nature and magnitude of the latent need that could be satisfied by personal computers.

Sometimes engineers accidentally stumble upon a revolutionary design while pursuing an unrelated project. One such example is the Post-It™ adhesive notepad. The easy-to-peel-off adhesive that makes this stationery product so popular was the result of a failed experiment. No one was looking for a new stationery product, and it was only after it was developed that the need was clearly identified from market research. Of course, there are also many contrary examples of clever new designs that were unsuccessful because the need never did develop.

In many situations, innovative engineering design can be successful in a competitive market only if the technology is complemented by skillful marketing techniques. A classic example of this phenomenon is the competition since the mid-1980s between the Macintosh and IBM clone for control of the personal computer market. It is virtually universally agreed that the Macintosh has been technologically superior, but it has been confined to a marginal role in the market because of smarter marketing decisions by Microsoft and the computer hardware manufacturers.

2.3 DEFINING THE PROBLEM

After clearly defining the need, we turn our attention to describing how we envision meeting that need. This step requires achieving a delicate balance between establishing the general scope of our design efforts, and avoiding being so specific that we unnecessarily narrow the opportunities for creative design solutions. To achieve that balance, we will use a structured framework to help us discipline our thinking and organize our approach to defining the problem. In particular, we will construct a problem statement that consists of three components: goals, objectives, and constraints. Let's examine each of these three components separately

Goals

A goals statement is a brief, general, and ideal response to the needs statement. We can think of it as the answer to the question, "How are we going to satisfy this need?"

The goals statement is usually so ideal that it could never be accomplished or so general that it would be difficult to decide when it was achieved. However, it does establish a general theme or direction for the design mission.

To illustrate what a goals statement might consist of, and how it differs from a needs statement, let us continue with the example of Jane's assignment to design an automobile bumper. For convenient reference, we'll repeat here the already agreed-upon needs statement from page 9: *There is too much damage to bumpers in low-speed collisions.*

As part of defining the problem in Step 2, let's say that Jane and Sandra agreed that a reasonable goal would be to *design an improved automobile bumper*. Note that this is very general; it allows for both the stronger bumper and the more flexible bumper mentioned in Jane and Sandra's initial conversation. It is also pretty vague (how do we measure 'improved'?). On the other hand, it does rule out other approaches to meeting the need, such as designing a radar-based collision avoidance system, or reducing the amount of automobile use. This choice for the goal strikes a good balance between establishing a general design approach while preserving many design options. A good analogy to consider is that the goal is like the first big branch coming off a tree trunk; it sets you on a path along which there are many smaller branches, each one of which represents specific design options. However, in doing so, it eliminates alternate goals and their associated design options (Fig. 2–2).

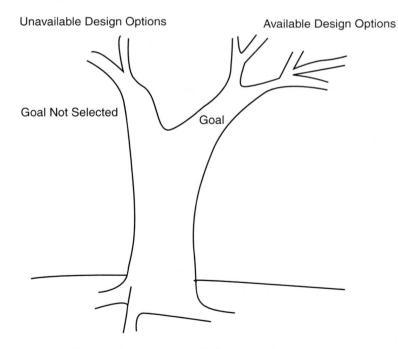

Figure 2–2. Goals and Their Associated Design Options

It is vital that we appreciate the difference between the need and the goal. The need describes the current, unsatisfactory situation; the goal describes the ideal future condition to which we aspire in order to improve on the situation described by the need. It may take several iterations before settling on a set of needs and goals statements that both you and your client are comfortable with. It is a good idea to keep them as simple as possible.

Objectives

The second component of a problem statement are the objectives. The objectives (there may be more than one) are quantifiable expectations of performance. They identify those performance characteristics of a design that are of most interest to the client, and they describe those characteristics in a manner so you and the client can decide how well a design meets the client's expectations. The objectives should also include a description of the conditions under which a design must perform. Specifying the operating conditions will allow us to evaluate the performance of different design options under comparable conditions.

Returning to Jane's current job assignment, let's say that she and Sandra agree (or alternately, that Sandra tells Jane) that the objectives for this project are as follows: *Design an inexpensive front bumper so the car can withstand a 5 mph head-on collision with a fixed concrete wall without significantly damaging the bumper or other parts of the car, or making the car inoperative. In addition, at the end of the useful life of the bumper, the bumper must be easily recyclable.*

We are really making progress in defining this problem. We now know that we are interested only in a front bumper and its behavior in a specific kind of collision. We also know that the client is particularly interested in low initial cost, low allowable damage level, and no impairment of drivability. The disposition of the bumper at the end of its useful life is also an important concern. We can envision how each of these objectives can be quantified.

As shown in Figure 2–3, an unambiguous description of the operating environment is necessary to clearly define the objectives and to establish the conditions under which all proposed designs can be tested and evaluated. Note that the emphasis of the objectives is on the behavior, or performance, of the design. We still have a lot of flexibility in selecting the material from which the bumper is made, its shape, and how it absorbs the energy from the impact. These features, which are under the control of the design engineer, are called design parameters. The best statement of objectives is one that includes all relevant performance characteristics without referring to any design parameters.

The advantage of separating the goal from the objectives is now clear. At some later step in the design process, Jane may find that it is not possible to satisfy all her client's expectations. Or, new expectations may be recognized. Either of these situations might cause Jane to return to this step and refine the objectives, but she can still retain the more general goal.

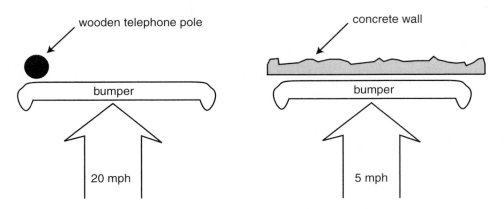

Figure 2–3.   Two Different Operating Environments for an Automobile Bumper

Constraints

The third key component of a problem definition are the constraints. Constraints define the permissible range of the design and performance parameters. They are essentially go/no-go conditions that all designs have to satisfy in order to be eligible for consideration. If a design does not satisfy a constraint, we reject it regardless of how well it performs with respect to one or more of the objectives. This is in contrast to objectives, which serve as the basis for comparing those designs which satisfy the constraints.

Constraints generally take one of two forms: equality constraints or inequality constraints. An equality constraint specifies that some characteristic of the design must take on a specified value, with no leeway for variation from that value. An example of an equality constraint that Jane's design might have to satisfy is as follows: *In order to prevent over-riding bumpers in collisions between automobiles, the federal government requires that all bumpers be installed 18″ up from the ground.* It doesn't make any difference how well a candidate design satisfies the objectives—if it doesn't satisfy this constraint, it is an unacceptable design.

Another type of constraint is a one-sided inequality constraint. This is a constraint that requires some characteristic of the design to be greater than, or less than, some specified value. Jane might be faced with a one-sided inequality constraint of this form: *The weight of the bumper cannot exceed 50 lb.* Note that both a 49 lb. bumper and a 20 lb. bumper satisfy this constraint. And since low weight is not explicitly identified as an objective, both of these bumpers are just as good as far as the problem statement is concerned. However, if in the process of considering this constraint, Jane thinks that a lighter bumper should be considered a *better* design than a heavier bumper, she can approach Sandra with the idea and propose that the objectives statement be modified accordingly. This is a good example of the interaction between different elements of a problem statement and how consideration of one element may lead to modifications in another.

The final kind of constraint we will consider is a double-sided inequality constraint. This constraint requires that a design or performance characteristic be greater than one specified value and less than another specified value. For example, Sandra might give Jane this constraint: *"The mounting brackets on the bumper must be between 8.0″ and 12.5″ from the center so they match with the brackets attached to the automobile frame.*

Recapitulation

We now have completely formulated the automobile bumper problem by specifying the need and specifying a three-part problem statement consisting of a goal, objectives, and constraints. Consider the summary of these components:

NEED

There is too much damage to bumpers in low-speed collisions.

PROBLEM DEFINITION

Goal: Design an improved automobile bumper.

Objectives: Design an inexpensive front bumper so the car can withstand a 5 mph head-on collision with a fixed concrete wall without significantly damaging the bumper or other parts of the car, or making the car inoperative. In addition, at the end of the useful life of the bumper, the bumper must be easily recyclable.

Constraints: In order to prevent over-riding bumpers in collisions between automobiles, the federal government requires that all bumpers be installed 18″ up from the ground.

The weight of the bumper cannot exceed 50 lb.

The mounting brackets on the bumper must be between 8.0″ and 12.5″ from the center so they match with the brackets attached to the automobile frame.

The keys to a successful problem formulation exercise are: distinguishing between the need and the problem definition; composing an appropriately abstract goal; listing unambiguous, performance-oriented objectives; and including all relevant constraints. There generally are opportunities to exchange objectives with constraints. For Jane's project, the objective of 'not causing significant damage to ... the car' could be recast as a one-sided inequality constraint that 'the maximum permissible damage is $200.' On the other hand, a constraint that 'the time for removing and replacing a damaged bumper must be less than 30 minutes' could be transformed into an objective of 'reducing the time for removing and replacing a damaged bumper.'

Since the particular formulation selected can affect the type of analysis conducted and the final form of the design (see Sec. 10.8 for an example of this), it is important to take care when specifying the objectives and constraints. In summary, the objectives establish the parameters by which alternate design concepts can be compared, without specifying allowable or required values of those parameters. On the other hand, if a design does not satisfy a constraint, it is not acceptable.

Finally, consider this cautionary note on semantics. As discussed at the end of Section 2.2, the client may not be familiar with the specialized terminology we have adopted for the various aspects of a problem formulation exercise. So when Sandra tells Jane that, "We have a problem with the amount of damage suffered by our automobile bumpers," Jane recognizes that Sandra is using the term 'problem' in a conversational way, not in the formal way it is used as part of this phase of design.

2.4. ALTERNATIVE TERMINOLOGY

In both Sections 2.2 and 2.3 we encountered conflicts with the way the words 'need' and 'problem' are used in conversational English and the way they are used within a formal problem formulation framework. It turns out that this is just the tip of the iceberg (OK, so swamps don't have icebergs . . .), and we should clarify our terminology before we go any further. Key words we have used so far are: need, goals, objectives, and constraints. Other terms used later in this book or by other authors in their discussions of engineering design problem formulation include: characteristics, attributes, criteria, requirements, functions, specifications, design parameters, performance parameters, and variables. Unfortunately, different writers on this subject use different words to describe the same thing. Also, the same word may have different meanings when used by different authors. This situation arises because different terminologies for problem formulation have historically evolved out of different contexts, and those terms conflict with each other when they are used in a new context. We would dearly prefer to be 100% consistent within this book, but historical usage of certain terms makes that difficult.

This situation is not all that rare in the English language, which is full of words with double meanings and multiple definitions. As one example, consider the word 'lead.' Its meaning and its pronunciation can only be determined from the context in which the word is used. Or, perhaps more relevant to the material in this chapter, is the word 'objective.' One meaning is 'unbiased,' a second is 'a target.' Again, we have to rely on the context to indicate which meaning is intended.

Recall the discussion in Chapter 1 of the lack of consensus regarding the definition of engineering design and on a model of the engineering design process. Under these circumstances, the lack of consistent terminology for the problem formulation aspect of design is not very surprising. There is no way we can enforce uniformity (unless you want to propose that a profession-wide standard terminology be adopted; see Chapters 3 and 4 for more information on standards and the standards making process). Until that happens (don't hold your breath) the best we can do is to try to use consistent terminology within this book, while recognizing that the same word used in a similar context by another author, or another engineer who studied design under another faculty member, may mean something different.

The lesson to be learned from this is that every time you find yourself in a new environment for formulating design problems, it will behoove you to ensure that everyone involved is speaking the same language. The specific terminology adopted is not nearly as important as making sure that everyone involved understands the terminology as it is being used in that environment. For convenience and completeness we'll start our terminology list by repeating the definitions of some of these words from Sections 2.2 and 2.3.

Need

This describes a current situation that is unsatisfactory. By doing so, the need thus establishes improvement in that current situation as the ultimate purpose of the project.

Goals

This is one of three components of a problem statement. It is a brief, general, and ideal response to the need. A goal can frequently be thought of as answering the question, "How are we going to address this need?" The goal is usually so ideal that it could never be accomplished, or so general that it would be difficult to determine when it was achieved. However, its selection establishes the general direction of the design effort.

Objectives

Objectives are quantifiable expectations of performance. They are the second component of a problem statement and serve as indicators of progress toward achieving the goal. Objectives define the performance characteristics of the design that are of most interest to the client, and describe those characteristics in a manner so you and the client can determine which of the alternative designs best meets his/her expectations. For example, one of the objectives for the automobile bumper project is: *keep the cost of the bumper low*.

Constraints

Constraints are the third component of a problem statement. They define the permissible range of the design and performance parameters. For example, one of the constraints on the new automobile bumper design is: *the bumper's weight cannot exceed 50 lb*.

Attributes

Generally, an attribute is any characteristic of a design or system. Attributes are not necessarily associated with either the need or any of the components of a problem statement. For example, some attributes of an automobile bumper are cost, color, thickness, and expected lifetime. According to our formulation of the bumper problem in this chapter, the cost attribute is associated with an objective but the other three of these attributes are not. Clearly, we could reformulate the problem to include objectives that incorporate these other attributes, if that was appropriate in terms of the need and goal. In certain other contexts, the term 'attribute' has a different meaning. For example, in Chapter 9, the term is embedded in a particular decision making methodology called Multiple Attribute Utility theory. In that context, attribute is a synonym for criterion. Also, in the context of discussing the Quality Function Deployment method (see Sec. 2.6) Cross[1] distinguishes between attributes and characteristics.

Criteria

Criteria are attributes of designs that are the basis for deciding among design options. They differ from objectives because objectives describe what we expect from the criteria. For example, a criterion for the automobile bumper is: *cost*. An objective involving this criterion is: *keep the cost of the bumper low*.

Characteristics

Generally, the term characteristics is a synonym for attributes. However, in discussing Quality Function Deployment (QFD), Cross[2] uses 'characteristics' to refer to a system's engineering or physical properties, while reserving 'attributes' to refer to those aspects of a product that are important to the client or customer. With respect to our discussion of QFD in Section 2.6, Cross's 'attributes' is a synonym for our 'customer requirements,' and his 'characteristics' is a synonym for our 'engineering requirements.'

Functions

Functions refer to services that must be provided by the design or its components. What must the design do? Functional analysis is a problem formulation framework particularly well-suited for dealing with complex engineering systems; it is discussed in Section 2.5. A particular function may not be explicitly related to any of the objectives, but may be necessary in order to achieve an objective. For example, the function of the support bracket of an automobile bumper is to allow the bumper to be attached to the chassis.

Specifications

With respect to problem formulation, this term generally refers to performance specifications.

Performance Specifications

Generally, performance specifications describe our expectations for the performance of a design. In this context, the term is a synonym for 'objectives.'

[1]Cross, p. 91.
[2]Ibid.

Design Specifications

This term usually refers to the detailed description of the completed design, including all dimensions, material properties, and fabrication instructions. It usually is not used in the problem formulation context; we include it here to distinguish it from performance specifications.

Customer Requirements

This term, as used in the QFD technique, refers to objectives as articulated by the customer or client.

Engineering Requirements

Also used in QFD, this refers to the design and performance parameters that can contribute to achieving the customer requirements.

Design Parameters

These are the characteristics of the design that the engineer can directly manipulate in order to achieve the objectives or satisfy the constraints. For example, if the objective is to reduce the weight, a design parameter that is available to the engineer is the density of the material. By selecting a material with a lower density (e.g., aluminum vs. steel) the objective can be realized. Similarly, if there was a constraint on how much the object could weigh, the same design parameter could be used to satisfy that constraint.

The engineer also may be able to reduce the weight of the design by reducing one or more dimensions of the system, in which case each of those dimensions is also a design parameter. Sometimes the relationships between the objectives and the design parameters are readily quantifiable. For a very simple example, let's say you are designing a rectangular box and one of the objectives is low weight. The four design parameters are length, width, and height of the box and the density of the material from which the box is made. The relation between the weight and the design parameters is

$$W = \rho l w h$$

Performance Parameters

Performance parameters are indicators of the performance of the object or system being designed. Some of the performance parameters may be synonymous with objectives, or serve as indicators of the extent to which the objectives are being achieved. In many situations, the performance parameter can be affected only indirectly by the engineer's design decisions. Consider the automobile bumper design project. One of the objectives is to withstand a collision without incurring "significant damage." A performance parameter that might serve as an indicator of "significant damage" is the cost of repairs. But the engineer has no direct control over repair costs—costs may be strongly affected by labor charges and the cost of replacement parts. However, the design parameters that the engineer does have control over may have a strong effect on cost.

2.5. FUNCTIONAL ANALYSIS

Establishing the need and problem statements may not sufficiently formulate the design problem for large, complex systems. In that case we also need to describe the tasks or 'functions' to be performed by the engineered system and its components. The term 'functional analysis' is widely used to indicate the process for identifying and describing these functions as part of formulating design problems.

Functional analysis is best accomplished with block diagrams of the system. These diagrams are similar in format to the system block diagrams represented in Figures 1–7, 1–8, and 1–9 but with two major differences. First, the system block diagrams in Chapter 1 show how the system components and subsystems relate to each other and to the overall system: they are an organizational or structural representation of the system. Here, the diagrams will represent functions and subfunctions; they are a process representation of the system. Second, in Chapter 1, there was no implied direction or flow to the system block diagrams. In this section, the box in the system diagram has either an input, an output, or both to represent the direction of the process flow.

The simplest system flow diagram is shown in Figure 2–4; the system transforms inputs into outputs. The function analysis approach to formulating the design problem is to describe each of the inputs and each of the outputs.

A more sophisticated formulation shown in Figure 2–5 divides both inputs and outputs into three categories: energy, materials, and information. Many systems have inputs and outputs in all three forms. Sometimes one of the forms is associated with the primary function of the system, and the other inputs are needed for auxiliary functions. For example, an automobile engine is primarily a system for converting the chemical energy in gasoline to the kinetic energy of a rotating drive shaft. A bridge is a system in which the input materials (concrete, steel, and asphalt) are assembled into a structure that permits people and goods to cross a river. A fuel gauge is a system for converting one form of information (the level of fluid in a tank) into another (the po-

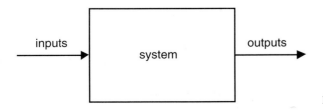

Figure 2–4.   System Flow Diagram

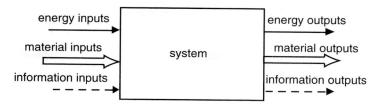

Figure 2–5. Energy, Material, and Information System Flow Diagram *(Adapted from Pahl and Beitz, p. 30, with permission from Springer-Verlag.)*

sition of a needle on a dial). The key characteristic of each of the three descriptions is that we focused on what the system does, without specifying how it does it. That is the crux of the functional analysis approach to problem formulation.

Subfunctions

The distinction between a function and a particular approach to performing that function depends on the level of detail of the system description. Consider the problem of designing an electric power generation plant. At the most general level, the function of the power plant is to convert nonelectric energy into electricity. The first law of thermodynamics requires that the conversion losses plus the amount of electricity generated is equal to the input energy, as shown in Figure 2–6a. While we have not yet addressed the issue of how that conversion occurs, our formulation of the problem has eliminated from consideration cogeneration systems in which some of the output is in the form of useful thermal energy. The system model for a cogeneration power plant would look like Figure 2–6b. This illustrates the value of the functional analysis approach to help define the problem at the very early stages.

  Sticking with the conventional power plant, we can examine the system function more closely by dividing the function into subfunctions. One such dissection, involving the subfunctions of generating the steam, spinning the turbine, driving the generator, and condensing the steam, is shown in Figure 2–7. Figure 2–7 describes how the conventional power plant depicted in Figure 2–6 generates electricity. At that level, it

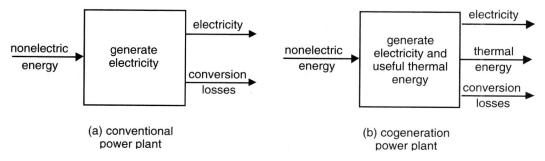

(a) conventional
power plant

(b) cogeneration
power plant

Figure 2–6. Functional Analysis System Diagrams for Conventional and Cogeneration Power Plants

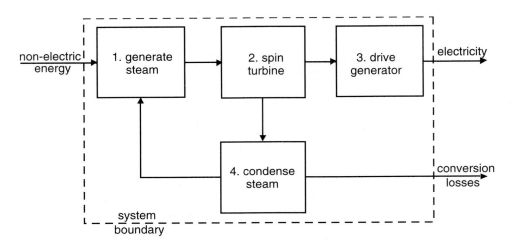

Figure 2–7.   Subfunction Diagram for a Steam Turbine Power Plant

represents the design decision to use a steam turbine, rather than a gas turbine, a wind turbine, photovoltaics, or hydroelectricity. This is the counterpart of the design decision made at the previous level to not use cogeneration technology. However, we have not said anything yet about how each of the four subfunctions in Figure 2–7 will be carried out. To do that, we can further subdivide each of the subfunctions. For example, one approach to using coal to generate steam is depicted in Figure 2–8.

The 1.x numbering system in Figure 2–8 is a reminder that each of the subfunctions in that diagram is, in turn, a subfunction of subfunction 1 in Figure 2–7. If you are considering the alternative of using natural gas rather than coal as the fuel, you would draw a slightly different diagram. Similar subfunction diagrams can be constructed for the other three subfunctions in Figure 2–7. In addition, any of the subfunctions in Figure 2–8 can be subdivided into its component subfunctions.

As with the prior diagrams, we continue to focus on the functions, as opposed to how those functions are accomplished. For example, in Figure 2–8 we don't specify how the coal is pulverized, or how the coal is burned, or how the hot gasses transfer their energy to the circulating water. Those decisions, involving design of pulverizers, burners, and heat exchangers, are rightfully reserved for subsequent stages in the design process.

2.6. QUALITY FUNCTION DEPLOYMENT

Quality Function Deployment (QFD) is a technique for identifying customer requirements and matching them with engineering design and performance parameters. QFD was first popularized in Japan as part of the customer-driven approach to product design. In fact, the term comes from a Japanese phrase that means "... the strategic arrangement (deployment) throughout all aspects of a product (functions) of appro-

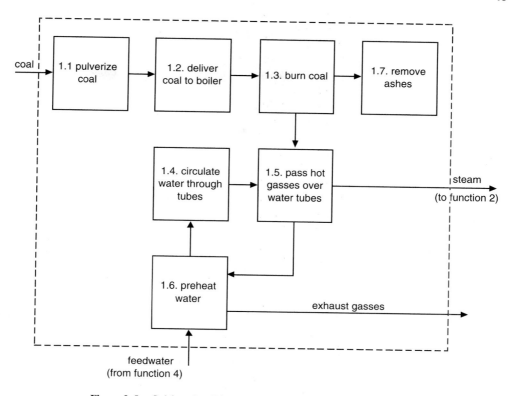

Figure 2–8.   Subfunction Diagram for Steam Generation (Function 1)

priate characteristics (qualities) according to customer demands."[3] It is an especially useful tool for formulating design problems for products in situations where several competing products are already on the market.

Constructing a QFD Table

The key element of a QFD exercise is constructing a chart that explicitly depicts the key relationships between customer requirements, engineering (or product) requirements, and the characteristics of competing products. The general arrangement of a QFD table (shown in Fig. 2–9) consists of the following five regions:

1. Customer requirements
2. Engineering requirements
3. Matrix of requirements relations
4. Competitive benchmarks
5. Engineering targets

[3]Cross, p. 92.

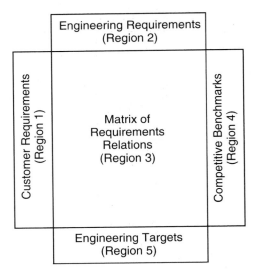

Figure 2–9. Structure of Quality Function Deployment Chart

We will examine each of these regions in the context of the automobile bumper design project. Then we will discuss some variations in the format of QFD tables.

Customer Requirements

Region 1 in Figure 2–9 contains a listing of customer requirements. This list consists of the features or characteristics of the design that the customer indicates are relevant. The list may be obtained from direct conversation with the customer, from observations of customer usage patterns with similar products, from customer response to a survey, or from a source such as Consumer's Reports that lists product features that its readers are interested in. An important aspect of Region 1 is that the list, to the extent possible, is in the customer's own words, and has not been "filtered" or otherwise interpreted by engineers or members of the marketing staff. An example of characteristics of an automobile bumper that might be articulated by the customer are displayed in the Region 1 part of Figure 2–10.

Engineering Requirements

Region 2 in Figure 2–9 contains the list of engineering requirements. This list is generated by the engineering staff and contains those quantifiable aspects of the system that can contribute to satisfying the customer requirements listed in Region 1. The system has not been designed yet so the engineering requirements may be a mixture of performance parameters and design parameters. The idea at this point (as it is when developing the list of customer requirements for Region 1) is to include as many requirements that you can think of. Some of the customer requirements articulated by

Customer Requirements	Engineering Requirements: yield strength	Young's modulus	mounting hole separation	plating thickness	effective spring constant	cross-section moment of inertia	weight	max. deflection	cost	Bench marks: Competitor A	Competitor B
looks good				x							o
holds license plate			x							o	
resists dents	x	x								o	
protects lights	x	x			x	x		x			o
doesn't rust				x						o	
lasts a long time				x						o	
inexpensive	x			x			x		x	o	
protects fender/hood	x	x			x	x		x			o
Units	psi	psi	in.	in.	lb/in.	in.^{4}	lb	in.	$		
Engineering Targets				0.05			50		100		

Figure 2–10. QFD Chart for Automobile Bumper

a technically sophisticated client might already be in the form of engineering requirements. No problem—include such items both in Region 1 and in Region 2. We have listed nine engineering requirements for the automobile bumper in Region 2 of Figure 2–10.

Matrix of Requirements Relations

We can think of Region 3 of a QFD table as a matrix with rows consisting of the customer requirements and columns consisting of the engineering requirements. Each relationship between an engineering requirement and a customer requirement is indicated by an "x" mark in the appropriate cell in the matrix. For example, the yield strength of the bumper material affects the ability of the bumper to resist dents and to protect the headlights, fenders, and hood. Also, high yield strength may adversely affect the requirement that the bumper be inexpensive. These relationships are reflected by the corresponding "x" marks in Region 3 of Figure 2–10.

Benchmarking

A key feature of the QFD approach is the opportunity to explicitly compare your design to that of competitors. This is called 'benchmarking' and is done in Region 4. The competitor that does the best job of satisfying each customer requirement is identified by a mark in the appropriate box. For the sake of this introductory application of QFD to the design of an automobile bumper, we assume that there are only two competitive bumper designs, A and B. We use an "o" as an indicating mark in Region 4 to provide visual separation from the "x" marks in Region 3. As depicted in Figure 2–10, competitor bumper A sets the pace for five of the customer requirements, while competitor bumper B is superior for the other three customer requirements. While we selected two bumpers for the purposes of benchmarking, we could have used a wider class of low-speed collision protection/avoidance systems for this purpose. The choice of systems to use for benchmarking has to be made in the context of earlier discussions that defined the need to be addressed (see Sec. 2.2) and the scope of the design project.

Engineering Targets

We list the units associated with each of the engineering requirements in Region 5, along with numerical values of the targets we establish for those requirements. The target for each engineering requirement may be the value that the requirement must achieve in order to compete with the alternative products that are used as the benchmarks. As shown in Figure 2–10, we establish the target for bumper cost to be less than $100 in order to match the cost benchmark established by competitive bumper A. However, it usually is not necessary that every engineering requirement achieve a target that is superior to the best of the competing products. A more holistic (and realistic) approach is to identify a set of targets involving a key combination of engineering requirements.

Variations to QFD Tables

Some authors suggest using numerical values in Regions 3 and 4 instead of just the identifying marks that we use in Figure 2–10. Such numerical values go beyond identifying the relationships between the requirements; they quantify the strength of those relationships. This helps to identify which engineering requirement has the biggest effect on each customer requirement, and thus helps establish priorities for the design activities. Also, quantitative benchmarks in Region 4 are useful guides for establishing the magnitudes of the engineering targets in Region 5.

Finally, some authors use an additional region next to Region 1 for weighting factors that reflect the relative importance the customer places on the various requirements. We postpone discussion of weighting factors until Chapter 9, where they are discussed in considerable detail.

House of Quality

One variation of the QFD chart is the addition of a triangular region at the top of Region 2. This region is used to keep track of relationships between the engineering parameters. The roof-type appearance of this region, on top of the rectangular base regions of the chart, gives rise to the term House of Quality for the chart shown in Figure 2–11. The "+" signs in the three locations of the triangular region in Figure 2–11 indicate that there is a positive correlation between yield strength and cost, between plating thickness and weight, and between plating thickness and cost. The three

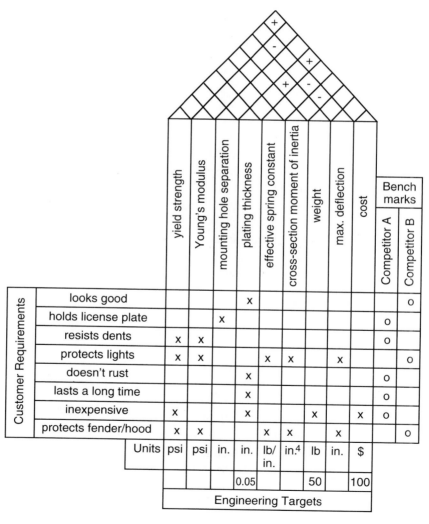

Figure 2–11.   House of Quality for Automobile Bumper

"-" signs indicates that the maximum deflection of the bumper decreases if Young's modulus, the effective spring constant, and the cross-sectional moment of inertia increase. Depicting these relationships on a House of Quality chart helps to identify the reinforcing effects as well as the opportunities for trade-offs between different engineering requirements.

2.7. DEVELOPING DESIGN CRITERIA

It may not always be obvious what the design objectives or criteria should be for a given project. Sometimes it isn't until several design options have been identified that we develop a clear understanding of the criteria. Consider the problem of designing a system to transmit power between parallel rotating shafts. The three alternatives under consideration are belts, chain drives, and gears. One approach to developing criteria is to list all the advantages you can think of for each option. Such a list is given below.[4]

Advantages of Belts

b1. Electrical insulation is provided because there is no metal to metal contact between driver and driven units
b2. Less noise than chain drive
b3. Can be used for extremely long center distances where chain weight would be excessive
b4. Can be used at extremely high speeds where chain inertia must be considered as influencing chain fit at the sprocket and chain tension
b5. No lubrication required
b6. Shaft center distance variation and shaft alignment is much less critical than for gear or chain drives

Advantages of Chains

c1. Variations in distance between shaft centers can be more easily accommodated than with gear drives
c2. Easier to install and replace than belts because the center distance between driver and driven units need not be reduced for installation (splice and link belts overcome this objection at the cost of lower power ratings)
c3. Require no tension on the slack side, so loads on bearings are reduced
c4. Do not slip or creep as do belts (except for toothed belts)
c5. More compact because sprocket diameters are smaller and chains are narrower than sheaves and belts for the same power transfer

[4]Adapted from Orthwein, p. 711 (© 1990), with permission from PWS Publishing Co., Boston, a division of International Thomson Publishing, Inc.

c6. Do not develop static charges
c7. Do not deteriorate with age, heat, oil, or grease
c8. Can operate at higher temperatures than belts

Advantages of Gears

g1. More compact because center distances are minimum
g2. Greater speed capability
g3. Greater range of speed ratios than chains
g4. Can better transfer high power at high speeds
g5. Do not deteriorate with age, heat, oil, or grease
g6. Do not develop static electric charges

We now transform each of these twenty items into a criterion and consolidate those items that are redundant. As a result, we end up with the following list of fifteen criteria. The statements from the above three groups upon which each criteria is based are indicated in parentheses.

- Shock protection (b1, c6, g6)
- Noise (b2)
- Large separation distance (b3)
- High speed capability (b4, g2)
- Lubrication requirement (b5)
- Misalignment (b6)
- Separation distance flexibility (c1)
- Installation/replacement ease (c2)
- Bearing loads (c3)
- Slippage/creep (c4)
- Size (c5, g1)
- Life expectancy (c7, g5)
- Operating temperature (c8)
- Speed flexibility (g3)
- High torque capability (g4)

Clearly this list of fifteen criteria is not unique, as the original list of twenty statements was not unique. For many purposes, this list meets our needs and no further refinement is needed. However, several refinements may be desirable to help clarify formulation of the problem. First, the one-, two-, and three-word descriptors of each criterion may be inadequate to unambiguously convey the objectives. Hence, we can add additional narrative to each descriptor. Second, we may gain additional insight into the problem by looking for patterns among the different items on this list. In particular, are there any natural groupings of any subsets of the fifteen criteria? Both of these techniques are discussed below.

Unambiguous Criteria

An important part of formulating design problems is getting the client and all members of the design team to agree on the need and each component of the problem definition statement. This is a non-trivial communication challenge (Step 8) but it is vital to ensure that we are all working on the same problem. It would be counterproductive, frustrating, and professionally embarrassing, to launch a major design effort only to find out halfway through that there was a big misunderstanding resulting from inadequate attention to defining the criteria.

Consider the criterion of life expectancy in the list we created. Are we talking about total life or operating life? That distinction may be important if there are very long idle periods between use and exposure to a harsh environment (heat, dust, or corrosion) during those periods. The equipment may be portable and spend as much time in transit as it does in use, and the loads it receives in transit may be different, but just as severe as, operating loads. If we are talking about operating lifetime, under what conditions should the lifetime be measured or calculated? Lifetimes during steady-speed operating conditions could be substantially different than during operating cycles that include impulsive loadings or many rapid start/stop operations with high acceleration/deceleration.

To remove ambiguity, you should develop longer, more precise definitions of those criteria whose short definitions are not crystal clear. These definitions should be documented so that all parties involved have an opportunity to suggest improvements and to ultimately agree on the final versions. In many cases, you might feel that this process is taking too long and you may be tempted to move on to the more exciting activity of actually designing the system. But the fact that it is taking so long may be an important clue that there are ambiguities that have not yet been resolved. Be patient. . . . extra time spent in this stage will be well worth it.

Objectives Tree

One approach to forming and organizing groups of criteria is to use a graphical tool called an objectives tree.[5] An objectives tree is a graphical display of criteria in a multilevel tree format. At the lowest level in the tree, subsets of criteria are clustered under groups displayed at the next level in the tree. In turn, some of these groups within a given level can also be clustered under more encompassing groups at the next highest level. The fifteen criteria for the shaft power transmission problem have thick borders and bold font in the objectives tree shown in Figure 2–12. In this case we've taken the *Life Expectancy, Lubrication Requirement,* and *Install and Replace Easily* criteria and grouped them together under the heading of *Maintenance.* Similarly,

---

[5]We use the term 'objectives tree' instead of 'criteria tree' to be consistent with the terminology used by several other authors (see Cross, p. 56; Dieter, p. 152, Ertas and Jones, p. 53).

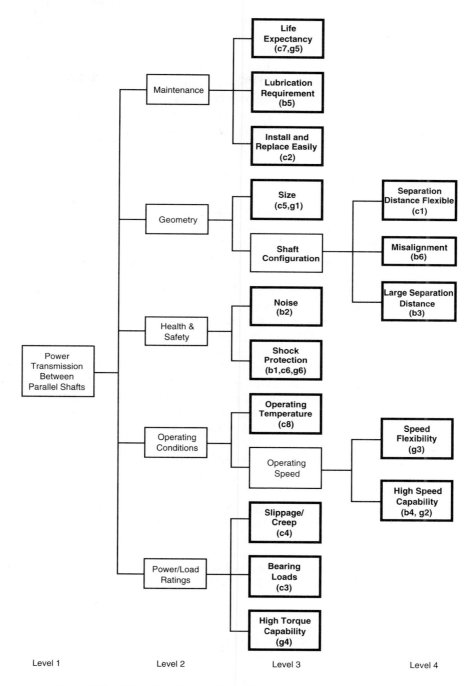

Figure 2–12.   Objectives Tree for Power Transmission Between Parallel Shafts

we've grouped the *Noise and Shock Protection* criteria under a *Health and Safety* group. Wherever it seemed reasonable to do so, we formed these groups of criteria. These groupings can, and in this case do, occur at several different levels. For example, we grouped three criteria under the *Shaft Configuration* group and in turn combined *Shaft Configuration* with the *Size* objective to form another group labeled *Geometry*.

The process of organizing the criteria into groups for inclusion in an objectives tree can help identify redundancies and gaps in a list of criteria. We will see in Chapter 9 how the objectives tree format can also facilitate selecting the best design option from a list of several alternatives.

2.8. CLOSURE

We have examined Steps 1 and 2 in the design process very carefully in this chapter. This attention to the early phases of design is justified by our observation that otherwise this can be a very difficult and frustrating experience. It may not be possible to identify all the objectives and constraints and to express them in an unambiguous manner at the first attempt. Additional information may be required (Step 3 in the design process), or analysis may have to be conducted (Step 5 in the design process) before the problem formulation can be finalized. Also, efforts to define the problem may cause you to re-examine the needs statement (Step 1). This is what we mean by the iterative nature of design: the need to revisit other steps as the result of insights gained in the step currently being pursued.

We have discussed several techniques for organizing the problem formulation process. Think of these techniques as tools in a tool chest. Use the ones that best fit the task at hand. In some cases, a combination of several of the tools should be used. An important key for success is to separate the problem formulation effort from the concept generation aspect of design (Step 5—covered in chap. 6). Another rule to follow is to separate the need (what's wrong with the current situation) from the problem definition (how are we going to address the need?).

Inexperienced design engineers will frequently run into trouble because they prematurely jump into other steps in the design process without devoting sufficient time to Steps 1 and 2. As a result, they end up solving the wrong design problem! A good rule of thumb is to spend ten times as much time on Steps 1 and 2 as you initially expect to. This part of the swamp is especially treacherous, with lots of alligators eager to devour naive design engineers. Devoting adequate time to carefully develop a needs statement and a problem statement is a vital survival skill.

2.9. REFERENCES

CROSS, NIGEL. 1994. *Engineering Design Methods: Strategies for Product Design, 2nd Edition.* Chichester, England: John Wiley & Sons.

DIETER, G. E. 1991. *Engineering Design: A Materials and Processing Approach, 2nd Edition.* New York: McGraw-Hill, Inc.

ERTAS, ATILA, and JESSE JONES, 1993. *The Engineering Design Process.* New York: John Wiley & Sons, Inc.

ORTHWEIN, W. 1990. *Machine Component Design.* St. Paul, MN: PWS Publishing Co.

PAHL, G. and W. BEITZ. 1996. *Engineering Design: A Systematic Approach, 2nd Edition.* Translation edited by Ken Wallace, Berlin: Springer-Verlag.

PUGH, STUART. 1991. *Total Design: Integrated Methods for Successful Product Engineering.* Wokingham, England: Addison-Wesley Publishing Co.

ULLMAN, DAVID G. 1992. *The Mechanical Design Process.* New York: McGraw-Hill Book Co.

2.10. EXERCISES

1. Your client asks you to design a cleaner burning wood stove. Describe a plausible "need" that might have triggered that request. Then identify two approaches to satisfying that need other than designing a cleaner burning wood stove.

2. Your client tells you, "We need to increase the capacity of urban transit busses." Describe a plausible real "need" that might be the basis for your client's statement. Identify two approaches to satisfying that need other than increasing the capacity of urban transit busses.

3. Your client asks you to reduce the noise made by the jet engines on their commercial aircraft. Describe a plausible "need" that might have triggered that request. Identify two approaches to satisfying that need other than reducing jet engine noise.

4. Your client asks you to extend the life of incandescent light bulbs. Describe a plausible "need" that might have triggered that request. Identify two approaches to satisfying that need other than designing a longer-lasting light bulb.

5. Consider the goal of designing a world class facility to house a professional sports team in an urban community. Identify a reasonable set of objectives and constraints for this project.

6. Identify a reasonable set of objectives and constraints for the design of a screen door.

7. Prepare a functional analysis diagram for a three-speed oscillating desk fan.

8. Prepare a functional analysis diagram for a potato harvesting machine.

9. Construct a QFD table for a three-ring notebook binder.

10. Construct a QFD table for a shopping cart.

11. Construct a House of Quality table for a pocket knife.

12. Construct a House of Quality table for a power screwdriver.

13. Consider the sixteen criteria listed below for the design of an automobile horn. Use these criteria to construct an objectives tree and identify opportunities for refining the criteria list.
   1. Ease of achieving 105–125 DbA
   2. Ease of achieving 2000–5000 Hz
   3. Resistance to corrosion, erosion, and water
   4. Resistance to vibration, shock, and acceleration
   5. Resistance to temperature
   6. Response time
   7. Complexity: number of stages
   8. Power consumption
   9. Ease of maintenance
   10. Weight
   11. Size
   12. Number of parts
   13. Life in service
   14. Manufacturing cost
   15. Ease of installation
   16. Shelf life

   ◆ *Source*: Pugh, p. 83.

14. As fleet manager for the Municipal Engineering Department, you are responsible for purchasing sedans for use by departmental personnel who visit project sites throughout the city. The following factors may be useful in deciding which brand of automobile to purchase: interior trim, exterior design, workmanship, initial cost, fuel economy, maintenance costs, handling and steering, braking, ride, and comfort. Arrange the attributes in the form of an objectives tree involving at least two but no more than four categories.

♦ *Source:* Dieter, p. 670.

15. Consider the following lists of advantages and disadvantages of bolts and welds as alternative joining methods. Develop a consolidated list of criteria and organize them in an objectives tree.

Advantages of Bolts

b1. Accommodate wide range of thicknesses
b2. Easily disassembled
b3. Dissimilar materials may be joined
b4. Simple and inexpensive equipment
b5. Minimum skills
b6. Cheaper than welding
b7. No residual stresses or warping
b8. Heat treatment of objects unaffected
b9. Join laminated, composite materials

Disadvantages of Bolts

b10. Joint is weaker than the object being joined
b11. Introduces stress concentration at the holes
b12. Not fluid tight unless sealed.
b13. May have poor electrical conductivity
b14. May loosen and may be vibration sensitive
b15. May corrode at edge of nut or bolt head
b16. Requires overlap or backing plates
b17. Weaken under large temperature changes

Advantages of Welds

w1. Connections are fluid tight
w2. Joint as strong as materials being joined
w3. Can form complicated shapes
w4. Lightweight
w5. Good electrical and thermal conductivity
w6. Ground welds avoid stress concentrations
w7. Strength and rigidity unaffected by temperature changes

Disadvantages of Welds

w8. May change heat treatment of metals joined
w9. Difficult to disassemble and reassemble
w10. Joined objects may warp
w11. May introduce residual stresses
w12. Few dissimilar metals may be joined
w13. Requires skilled operator, expensive equipment
w14. Inspection requires special equipment

♦ *Source:* Adapted from Orthwein, pp. 375–376 (© 1990), with permission from PWS Publishing Co., Boston, a division of International Thomson Publishing, Inc.

CHAPTER 3

Information
and Communication

3.1. OVERVIEW

The design process model described in the previous chapter includes Step 4—acquiring the information needed to carry out the design activities and Step 8—communicating the results of the design efforts. We deal with both of these in this chapter because they are different aspects of a more general activity—the transfer of information. We discuss information collection in the first part of this chapter and information dissemination at the end of the chapter. We connect these topics with a transition discussion where we demonstrate that information collection and dissemination overlap in interactive information exchange activities.

We begin with a brief treatment of different ways to organize our thinking regarding the information needs of design engineers. We then focus on different forms of technical information and we identify the private sector and government organizations that produce technical information. Next comes the transition portion of the chapter where we discuss the simultaneous roles of engineers as both consumers of information and providers of information. That discussion will address personal networks, group dynamics, and oral communications. The concluding sections of the chapter cover guidelines for effective written and graphical communications.

3.2. INFORMATION TYPOLOGIES

Engineering students get a great deal of practice solving the traditional math, science, and engineering science homework and examination problems that pervade engineering curricula. Most information needed to solve such problems is prepackaged, compact, highly compartmentalized, and likely to be found in a textbook. On the other hand, information needed for a design project is more likely to come from a variety of non-textbook sources, and has to be discovered, distilled, and synthesized to extract the key ingredients. Since the success of a design project can hinge on proper incorporation of relevant information, design engineers must be proficient at collecting information.

Because the engineering profession is so dynamic, the half-life of technical information is very short. Some estimates put it at about seven years, on the average. That means that half of the technical information you possess at any time will be obsolete within seven years. In some engineering fields, like computer hardware and software design, the half-life is substantially less. As you embark on an engineering career, you should develop a long-term information acquisition strategy that will enable you to keep up to date in your field.

We will examine information gathering from several crosscutting dimensions: the type of information; the physical and electronic location at which it is available; the form in which it is available; and the dynamics of acquiring it. We defer a detailed discussion of the forms of technical information until Section 3.3 and devote the remainder of this section to examining information types, location, and acquisition dynamics.

Type of Information

Design engineers may need as many as four distinctly different types of information over the course of a project; technical, stimulation, economic, and acquisition. We will examine each of these in turn in this subsection.

Technical.   This is the most familiar and frequent type of information used by design engineers. Whether we are talking about charged particles moving through an electromagnetic field, the laws of thermodynamics, the properties of materials, or the heat released in a chemical reaction, this is what engineers spend most of their time learning in college. In professional life, a grasp of this kind of knowledge and the ability to use it effectively in a design effort is what distinguishes engineers from other professionals, whose specialized fields of knowledge lie elsewhere. We devote much of the next section to discussing the different forms in which technical information is available and the techniques for acquiring it. Let us focus here on three other types of information.

Stimulation.   We dedicate Chapter 6 of this book to the creative aspect of design (Step 5 in our design process model). The focus there is on understanding the nature of creative thinking, barriers to being creative, and techniques for overcoming those barriers. Here we want to concentrate on the type of information that can help stimulate your creative juices.

You don't have to sit alone in a dark corner or walk along an isolated beach in order to generate creative design concepts. As we will discuss further in Chapter 6, a lot of creative ideas are modifications of existing ideas or existing ideas that have been applied to a new setting. Hence you can assist your search for innovative approaches by examining other peoples' creative efforts. A good place to look is the patent literature on the same or related topics. In the next section we will discuss the patent literature and how to use it effectively. There are several other types of information that you may want to scan for ideas on creative thinking.

You can read biographies of Thomas Edison, Benjamin Franklin, Thomas Jefferson, and other prolific inventors to gain insight into their approaches to idea generation. Entire subfields of psychology and business management are concerned with understanding and stimulating creative thinking. The advertising and marketing literature in particular are replete with discussions of the creative process. Also, there is a subculture in the United States of amateur (and, in some lucky cases, professional) inventors who spend a great deal of their energy dreaming up, and trying to sell, ingenious devices.

Economic.   There are two kinds of economic information that design engineers find useful. The first kind is general information about the market prospects for your design. Is the industry or activity to which you are targeting your design expanding? What other approaches are currently being used and at what cost? How much of the market can you expect to capture? How many devices or systems will you be able to sell?

Many private consulting firms conduct this kind of market research. Some companies develop this information for customized markets upon request from a specific client and provide the results to that client on a confidential basis. Other firms, like Frost & Sullivan, identify key technological and market trends, develop their own market forecasts, and sell them on the open market.

Another good source of market prospects for new technologies are reports published by the U.S. Department of Commerce that assess the opportunities and challenges facing various industries. For example, a 1994 report contains sixteen pages on chemicals and allied products, including a summary of major federal regulations affecting the chemical industry. It includes sections on international competitiveness of organic and inorganic chemicals, paints and coatings, adhesives and sealants, and fertilizers.[1]

The second kind of economic information needed by engineers are detailed cost estimates for specific project elements. Chapter 8 of this book deals with approaches to comparing the economic merits of alternative design concepts. Such analyses require access to cost data in the same way that a technical analysis requires access to data on voltage, mass, temperature, velocity, or other physical characteristics of the object or system. There are two ways in which this cost information is used. First, your supervisor or client may ask you to estimate the cost of building the system you are designing. Second, you need cost estimates for those components you plan to purchase from suppliers. You need to be aware of the sources of information that can assist you in making these cost estimates.

Obtaining and making cost estimates is such an important part of many design projects that some engineers develop specialties in that area. In fact, there is a professional engineering society serving the needs of those people—the American Associa-

---

[1]International Trade Administration, January 1994.

tion of Cost Engineers (AACE). There are some inherent differences among the various engineering disciplines that affect the needs for obtaining and the techniques for generating cost information. In general, mechanical and electrical engineers tend to design systems that are manufactured in large numbers, such as computer terminals, gears, automobiles, and voltage regulators. On the other hand, chemical and civil engineers tend to design systems that are constructed one at a time, such as a bridge, an oil refinery, a wastewater treatment plant, or an irrigation canal. However, even a radically new bridge or chemical process design may still use many common engineering components and materials. For example, the cement for the bridge and the piping in the oil refinery are mass produced items.

Without going into details (see Chapter 8 for more), there are several excellent sources of cost data. At a gross, or overview level, technical handbooks may contain useful cost information. A good example is an eighty-page chapter on process economics in *Perry's Chemical Engineering Handbook*. This chapter includes estimates of capital costs for several dozen types of process plants and cost estimates for large pieces of process equipment. One table focuses on materials and labor costs, while another one provides adjustment factors for the country in which the plant will be built to account for local material and labor costs.[2]

A large commercial or industrial facility is a very complicated system and consists of communications, power, lighting, ventilation, and structural subsystems that are designed by electrical, mechanical, and civil engineers. Many engineers work for firms that specialize in designing and/or building such facilities. Detailed estimates of facility construction costs are frequently necessary when a firm bids on a job. Too high of a cost estimate, and your firm won't be hired because a competitor submitted a lower bid. Too low of an estimate, and you might get the job but lose money on it. Too many "too-low" cost estimates and your company will quickly go out of business. Clearly, engineers in this field require dependable information sources for such cost estimates.

Several commercial firms publish cost estimates on a regular basis. The *National Construction Estimator* is one source of detailed cost estimates for construction work. More detailed cost estimates are available from a set of about a dozen books published by R. S. Means. As an example, Table 3–1 shows the Means cost estimate for a

TABLE 3–1.   COST ESTIMATE FOR ROOFTOP MULTIZONE AIR CONDITIONER FOR 3,000 FT² APARTMENT CORRIDORS

Components of System 8.4-221-1280 for 3,000 ft²	Quantity	Unit	Cost Each		
			Material	Installation	Total
5.5 ton rooftop multizone unit	1	Ea.	14,834.60	2,413.40	17,248.00
Ductwork package	1	System	2,985.40	6,227.10	9,212.50
Total			17,820.00	8,640.50	26,460.50
Cost per ft²			5.94	2.88	8.82

[2]Holland et al., 1984.

rooftop multizone air conditioner to serve apartment house corridors.[3] A continuation of this same table provides cost estimates for similar air conditioners installed in other types of buildings. This kind of generic information is valuable early in the design effort when approximate cost estimates are appropriate. When the design progresses to a more detailed stage, you have to contact a vendor to get a specific price on a specific model of a specific brand multizone air conditioner.

Acquisition.   Once you have decided what components and materials you wish to use in your design, the next phase is ordering the needed products. There are several sources of information that can help you identify a vendor who can satisfy your requirements. One such source of vendor-specific information is advertisements in technical magazines.

Generally those advertisements provide phone and fax numbers for you to request price sheets, product catalogs, or other literature (see Fig. 3–1 for an example).

(Circle 22 Reader Service Coupon)

Figure 3–1.   Advertisement for Technical Equipment. *(Civil Engineering, Jan. 1995, p. 37. Reprinted with permission from Richard Dudgeon, Inc.)*

---

[3]R. S. Means Co. Inc., 1991, p. 807.

To make it easier for readers to obtain more information, most of these magazines contain a reader service card that contains a number for each advertisement in that issue. The reader simply circles the numbers on the card corresponding to the relevant advertisements (see the bottom line in the advertisement displayed in Fig. 3–1) and mails in the card.

Many manufacturers of engineering materials, components, and equipment have local agents and representatives in major metropolitan areas. These information sources are listed under the product heading in the local phone company's yellow pages directory. As an illustration, there are twenty-one listings under the heading "bearings" in the *Seattle Yellow Pages*.

The *Thomas Register of American Manufacturers* is similar to a national yellow pages phone directory. Published annually, the *Thomas Register* lists suppliers of various components, systems and services of interest to design engineers. It consists of three main sections. First is the eighteen-volume listings of products and services, with over 50,000 entries. There is also the two-volume company profiles that list key contact personnel for almost 150,000 companies in the U.S. and Canada. The third section of the *Thomas Register* contains reproductions of 1,700 vendor catalogs in an eight-volume catalog file section.

Information Repositories

We now look at information from the perspective of where that information is stored and how design engineers access it. Libraries are the traditional repositories of technical information. Many engineers develop and maintain their own personal library. You've probably already started your own by retaining textbooks from some prior courses. During the course of your undergraduate career, you may also buy technical handbooks and software manuals. Many students retain their lecture notes, homework assignments, and exams from courses they deem relevant to their career plans. Add one or two technical magazine subscriptions and you've got a valuable and convenient library right on your desk.

Professional libraries. A major advantage of a professional library over a personal library is that the professional library has the resources and the professional staff to select and organize additions to the collection. The most common forms of professional libraries are general purpose public and university libraries. In addition there are many specialized libraries. Some large companies and government agencies have their own specialized in-house libraries and some large universities have separate libraries for their engineering, medical, business, and law schools. If you are a member of one of the major engineering societies, you automatically have borrowing privileges at the Engineering Societies Library in New York City; you can borrow books and documents over the telephone.

You should invest some time to learn what resources are available to you through the libraries that you have access to. Probably the single most important library resource is the reference librarian. Reference librarians are professionals who devote their careers to knowing the contents of important reference works, how to access them, and satisfy the information requirements of their customers. The more you interact with them, the more impressed you will be by their familiarity with very arcane and specialized information sources. Of course, a specialized library, such as an engineering library, will have a reference librarian who is more familiar with engineering literature than a reference librarian in a general purpose library. Whenever your search for technical information becomes frustratingly unproductive, or is hard to even get started, seek help from the reference librarian. Depending on how busy they are at the time, many reference librarians can respond to phone inquires.

We are in the middle of an information technology revolution. Information that was at one time available only in written form is now available electronically. These changes will greatly impact the look, feel, and mission of traditional libraries. Most libraries have replaced their traditional card catalog with on-line computer versions that also indicate whether the item is currently checked out, and if so, when it is due and how long the waiting list is. On many campuses, local networks connect faculty offices, dormitory rooms, and student computer laboratories and provide access to the library's electronic card catalog. Via the Internet, electronic versions of catalogs of many other libraries are also available.

Many campus libraries use e-mail to provide other services by (free) subscription. For example, the engineering library at the University of Washington e-mails its new books list to subscribers every week. What if you wanted to be notified only about new acquisitions in a certain topical area? You can subscribe to an individualized monthly new acquisition notice based on your choice of key words. More and more campus libraries are also providing on-line versions of dictionaries, encyclopedias, and government documents such as the *Federal Register* accessible from your desk-top computer.

Many documents are available directly to you through the World-Wide Web without having to go through a library. More and more information is becoming available on the Web before it is released in hard copy form. Some providers of technical reference information and databases are even abandoning their traditional hard copy format and publishing "documents" only in electronic form. In addition, the Web provides opportunities for interactive technical chat-rooms and forums where engineers can post questions and respond to questions posed by others. Some of the professional engineering societies host these forums on their web sites.

As a regular user of the Web, you can build up your own collection of bookmarks that provide you with virtually instant access to your favorite sources of technical information. Using the Internet to gain access to libraries and other sources of technical information is a real boon to small companies without their own library, or companies that are located in a small town far from a major library.

Having all these wonderful resources available to you on your desktop permits you to significantly increase your productivity. No more frustrating treks to the library and through the stacks only to learn the book you desperately need is checked out, or that the specific issue of the magazine with that crucial article is at the bindery. On the other hand, these marvelous new information technologies come at a real, but subtle, cost. It is a well known phenomenon that much valuable information on Topic A is discovered in the course of searching through the library stacks or periodical section for information on Topic B. Something just happens to catch your eye that, while not directly connected to the focus of your current search, is that sparkling gem of information that you had looked high and low for during a previous search. Or you may realize that this gem will vastly simplify an upcoming search. Almost by definition, electronic searches are very focused and preclude the opportunity to discover these sparkling gems. To overcome this limitation, it is a good idea to set aside some regularly scheduled time for nonfocused browsing through the library. It will be well worth your time.

Limitations of library sources.   A major drawback of relying on libraries for your primary source of technical information is that even the most recent additions to a library contain information that is several years old. Consider the lead time for a book. Most authors of technical books decide to write a book only after accumulating lots of research or teaching experience with the subject matter of the book. From the time the decision is made to write the book, it usually is two to four years before the book finds its way onto a library shelf. The lead time for a conference paper or research article is somewhat shorter, but can easily be one or two years. Section 3.5 of this chapter deals with ways to overcome this time lag.

Information Acquisition Dynamics

It is important to approach the information acquisition task with the proper attitude. Many engineering students, having been groomed to solve self-contained problems from their textbooks, tend to underestimate both the amount of information available and the benefits of investing time in acquiring it. It's the familiar notion learned from traditional analysis-oriented courses—if the information was really needed, it would have been given to me in the statement of the problem. It takes discipline to overcome this extremely narrow perspective. A good slogan to post prominently in your dormitory room or office is: Never underestimate the amount of information available on any subject.

Even engineers with decades of experience continue to discover new avenues for obtaining important design information. No matter how specialized or arcane the information is that you are looking for, never fall into the trap of thinking that there can't be anything available on your topic. In fact, always assume the opposite to be true! Armed with this "go-get-'em" attitude, you are now ready to begin a search.

Stages of information acquisition.   The first stage of gathering information is identifying the kind of information required and determining if and where it is available. The second stage involves physically or electronically gathering the information. In stage three, the engineer must determine how reliable and credible the information is, assimilate and interpret the information, and determine how to apply it to the problem at hand. Finally, the engineer must decide when to stop looking.

The engineer must make these decisions based on experience and personal judgment. One helpful approach is to rephrase questions as quantifiable questions. "Do I have the right amount of information?" can be phrased as, "Will the expected value of additional information exceed the expected cost of obtaining it?" This does not answer the question, but it provides a useful rule of thumb by which to compare options. Often we collect too much information at the expense of leaving time to properly understand, interpret, and apply it. Information that cannot be digested has a negative value because of the time and energy devoted to gathering it and attempting to use it.

Passive versus active.   Another perspective on information acquisition dynamics is the notion of passive versus active information gathering. Passive means there is a piece of information that someone has already prepared and all you need to do is find out where it is and retrieve it. On the other hand, active information gathering means the real-time interaction of two or more people involving the transfer of information between them. Engaging a colleague or a professor in a dialogue about the information you are seeking is an example of an active information search. This notion of active information exchange suggests that sometimes you may be on the receiving end of such a request. Now the distinction between being a consumer of information and a provider of information gets blurred. For the moment, we will confine our discussion to information seeking skills, and wait until Section 3.6 before crossing over the boundary into information communication skills.

3.3. FORMS OF TECHNICAL INFORMATION

As discussed in Section 3.2, technical information is one of four types of information that design engineers may need over the course of a design project. Because of its central role in engineering design, we devote this section to examining the various physical and electronic forms that this information may take.

Each new design problem you tackle is different from any previous one (if it wasn't, you would use the existing design). Virtually by definition then, you will require a different combination of information resources. When designing in an area that is well developed, a wealth of information and design examples may provide you with considerable insight and save you much time. This might be the case, for example, of drawing upon your prior design of an air conditioning system for apartment house corridors as background for your new assignment to design an air conditioning system for a hotel ballroom. You may be able to satisfy your new information requirements by extracting information from the same source previously used, but from a different table, or a different part of the same table.

On the other hand, you may have very little information already at your disposal if, after designing several air conditioning systems, your next assignment is to design a solid waste disposal system. Your information collection task is now a lot more difficult, although you are confident that there is an extensive body of literature on solid waste disposal. It's just a matter of locating it and tapping into the portions relevant to your assignment in an efficient manner.

You also could face a design problem in an area which is radically different and for which very little information is available. Even if there is some information on the topic, it may be theoretical research and not directly applicable to design situations.

TABLE 3–2. HIERARCHY OF LIBRARY INFORMATION SOURCES

most general	Technical dictionaries
	Encyclopedias
	Handbooks
	Textbooks
	Indexes and abstracts
	Advanced technical books
	Technical magazines
	Conference papers and proceedings
	Research journals
	Technical reports
	Translations
most specific	Company catalog and brochures

Source: Adapted from Dieter, p. 619 with permission from The McGraw-Hill Companies.

The engineer may need to perform experiments or develop new models in order to obtain the necessary information. Such a situation arose when air conditioning system designers had to cease using most common refrigerants because they adversely impact Earth's protective ozone layer.

Table 3–2 lists twelve different forms of technical information and ranks them roughly in the order from the most general to the most specific. Where on this list you start a particular information search depends on how familiar you already are with the topic. Regardless of the starting point, any given search will involve collecting information at several different levels. To illustrate, we will retrace the path of a specific search for information on designing a particular kind of electric circuit. Then we will look closer at many of the information forms listed in Table 3–2.

Case Study of Band-Pass Filter Circuit

Suppose you are asked to design an electric circuit that takes an incoming electric signal and only allows that part of the signal within a certain band of frequencies to reach the output part of the circuit. Such circuits are called band-pass filter circuits. They are used extensively in communications and control systems. For example, your radio receives signals of many different frequencies. When you select a particular station, you are using a band-pass filter to filter out all signals except for the one with the desired frequency. See Figure 3–2 for a simple form of a band-pass filter circuit.

It can be shown that if the frequency of the signal is related to the inductance L and the capacitance C according to

$$f = \frac{1}{2\pi\sqrt{LC}}$$

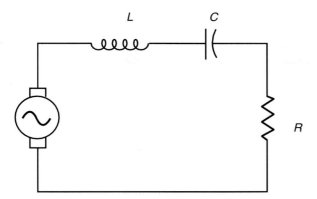

Figure 3–2. Simple Band-Pass Filter Circuit

then a large amount of current will flow through the load R. For a radio, the amount of current passed through should drop off very rapidly for any signal not of the selected frequency. The design problem is to select L and C in order to establish a range of frequencies within which the current passed through is more than a specified amount (strong signal) while keeping the amount of current passed through by unwanted frequencies below an acceptable level.

Suppose we remember studying this topic in school, but it was a few years ago, and we can't even recall what the term band-pass means. Our first step is to refresh our memory as to what a band-pass filter is. Our solution is to go to the reference section of the library and find the *IEEE Standard Dictionary of Electrical and Electronic Terms*. The entry for "band-pass filter" is, "see optical filter." Kind of strange, we think, since we're designing an electrical circuit, not an optical device. Nevertheless, since the book is already open, we find the following entry, "optical filter: an element that selectively transmits or blocks a range of wavelengths." This is somewhat helpful, though the use of wavelengths instead of frequencies sets us back momentarily until we recall the relationship between the two.

Our memory is refreshed, but since we have never designed such a circuit before, we want more information about real filter circuits and signals, not the idealized version shown in Figure 3–2. We try the *Encyclopedia of Science and Technology*, a twenty-volume set found in many general purpose libraries. Looking up "band-pass filter" in the Index (Vol. 20) refers us to a six-page article in Volume 6 entitled "Electric Filters." The article was written by Dr. Norman Balabanian, a prominent electrical engineering professor at Syracuse University. In relatively simple terms (the article is aimed at a technical audience, but not necessarily electrical engineers) he discusses design considerations and applications. The article describes the trend towards miniaturization in electronic components and systems. In particular, it discusses the dwindling use of inductors and their replacement by devices called active RC filters. This facilitates designing filters in the form of tiny semiconductor chips. The article concludes with a list of six references for further information.

Since we're already at the library, we decide that we should check out some similar sources. We go to the *Encyclopedia of Physical Science and Technology*. This fifteen-volume set is more specialized than the *Encyclopedia of Science and Technology* and the articles are much more mathematical in nature. There is no "band-pass" entry in the index, but an entry for "active, in circuit theory" leads to a twenty-one-page article on circuit theory. In that article, they use the terminology "pass-band." This is a good example of how slight differences in terminology can complicate an information search. It is a good idea to use several variations of the terminology to make sure your search is capturing the needed information.

With this basic understanding of band-pass filters under our belts, we feel we're ready for a more in-depth treatment. The *Standard Handbook for Electrical Engineers* seems like a likely source. There, in Section 2 under the general heading of "Electric and Magnetic Circuits," we find a thirty-seven-page article entitled "Filters." Section 201 on p. 2-84 is devoted to "The Band-Pass Filter." It includes a worked out example of designing a band-pass filter with a center frequency of 4 kHz and a bandwidth of 900 Hz.

Our confidence has grown to the point that we decide to examine the state-of-the-art in band-pass filter design. We don't see the *Engineering Index* on the shelf, so we ask the reference librarian. She says they don't have it but suggests we check the *First Search© Catalog*, an on-line system that provides access to many different databases. We find the *Concise Engineering and Technology Index* among the databases in its inventory. We conduct a subject search, and input the subject as "band pass filter." The computer screen shows a list of fourteen items, giving title and author, from 1988–1994. Scanning through the list, the only item that appears relevant to our interest is an article entitled, "New Filter Synthesis Technique—The Hourglass," written by Byron Bennett. We decide to take a closer look at that item. The next screen displays the full title, the name of the research journal in which the article appeared, when the article was published, and the author's affiliation (Montana State University). The following four screens contain the abstract, other descriptors, and an indication that the article contains nine references. We did this search at a library in the state of Washington and want to know which libraries in the region subscribe to that particular research journal. The computer not only yields a list of five libraries in Washington, five in Oregon, and one in British Columbia that subscribe to the journal, but also indicates which of those libraries can supply it via inter-library loan. With this excursion successfully completed, let us examine some of the key generic forms of technical information.

Books

Most undergraduate engineering students think of textbooks as the primary source of technical information. There are generally at least several textbooks available for every undergraduate engineering course. If a specific topic is omitted or not explained very clearly in one textbook, you can often find that material in another textbook on the same subject. Even after graduation many design engineers continue to use undergraduate textbooks as reference works.

Textbooks are also published for advanced undergraduate and graduate level courses. These tend to treat more specialized topics in a more sophisticated manner. It is not unusual for practicing design engineers to refer to these advanced textbooks for specialized information. In addition, many technical books are published primarily for use by practicing engineers and are not intended for use as a text. One popular book format of interest are books in series that are usually annual volumes containing specially commissioned state-of-the-art review articles. One of many such series of interest to engineers is the *Annual Review of Energy and Environment.*

One of the major advantages of technical books (whether or not they are primarily marketed as textbooks) is their exhaustive and detailed coverage of the subject. At the same time, this can limit their usefulness in helping engineers make design decisions under the press of an imminent deadline. Many times, you'll find that you need a quick, concise explanation, and don't have the time to digest a thorough treatise.

Dictionaries, Encyclopedias, and Handbooks

You may be presented with a design problem in a field in which you have no previous experience. Of course, this is more likely to happen early in your career, when you have very little experience in any field. Later on in your career, when you have established a reputation in a particular field, most of your work will come from within that field. For example, if you've made a name for yourself as an excellent designer of suspension bridges, most of your work will be designing suspension bridges. It is unlikely that a client will approach you with a request to design a computer chip. However, there are always circumstances in which you venture beyond familiar territory. In those situations, you may need to quickly familiarize yourself with some basic concepts, principles, and processes in a very concise way. You will need a quick overview, just enough to get you started. In such a situation, a technical encyclopedia may be able to bring you up to speed in a few paragraphs.

In some cases, you may not even be familiar with the vocabulary and terminology. In these circumstances, a technical dictionary can come in very handy. There are many technical dictionaries, and some of them are quite specialized. Consider your situation as a newly hired engineer with a consulting firm to the paper industry. You are assigned to a project involving the design of a new, more environmentally benign, paper pulping process. As part of your background reading on pulp making, you come across the term 'alpha and dissolving pulp'. You never heard of this before, so you walk down the hall to your company's small reference library and find definitions for both alpha pulp and dissolving pulp in the *Pulp and Paper Dictionary.* That didn't take long!

On the other hand, you may have a passing or even working knowledge of a technical field, but don't have many of the details at your disposal. A handy desktop reference is a handbook. Handbooks tend to be quite voluminous, but each chapter contains key equations, concepts, and design tools in a compact fashion. Frequently each chapter in a handbook is written by a different author; the individual contributions are edited and integrated by a technical expert who serves as the editor. There are standard handbooks for each of the major engineering disciplines, such as *Mark's Standard Handbook for Mechanical Engineering.* Examples of more specialized engineering handbooks are: *Electromechanical Design Handbook, International Earth-quake Engineering Handbook, Industrial Thermal Processing Handbook.* As with other forms of information, it is good policy to assume that there is a technical hand-book that covers the subject you are interested in.

Periodicals

The engineering profession is replete with technical information published in periodi-cals that have publication schedules ranging from weekly to quarterly. The two most common formats are research journals and magazines.

Research journals.   Results of engineering research are often published as pa-pers in scholarly research journals. A key characteristic of most engineering research journals is that they publish unsolicited papers, but only after subjecting them to an independent, confidential peer review process. Many papers get rejected, or are pub-lished only after extensive revision. Typically, the editors and reviewers are unpaid volunteers who are distinguished researchers in the field, and who serve out of a sense of professional responsibility and desire to advance the state of the art. Though some journals include book reviews and announcements of conferences, they generally do not accept paid advertisements. Their revenue is derived entirely from subscriptions and sales of reprints.

Much of the material published in engineering research journals is very theoreti-cal with little immediate application to design. However, there are some articles that have direct relevance to design projects. In fact, some journals publish applied-type research that may be quite useful to design engineers. Several relatively new journals are devoted exclusively to engineering design topics. For example, *Materials and De-sign* is a bimonthly research journal whose focus is, ". . . high quality refereed papers on all types of engineering materials, with an emphasis on design and product perfor-mance. The practical combination of materials development and engineering design aspects . . . is . . . critical for practicing engineers needing an awareness of the capabili-ties and opportunities afforded by all modern materials." An example of a paper pub-lished in *Materials and Design* is, "Reliability-based Design of Ceramics." Like most research journals, the editor is a distinguished researcher in the field, and a group of experts serve on the journal's editorial advisory board.

Another research journal is *Design Studies*, described as, "The international journal for design research in engineering, architecture, products, and systems." It fur-ther describes itself as, "The journal reports on new developments, techniques, knowl-edge and applications in the practice of design, as well as design education: how de-sign techniques may be taught, the approach to ill-defined problems and the impact of new technologies." A recently published paper in this journal is "New Insights into Computer-Aided Conceptual Design."

Magazines.   Technical magazines differ from scholarly research journals, in that they cater more to the needs of practicing engineers. They are more commercial in tone and appearance, typically containing extensive advertisements for products from manufacturers and services from design consulting firms. Many advertisements in these magazines are from software vendors. These magazines usually also contain announcements of job opportunities, and also allow job-seekers to place announcements. Some of these technical magazines derive enough revenue from advertisements and library subscriptions to allow them to offer free subscriptions to individual qualified engineers.

Technical magazines employ paid professional editors and writers, and generally do not publish unsolicited articles. Their articles tend to be more applied in nature, frequently describing design and performance of actual products and systems. In addition, they may have short articles discussing economic trends and government activities of interest to their readership.

Most of the professional engineering societies publish a monthly magazine that is distributed to all its members. As you would expect, these magazines are oriented toward the particular engineering discipline, in keeping with the makeup of the sponsoring society's membership.

Many specialized technical magazines are also published by commercial publishers. These magazines tend to focus on an industry, rather than an engineering discipline. Engineers working in those industries or providing products and services to those industries are the primary audience for this group of magazines. To give a flavor for the scope of this segment of the technical information enterprise, a few of the titles of the many dozens of commercial technical magazines are listed here:

Aviation Week and Space Technology
Ceramic Industry
Concrete Construction
EDN—The Design Magazine of the Electronics Industry
Glass Industry
Machine Design
Microwaves & RF—for designs at higher frequencies
Power Engineering
Railway Age

Many of these magazines publish an annual yearbook and product directory, summarizing trends in the industry and providing comprehensive lists of products and services, and the companies that provide them to the industry.

Conference Proceedings

As described in Section 3.5, hundreds of large technical conferences are held each year. At a typical conference, hundreds of engineers present papers describing their recent research and design efforts. These collections of papers are frequently published by the conference sponsor in the form of bound, often multivolume, conference proceedings. Since many of these conferences are held annually or biennially, many technical libraries have standing subscription orders for the proceedings. If you are aware of a regularly scheduled technical conference covering a topic relevant to your design assignment, you should examine recent editions of that conference's proceedings for useful information.

Technical Reports

Not all important contributions to the advancement of the state of the art are published in research journals or technical magazines, or presented at conferences. Much valuable information is written in the form of technical reports issued by the company, research institute, or government agency at which the work was conducted. Sometimes these technical reports are simultaneously or subsequently submitted for presentation in a conference or publication in a research journal, but in many cases technical reports are the final form in which the information is published.

Government agencies like NASA conduct a lot of their own engineering research and design studies in-house and publish their results in a regular series of technical reports. Other government agencies have much of their engineering studies conducted under a grant or contract with a university or private firm. As part of the conditions for the contract, the agency may require the grantee or contractor to issue one or more technical reports.

Indexes, Abstracts, and Reviews

Until now, our discussion of forms of information has focused on documents that contain substantive information. In this subsection we will focus on a class of reference periodicals that serve as locators and summaries of substantive sources. These publications come in several different forms, and virtually all of them are available either on-line, on CD-ROM, or in hard copy. A word of caution about terminology: I distinguish between generic forms of locator and summary periodicals according to the following criteria. If articles and documents are summarized only by title and author, I call it an index; if it includes a short summary of the article or document, I call it an abstract; and if the summary is an interpretive critique by someone other than the author, I call it a review. While this distinction may be helpful when discussing generic forms, many of these periodicals use the terms index, abstract, and review in their title without regard to this generic distinction. For example, you may very well find a periodical that calls itself an index but fits our definition of an abstract.

One family of indexes is *Current Contents,* a weekly that publishes the tables of contents of scholarly research journals. The version we are most interested in is *Current Contents/Engineering, Technology and Applied Science.* This covers more than 800 journals and 400 books published in regular series. Organized by journal and book series title, you can also search by key words in article titles and by authors.

Abstracts in the field of engineering generally cover research journals, technical magazines, books, conference proceedings, and technical reports. The most comprehensive engineering abstracts are the *Engineering Index* and the *Applied Science and Technology Index* (note the aforementioned terminology problem). The latter has a broader scope, covering computer science and other applied sciences as well as engineering. Both are annual publications with weekly, monthly, bimonthly, or quarterly updates (depending on the format).

The CD-ROM version of the *Engineering Index* is called *Compendex Plus®*. The *Engineering Index* contents without the abstracts (that makes it a true Index in our terminology) is available on CD-ROM as *Ei Page One™*. For those engineers interested only in information within a particular discipline or topical area, *Compendex Plus®* is available in seven specialized versions covering the following fields: chemical engineering, civil engineering, electrical engineering, mechanical engineering, energy and environment, advanced materials, and manufacturing.

Many other specialized technical databases are available. Some self-explanatory examples are *Applied Mechanics Reviews, Corrosion Control Abstracts, Computer Select, Food Science and Technology Abstracts, Fuels Abstracts, Highway Research Abstracts, Metals Abstracts,* and *Selected Water Resources.*

Many contemporary engineering design issues and developments are covered in the popular mass media. These materials are covered by *InfoTrac®*, whose two main components are *Magazine Index Plus™* which provides an index and abstracts for over 400 different periodicals; and *National Newspaper Index™* which covers the *New York Times, Wall Street Journal, Christian Science Monitor, Los Angeles Times,* and *Washington Post.* The aforementioned newspapers are written for a sophisticated nationwide audience, and have excellent coverage of science and technology issues.

You may want to follow up on a particular design innovation. In the case of the band-pass filter discussed earlier, we were led to a design innovation described in a paper from Professor Bennett from Montana State University. But since the date of that paper is 1988, we wonder whether Professor Bennett or anyone else has followed up on that work. Here's where the *Science Citation Index* may be helpful. This resource, issued annually, lists all the subsequent articles in research journals that cite Professor Bennett's paper as a reference.

Codes and Standards

A complex technological society requires standardization. Nuts have to fit bolts. Staples have to fit staplers. Can you imagine if the only company from whom you could buy staples is the one that manufactured your stapler? Suppose there are twenty companies that manufacture staplers and they each used a slightly different size staple. Stationery stores would have to stock all the different size staples. The cost of both staplers and staples would increase, not to mention the inconvenience when you could not find the correct match. What about the location and size of the holes in three hole notebook paper and the spacing of the rings in three-hole ring-binders?

Design engineers need to recognize the existence of standards so that they can utilize them in their designs. If you specified a nonstandard size wire for a circuit you were designing, your design will take longer and cost more to make, because you would have to have the wire custom made for you by a vendor rather than ordering the standard size wire available immediately from a warehouse.

Lets now consider several applications of standards to engineering designs. A more comprehensive examination of the standards system and processes is contained in Chapter 4.

Case study of Uniform Building Code.   One of the most widely used standards is the *Uniform Building Code* (*UBC*), published by the International Conference of Building Officials (ICBO). Consisting of three volumes, the *UBC* deals with fire and safety issues, structural design, and materials and testing. The *UBC* is revised every three years. ICBO also publishes a series of other codes, such as the *Uniform Mechanical Code, Uniform Fire Code,* and *Uniform Housing Code* that are compatible with the UBC.

Suppose you are working for a consulting engineering firm that is designing a new hospital complex. You are a member of the team responsible for the fire safety aspects of the design. In particular you are designing a system to control the spread of smoke in the event of a fire so that building occupants can be relocated or evacuated. You are considering using the flow of air to prevent smoke from migrating through fixed openings between smoke-control zones. The design parameters you are working with are openings of height $h = 12$ ft, smoke temperature $T_f = 150°F$, and ambient air temperature $T_o = 70°F$.

You turn to UBC Section 905.4.2 to determine the required minimum average air velocity v(ft/min) through the opening as

$$v = 217.2 \left[\frac{h(T_f - T_o)}{(T_f + 460)} \right]^{\frac{1}{2}}$$

which leads to

$$v = 273 \text{ ft/min}$$

However, you notice that UBC Section 905.4.3 prohibits airflows toward a fire that exceed 200 ft/min. Hence the airflow method for smoke control cannot be used in this situation. This is an excellent lesson on the importance of understanding the context of any code provision. If you had not looked any further than the formula, you would not have realized that the 200 ft/min limit existed.

Prudent design engineers will read the broad provisions of any code language to make sure that the specific provision is being properly interpreted. It is also a good policy to always assume that a standard exists that will affect your design.

The most comprehensive guide to standards developed in the private sector is the *Index and Directory of Industry Standards.* Volume I of this five-volume work is the *U.S. Standards-Subject Index.* Volume II is a *Numeric Listing of U.S. Standards.* Volumes III, IV, and V cover standards issued in non-U.S. countries and international standards. This publication is a standard reference work in technical libraries.

Case study of children's bicycle helmet.   Let's go back a few years when you first came up with a clever idea for designing a bicycle helmet for young children. One of the main features of your design is an innovative restraint mechanism that prevents the helmet from being dislodged in a collision. While chatting with a colleague at a local meeting of your professional engineering society, she mentioned that the federal government was about to issue a new standard for bicycle helmets. To check into it, you went to the *Federal Register Index* in the reference section of your local public library. There, in the monthly update for August 8–September 9, 1994, you see that a proposed rule on bicycle helmets was published on page 41719 in the August 15, 1994 *Federal Register.* Armed with that information, you accessed the electronic version of the *Federal Register* posted on the Internet by Counterpoint Publishing. There you found the federal Consumer Product Safety Commission's (CPSC) proposed new regulation entitled "Safety Standard for Bicycle Helmets." If adopted, this regulation would add a new section to Title 16 of the *Code of Federal Regulations (CFR)*. The proposed rule takes up twenty pages, including twelve figures regarding helmet geometry and testing apparatus. The three pages of background information prior to the text of the proposed rule indicated that this rule was being proposed as mandated by the Children's Bicycle Helmet Safety Act of 1994. You were particularly interested in the procedure and apparatus for testing the retention system. Since this electronic version of the *Federal Register* only displayed the text of the proposed rule, you went to the reference section of the library to get a hard copy of the diagrams. Figure 3–3 shows the diagram from p. 41736 of the *Federal Register* for the apparatus to be used in testing the retention system. You also note that the CPSC deadline for receiving comments on their proposed rule was October 31, 1994. Although you missed the deadline to provide input, you noted the name, mailing address, and phone number for the project manager from the CPSC Directorate of Engineering Sciences as the primary contact on this proposal.

Federal Register.   The *Federal Register* is a daily announcement of new federal regulations, proposed regulations, meetings, and other announcements. All federal agencies must use the *Federal Register* to disclose their regulations and regulatory

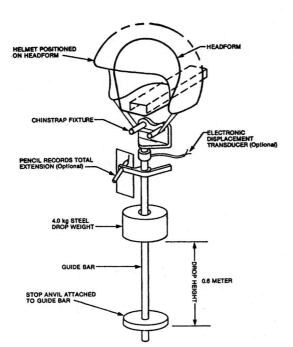

Figure 3–3.   Apparatus for Testing Bicycle Helmet Retention System

intentions. By law, federal agencies are required to announce their proposed regulations and provide fair opportunity for interested parties to provide written input before final decisions are made. This may include scheduling meetings in different parts of the country to receive comments on the proposals. When the final regulatory decision is published in the *Federal Register*, the agency must summarize the nature of the comments it received on the draft regulation and explain how the form of the final rule is responsive to that input. As illustrated by the case of the bicycle helmet design discussed above, the *Federal Register* is the place to seek information regarding current federal regulatory activities that may affect your design efforts.

Code of federal regulations. Once a federal regulation has been officially adopted, it has the force of law. It becomes codified and entered into the *Code of Federal Regulations (CFR)*. Updated every year, the CFR is organized into fifty titles covering different aspects of federal rules and regulations and serves as the permanent record of federal regulatory actions. If you want to locate an existing federal regulation (at least one year old), then the *CFR* is the place to look. Because some federal regulations are very complicated, our earlier admonition (see the subsection on the *Uniform Building Code*) regarding understanding the full context of specific regulatory provisions is especially relevant here.

Patents

The adage that there is nothing new under the sun applies here. No matter how unusual you think your design task is, there is a good chance that someone else had a similar design problem and received a patent on a device to solve it. Approximately 5,000,000 U.S. patents have been awarded and about 100,000 new patents are issued each year. There is not much point to spending a great deal of effort on what you consider to be an innovative design concept until you determine whether or not it is already covered by a patent. The next chapter will discuss the process of applying for a patent and other issues associated with the patent system. Here we treat patents purely as a source of information.

Patent search. The patent system in this country is administered by the Patent and Trademark Office, a branch of the U.S. Department of Commerce. The headquarters office in a Washington, D.C. suburb includes a patent search room that is open to the public. For most of us who do not have easy access to D.C., the best place to search through the patent collection is at a Patent and Trademark Depository Library (PTDL).

The PTDLs are a network of about seventy libraries located in about forty states, usually as part of the library at a major engineering school or within the public library in a major metropolitan area. The extent of the patent collection varies among the PTDLs, but most have complete collections of recently awarded patents.

Each patent in the patent collection is assigned a patent number, and in addition is classified by subject according to a classification system that organizes the collection into subject categories in order to facilitate systematic patent searches. The classification system consists of about 400 classes, each one of which is divided into subclasses. Each class and subclass has a numerical identifier. When a patent is assigned to a class and subclass, the appropriate numerical identifiers are associated with that patent. It is not quite the same as the library assigning a call number to a book because classes and subclasses are not mutually exclusive. Thus, each patent may be assigned to more than one class or more than one subclass within a class.

Three documents and a CD-ROM database issued by the Patent and Trademark Office are the beginning point of most patent searches. The key documents are the *Index to Classification*, the *Manual of Classification*, and the *Official Gazette of the U.S. Patent and Trademark Office.* The CD-ROM patent search system is known as the Classification and Search Support Information System (CASSIS).

Normally, a search will start with a subject area or subject areas that the searcher thinks best describes the field of interest. The first stop is the *Index to Classification*, an alphabetical listing of descriptors of classes and subclasses, to see if the subject area is listed. If not, consider variations on the description of the subject until you find a version that is listed. Suppose you're designing a multiple-speed drive for manual wheelchairs to make it easier for manual wheelchair users to climb slopes and ramps. From the *Index to Classification* you find "wheelchair" as the major heading for class 280. The first and apparently most comprehensive subclass is 250.1.

Next you look under class 280 in the *Manual of Classification*. In addition to subclass 250.1, many other subclasses look interesting, but you choose at this point to focus on subclass 250.1. You then go to CASSIS to get a listing of 359 patents issued in class 280, subclass 250.1. Intimidated by this huge list, you realize that you need advice about narrowing down the scope of our search. However, since the reference librarian is busy at the moment, you might as well start at the top of the (chronological) list. A recently issued patent on this list is patent #5,356,172.

You find the title, background information, and abstract for patent #5,356,172 in the October 18, 1994 issue of the *Gazette*. Clearly this patent, entitled "Sliding Seat Assembly for a Propelled Wheel Chair" is not relevant to our interest in multiple-speed drives. So you cross it off your list—only 394 to go! If and when you reduce this list to a reasonable number of relevant patents, your next step will be to obtain the complete version of each of them from the microfilm files.

3.4. INSTITUTIONAL GENERATORS OF TECHNICAL KNOWLEDGE AND INFORMATION

In the prior section, we focused on the form of the technical information itself. In many situations, you may not be looking for a specific document, but just want to keep track of certain technological developments. You may want to talk to an individual who is doing cutting edge work in a field that could have major implications for your design efforts. It may be a specific project you are currently working on, or a more general situation that could have long-term implications for maintaining or upgrading your professional design skills. For these reasons, engineers should be familiar with the organizations that are the major sources of technical information.

Professional Engineering Societies and Trade Associations

Two major sources of technical information are professional engineering societies and trade associations. Both organizations exist to meet the technical and professional needs of its members. Their policies and programs are determined by their dues-paying members, and managed by a full-time professional staff. One fundamental difference is that the members of professional engineering societies are individual engineers while trade associations' members are companies who make products or provide services in a specific industry. Nevertheless, these organizations have many similarities. Both the professional engineering societies and trade associations typically hold conferences, publish newsletters and magazines, may develop codes and standards, and sponsor continuing education programs.

A good source of information about professional societies and trade associations is the *Encyclopedia of Associations*. A more specialized document that gives vital statistics on the engineering societies is the *Directory of Engineering Societies*. More discussion of the nature and activities of professional engineering societies is contained in Chapter 4. Here we will briefly examine some aspects of trade associations.

Consider the food processing industry. Many chemical, electrical, and mechanical engineers devote careers to designing machinery, control systems, and process flows for the food processing industry. A major trade association for this industry is the National Food Processors Association. Its membership consists of 500 companies; it has a $16 million annual budget and a full-time staff of 185 people. Parts of the food processing industry are also served by smaller, more specialized trade associations. Examples include the Frozen Potato Products Institute, Pickle Packers International, and the Tortilla Industry Association.

Suppose you are a design engineer for a firm that builds sophisticated temperature measurement and control systems. One of the biggest markets for this equipment is in the metals production and fabrication industry where precise control of temperature is needed in order to achieve the desired material properties. You have an idea for a new laser device that will significantly improve the accuracy of measuring the temperature of molten steel. In order to convince your boss to let you devote ten hours a week for the next month to develop the design concept, you decide to collect some information on key technological trends in steelmaking. While perusing the *Annual Statistical Report* published by the American Iron and Steel Institute (the trade association of the steel industry), you are particularly struck by how rapidly continuous casting technology has replaced traditional ingot pouring, cooling, and reheating as the first stage in steel fabrication (Fig. 3–4). This turns out to be extremely valuable information, because you believe that your laser device will be extremely effective in monitoring temperatures in the continuous casting process.

One of the key functions of trade associations is collecting, analyzing, and disseminating industry-specific data such as that shown in Figure 3–4. This data may include estimates of employment, production costs, sales, consumption of raw materials and energy, and imports and exports. Typically this data is published in an annual report issued by the trade association. Many companies are reluctant to release certain data about their operations because it could affect their competitive advantage if it

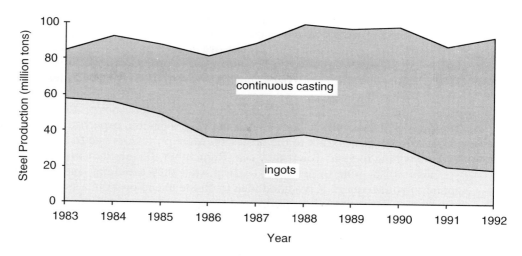

Figure 3–4.   Trends in Steel Casting. *(American Iron and Steel Institute, p. 71.)*

fell into a competitor's hand. These same companies may be willing to give this data to their trade association provided that the association will only publish industry-wide totals that could not be deciphered to reveal company-specific data. In this way, all companies in the industry benefit from having access to the industry-wide totals without compromising the confidentiality of their company-specific data.

Industrial Research Institutes and Consortia

Some industries have established affiliated entities which carry out the research agenda of the industry. Examples are the Electric Power Research Institute (EPRI) and the Gas Research Institute (GRI) which are the research arms of the electric power and natural gas industries respectively. Also, the National Food Processors Association operates three scientific research laboratories, which examine such areas as new food processing technologies, quality control, and sanitation techniques. Another example is the cooperative consortium between the federal government and the automobile industry to develop battery technology for electric cars. These and similar institutes typically publish a monthly magazine highlighting some of their research activities. They also issue many dozens of technical reports each year on specific topics. Some technical libraries subscribe to the technical reports series, and microfiche versions are available in the reference section.

Colleges and Universities

Engineering faculty members are excellent sources of technical information. But most undergraduate engineering students see their professors only within the context of the classroom. Some students may be intimidated by a professor who "knows so much," or is an authority figure. As a result, they might not even entertain the possibility of approaching a professor outside a classroom.

However, most engineering faculty are actively involved in research or consulting. Even if they are not currently working on a project related to your needs, they have a vast reservoir of prior experience to draw upon. It is a good idea to familiarize yourself with the technical and professional interests of the faculty in your department. Many departments have a list or brochure that describes professors' interests. If you find what looks like a good match between available expertise and your current information needs, go talk to the professor. Most professors like to discuss their work and are willing to assist you if they can. Remember though, they are busy people and you may not be able to get them to drop what they are doing and immediately respond to your request. It is a good idea to phone ahead or write a short note, requesting an appointment. While prior contact with the professor, such as taking their course, gives you a headstart, don't be reluctant to contact any individual faculty member. Engineering faculty are a source of information even after you graduate; many of them enjoy interacting with alumni on technical matters.

Federal Government

The federal government is the source of many different kinds of information. Regardless of your political persuasion or opinions of federal government bureaucracies, the United States federal government is one of the main collectors, analyzers, and disseminators of technical information that is valuable, if not absolutely necessary, for the success of many engineering design activities. The main forms are statistical data, research results, and laws and regulations. Many libraries have separate sections for government documents. While we confine our discussion here to the United States, all industrial countries have their own government-based technological infrastructure that are vital sources of technical information.

Executive branch.   Many federal agencies are essentially engineering research and development agencies. They publish regular reports on their various projects, including work done in-house by federal employees and work done at universities or private firms under contract to the agency. Some of the premier engineering test facilities in the country are at federal agencies. The National Aeronautics and Space Administration (NASA) is one such agency, whose aeronautics and aerospace research facilities such as Ames, Goddard, Langley, and Lewis approach legendary status. NASA's role in the manned space flight program makes it the client for one of the largest and longest running engineering design efforts in this country. Another premier federal engineering agency is the National Institutes of Standards and Technology (NIST) [formerly the National Bureau of Standards (NBS)], an arm of the U.S. Department of Commerce.

The national laboratories originally established in World War II to develop nuclear weapons are now overseen by the U.S. Department of Energy (DOE) and have broad mandates in the environmental and energy technology areas. Like their counterparts at NASA, the DOE national laboratories at places like Argonne,

Brookhaven, Oak Ridge, and Hanford are at the cutting edge of technological development. The newest member of that prestigious club is DOE's National Renewable Energy Laboratory (NREL) in Golden, CO.

Other federal agencies such as the Environmental Protection Agency (EPA) and the Department of Transportation (DOT) also have high quality engineering development and design elements to support their missions. For example, DOT operates a fifty square-mile Transportation Test Center near Pueblo, CO that is used by both private industry and government for testing many different ground transportation systems.

Although small by comparison with many other government agencies, the National Science Foundation (NSF) is a crucial source of technical information because, through its Engineering Directorate, it funds much engineering design research and curriculum development projects. The Department of Defense (DOD) also has some of the best technology research and development facilities. While the size of their activities may have decreased since the end of the Cold War, the quality of their technical personnel and equipment remains excellent. Included in this category are places such as the Naval Research Laboratory in Washington, D.C. and the Air Force Arnold Engineering Development Center in Tullahoma, TN.

Much of the engineering research and development effort conducted in-house at these government facilities is described in technical reports issued by the appropriate federal agency. A comprehensive guide to these reports is the *Monthly Catalog of United States Government Publications*. A typical issue may contain 600–900 pages.

The National Technical Information Service (NTIS), an arm of the U.S. Commerce Department, is the major distributor of technical reports written by federal agency grantees and contractors. A listing of new reports available from NTIS, *Government Reports: Announcements and Index*, is published twice a month.

In many cases, individuals can get on the mailing list for free subscriptions to agency announcements. For example, the Energy Information Administration of the U.S. Department of Energy issues *EIA New Releases* every two months. This document describes new reports and how to obtain them. Many of its products are available for purchase from EIA's National Energy Information Center. Other documents also available free from EIA include the *EIA Directory of Electronic Products* and *EIA Annual Report to Congress*. Data and model diskettes are available from NTIS or the DOE Office of Scientific and Technical Information (OSTI). Single issue EIA reports and subscriptions to regular series of reports can be purchased through the U.S. Government Printing Office.

Statistical and data sources.   One of the unique roles of the government is to collect, analyze, and disseminate lots of statistical data that no other entity can do. See our earlier discussion in this section regarding a similar role played by trade associations for specific industries. Since each trade association focuses on the data requirements of its members, and uses its own procedures and methodologies, only the federal government can apply a consistent approach across all industries to technical data requirements.

The major federal agency in this field is the U.S. Census Bureau, which is part of the Department of Commerce. In addition to the familiar population census conducted every ten years, the Census Bureau also conducts other, more specialized censuses. Of particular interest to engineers are the Census of Manufactures and the Census of the Construction Industry conducted every five years. The Census Bureau also administers data-collection efforts for the agencies. An example is the Manufacturing Energy Consumption Survey conducted for the U.S. Department of Energy every two years.

The results of federal agency statistical collection and analysis efforts are published in regularly issued reports. These reports are listed and summarized in the *American Statistics Index (ASI),* published annually. As with most data-bases, a CD-ROM version of *ASI* is available.

Congress.   Congress passes many laws of direct interest to engineers, and annually approves the federal budget; that includes the R&D budget. In the process of doing so, Congressional Committees hold hearings in which they examine the proposed budget and receive testimony from agency representatives and other interested parties. These hearing records, often with detailed appendices, are published as committee reports. It is fascinating how thoroughly some of the major technological issues and trends in this country are examined and discussed in Congress. Recent high-profile issues that were the subject of intensive congressional debate include the space station and the superconducting supercollider. To find out what reports are available, and who testified at the hearings, use the *Abstracts of Congressional Publications.*

Decisions on technical matters are made by almost every committee, but the locus of activities relevant to the engineering community is the Commerce, Science, and Transportation Committee in the Senate, particularly its subcommittee on Science, Technology, and Space. In the House of Representatives, the key committee is the House Science and Technology Committee.

General Accounting Office.   The General Accounting Office (GAO) is the investigative arm of Congress. This office conducts studies of how well executive branch agencies are implementing congressional directives. You can get a free subscription to their monthly listing of new reports. Among their topical areas are science and space, environment, and energy. Single copies of the reports are free.

State and Local Governments

Not all laws and regulations affecting engineers are at the federal level. States and communities have long been involved in public safety measures such as building codes, water supply and sewage disposal, and regulating operations of telecommunications and energy utility companies. Statistics published by state agencies, as well as trade associations and private research organizations are listed in the *Statistical Reference Index (SRI).*

There are state counterparts to the federal Code of Federal Regulations. In Washington state, it is called the Washington Administrative Code (WAC). Every state has an equivalent body of laws and regulations that potentially apply to many engineering design activities in that state. For a given technological issue, such as incineration of hazardous waste, it would be inefficient if each state government, in isolation from all other states, researched and analyzed the issue before adopting rules. There is a natural tendency for the states to keep in touch with each other and exchange information. Two of the mechanisms for doing that are the National Conference of State Legislators (NCSL) and the National Governors Association (NGA).

International Organizations

The engineering profession is becoming increasingly global in its outlook and activities. Engineering firms are accepting more design projects from clients in other countries, and opening up overseas branch offices. While our discussion of information requirements and sources in this chapter has been almost exclusively U.S. oriented, you may need to broaden your perspective to include other countries, or the international community at large. A good place to find global-scale statistics published by the United Nations, World Health Organization, and other international organizations is the *Index to International Statistics*.

3.5. PERSONAL NETWORKS

We now turn to active forms of information exchange in which the distinction between collection and dissemination becomes blurred. In this section we concentrate on the personal networking phenomenon.

Many design engineers develop their own network of other engineers working in the same field. They may see these colleagues once or twice a year at technical conferences and keep in touch by phone more frequently. Even though they may work for different companies, much technical information is exchanged via these informal networks. An adage worth keeping in mind is a variation of a familiar saying: It's not what you know, it's not who you know, it's what who you know knows. An example of this is relayed by the distinguished aeronautical engineer Walter Vincenti in his fascinating account of major breakthroughs in aircraft design. In his story about the development of flush riveting technology in the 1930s and 1940s, he explains the situation:

> As usual, generation of new knowledge depended in part on its interchange. Although the various companies operated essentially independently in the beginning (and maintained their individual programs later), communication grew as time went on. . . . presentations and discussion at technical meetings provided a useful if somewhat delayed channel. So also did technical journals and trade magazines. More immediate exchange took place by word of mouth; according to an engineer at North American, "When I got in a jam, I'd call my college friends at other companies."[4]

[4]Vincenti, p. 187.

Several more contemporary stories dealing with knowledge exchange are offered by Eric von Hippel.[5] At first glance, you might be tempted to question the loyalty of the engineer who provides this information to a friend who works for a competing firm. But as pointed out by von Hippel, this informal exchange is conducted only among peers that respect each other's professional skills, and there is an implication that a continuing relationship will result in information flowing in both directions. Both companies expect to gain from the exchange over the long term. Also, the sharing stops short of revealing closely guarded trade secrets that confer a competitive advantage.

Conferences

Engineers with similar technical interests get together to exchange information at technical conferences. Each of the large professional engineering societies sponsor dozens of technical conferences each year. For example, the December 1994 issue of *Mechanical Engineering*, the monthly magazine of the American Society of Mechanical Engineers (ASME), listed sixty-seven national and international conferences held in 1995 that were either sponsored or co-sponsored by ASME, or were of primary interest to its members. The smaller conferences may have fifty to one hundred attendees; the larger ones can involve several thousand engineers.

Conference Attendance

Some of these conferences last for several days. If you attended as many of these conferences as possible, you would spend so much time at conferences that it would be the equivalent of a full-time job. A more realistic scenario is to make attendance at one or two conferences a year an integral part of your strategic career development plans. Many companies are willing to provide at least partial financial support for your attendance at conferences that are relevant to your job responsibilities.

Five major kinds of activities occur at these technical conferences:

1. Each conference has a formal agenda that consists of oral presentations by individual engineers on specific topics. These presentations are usually grouped into sessions with common themes, with the program published in advance of the conference. Each session typically has several presentations lasting ten to twenty minutes each with opportunity for questions and discussion from the audience. In many conferences, the oral presentations are summaries of written papers that are available as separate documents or as bound conference proceedings. Depending on the conference, a copy of the proceedings may be included as part of the registration fee.

2. Many conferences also include vendor exhibits that display and demonstrate the vendor's products and services. Typical exhibitors at engineering conferences

---

[5]von Hippel, pp. 58–64.

are equipment manufacturers, software developers, book and periodical publishers, and consulting firms. Exhibit booths are usually staffed by the vendor's engineering personnel throughout the conference. Conference attendees can wander through the exhibit area at their own convenience, picking up samples and brochures and talking to vendor representatives. This is an excellent way to acquire information and expand your personal network.

3. Field trips to nearby manufacturing plants, research laboratories or construction sites give conference attendees a chance to see some of the outstanding technical activities and accomplishments of the host city. There are also one or more luncheon or dinner meetings with a keynote speaker or other ceremony.

4. Many conference organizers schedule formal seminars, workshops, and continuing education short courses as special events either just prior to or immediately after major technical conferences. These events, though held at the same location as the conference, may not be an official part of the conference program. The conference organizers hope that some conference attendees will either come a day or so early or stay an extra day or two to participate in the special event. Sometimes the dynamics operate in reverse—the course may be the major attraction for some engineers who then decide to take in a day or two of the conference. These events generally last from a half day to two days and have more of a classroom atmosphere than the technical paper sessions.

5. Some of the most useful information exchanges occur through informal interactions during the conference. You may be standing in line during a coffee break and strike up a conversation with the person next to you. Or you may be talking to an acquaintance at one of the exhibits when another party joins the conversation. A particular presentation at one of the sessions may have caught your attention, but you didn't have a chance to ask a follow-up question. You may run into the presenter several hours later in the hallway outside another session and take advantage of that opportunity to introduce yourself. Other opportunities may occur if you join a small group to go out for dinner at a local restaurant, or when you run into someone later that night in the hotel lobby. Many engineers feel that these interactions are the most valuable part of any technical conference. It is a good idea to make sure you have an ample supply of your business cards with you so that you can follow-up on the instant network expanders described above.

Conference participation.   In addition to attending technical conferences, there are opportunities for you to participate by presenting a paper or by assisting in planning the conference. With regard to the former, most conferences issue a "Call for Papers" from six months to a year or more prior to the conference. These announcements usually describe the scope of the conference, the special areas in which papers are being solicited, and the deadlines for submission of abstracts and complete papers. The normal process is to invite abstracts first, then the conference planning committee decides which abstracts to accept. The authors of the accepted abstracts are then invited to submit their complete manuscript by a certain date. A typical "Call for Papers" announcement is the following one that appeared in the October 1994 issue of *Mechanical Engineering* magazine:

> Fourth International Conference on Computer-Aided Optimum Design of Structures
> Sept. 19–21, 1995
> Key Biscayne, FL

> The objective of the conference is to bring together researchers and engineers in order to communicate recent advances in structural optimization and to demonstrate how optimization can be applied in engineering practice. Papers are sought on the following, or related, topics: . . . Submit abstracts by Dec. 19 to . . . Final papers are due May 26, 1995.

The technical program for most technical conferences is organized by volunteer engineers who serve on various committees. Much of the logistical work such as locating a conference facility (frequently a hotel) that can accommodate the conference needs for meeting rooms, exhibit space, and banquets is handled by professional staff.

Being involved in these conferences in a capacity other than a spectator is an extremely valuable way for you to strengthen and expand your personal network of peers. Also, many employers who may be reluctant to fund your attendance at a conference, may be favorably inclined if you are presenting a paper or are part of the organizing committee.

Continuing education courses. Continuing education courses generally are intensive treatments of very specialized topics. They usually consist of four to eight hours a day of class lecture/discussion for one to three days. An example is the two day course on *Water Hammer and Fluid Structure Interaction in Piping Systems* offered by ASME. This is clearly an atmosphere in which you can do some serious networking with the other attendees.

While some continuing education courses are scheduled in conjunction with major technical conferences, others are scheduled independently of conferences. The more popular courses may be offered several times a year in different locations around the country. Some engineering schools offer their own schedule of continuing education courses, and there are some for-profit companies in the business of offering continuing education courses for engineers.

3.6. GROUP DYNAMICS

Another form of active information transfer occurs within design teams. Most engineering design work is done as part of a project team, so engineers need to communicate with other engineers and with nonengineers who are part of many design teams.

At first, participation in group activities may be difficult for many engineering students. Outstanding students may feel that they have nothing to gain from being part of a group effort; they believe they can do the entire assignment better on their own. Students who don't have an assertive personality may feel awkward within a group structure, and upset because they are never involved in the real fun part of the project. And what about grades? How are you going to get credit for all the extra effort that you put into the project? One of the other group members may be freeloading, yet they may end up with the same grade as you. Or, you may feel that one or more of the other group members are just jerks!

Effective communication among team members is part of a broader relationship called group dynamics. This is an important field of psychology, and there are many books devoted entirely to the subject of group dynamics. As we shall see many times throughout this book, many of the topics covered are just the tip of the iceberg—introductions to fields of knowledge that in themselves are the subject of books and entire courses. One of the important characteristics of the study of engineering design is synthesizing information from many disparate fields. A good design engineer should be willing to venture out into those previously unexplored regions of the swamp to capture the more exotic creatures.

Personality Characteristics

One of the reasons that groups do not function well is that some individuals do not appreciate the differences in personality types among the group members. Many individuals assume that the other group members think and work the same way they do. There is a large and rich psychology literature dealing with different types of personalities. Almost everyone has heard of Sigmund Freud and Carl Jung, who were pioneers in this field. Included in this literature are various attempts to identify distinct personality types and devise tests by which individuals can be characterized according to those types. One of the approaches that has been used to provide insight into the group dynamics of engineering design teams is the Myer-Briggs Type Indicator. This approach uses all combinations of the following four personality dimensions to classify individuals into sixteen categories:[6]

1. Extroversion versus Introversion: whether a person's attention is directed to people and things or to ideas
2. Sensing versus Intuition: whether a person prefers to perceive information by the senses or by intuition
3. Thinking versus Feeling: whether an individual prefers to use logic and analytic thinking or feelings in making judgments
4. Judgment versus Perception: whether an individual uses judgment or perception as a way of life; that is, does the individual evaluate events in terms of a set of standards or simply experience them?

Our point here is to alert you to the fact that you are almost certainly going to be part of a design group in which there are a wide range of personality types. We don't expect engineers to become psychologists in these situations. But we can ask engineers

---

[6]Corsini, p. 447.

to be aware that there are distinctly different ways in which people think and act and interact with others. We can expect engineers to be sensitive to these differences, and to be flexible enough in a group situation to allow each person to feel comfortable, to contribute in their own way, and to feel as if their efforts are appreciated by the other members of the group. Engineering design groups are transitory. They usually are disbanded as projects terminate, or even modified as people enter and leave over the course of a project. In these circumstances, it doesn't pay to try to change people's personalities. A strategy that accommodates those differences is much more likely to lead to a successful group experience and product.

Group Structure

One of the purposes of group design projects in undergraduate engineering is to prepare you for professional design experiences. However, it is important to realize some of the ways in which a university-based design project team cannot capture the more complicated reality of professional design teams.

If you are taking a design class in a specific engineering discipline, say chemical engineering, then all members of your design team are also chemical engineering students. In a freshman or sophomore design class, students may not yet have identified with a particular engineering discipline, and their interests may be quite diverse. At least all the students are engineering students. In some situations, an upper division or graduate level interdisciplinary design course or project might be structured to engage engineering students from several engineering disciplines. In other cases, engineering students may be teamed up with nonengineering students, such as students from business administration, urban planning, or environmental studies, or law. Such projects are infrequent in a university setting because of the institutional barriers between disciplines, but they come closest to "real world" design situations. The more disciplines involved, the greater the communication barriers since each discipline has its own terminology, jargon, and paradigms.

Design groups can be formed in different ways. In a university setting, some professors allow the students in the class to organize themselves into teams. Others assign the individuals to groups on a random basis. Another approach is for the instructor to survey student interests and skills, and then make the group assignments. This comes closest to the way groups are formed in the professional world. Rarely do individuals in a professional design setting have much say about which project they get assigned, or who the other members of the group are. Those decisions are made by supervisors who are responsible for matching the firm's talent with their clients' needs.

Another characteristic of professional design groups is the different levels of experience and skills among the team members. Although the opportunities for doing so are rare, this can be modeled in a university by involving students at different levels of education, from freshman to graduate, in a group.

Both in school and in professional life, each individual member of the group may have different goals and priorities. Individual goals may not coincide with the group goals, or of other members of the group. To the extent that each group member is aware of each other member's agenda, the group activities can be more realistically planned.

Many engineering students complain that the workload expected of them by each faculty member is unreasonable because the faculty member does not fully appreciate the demands placed on them simultaneously by other faculty members. Many of these students indicate they can't wait to get out in the real world where they can focus on one thing at a time. Well, more often than not, engineers in the real world are participating in more than one design project at a time, with expectations and demands from each group leader similar to what they experienced back in the good old days in school.

Leaders and Followers

Any discussion of group dynamics is incomplete without addressing the issue of group leadership and the relation between a group leader and the other members of the group. Leadership is one of the more difficult attributes of human behavior to understand. Many studies have succeeded in debunking many longstanding myths about leadership, but they have been less successful in replacing them with conclusive evidence that explains individual and group leadership qualities. Nevertheless, it is worthwhile to discuss some of the key evidence, and some of the tentative indications of that evidence.

You've probably heard the expression "born leader." Well, there is no evidence to support this myth that leadership is a genetic trait. In fact, there is little evidence to suggest that leadership is associated with either certain personality, physical, or intellectual characteristics.

If a leader is not appointed when a new group is formed, it takes a while for leaders to emerge. The one behavior that seems to be the best indicator of a potential leader is the individual who talks the most. It's not necessarily what the person says, but how much they say. One possible interpretation of this phenomenon is that being animated indicates a strong motivation, interest, and desire to succeed.

Before we go too much further with this discussion, let us recognize something that most of us already know from experience—some people do not want to be leaders. They are more comfortable being a follower, and are very good at it. A good leader may be good primarily because they are in a group with good followers. A more sophisticated analysis is to determine what followers expect from their leader. If we could figure that out, then the individual that can best fulfill their group's expectations may be the best choice for a leader. Also, an individual's desire or willingness to be a leader may depend strongly on the situation. The same individual may be an effective leader in one situation but not in another.

What does a leader do? There appear to be two key functions of a leader. The first one is task oriented. The role of the leader is to organize the group, help define its goals, monitor its progress towards those goals, and make suggestions, decisions, and adjustments as necessary to achieve those goals. The second leadership function is relationship oriented. The leader should be responsive to and considerate of the needs of each group member.

Is a leader always needed? The need for a leader seems to increase as the size of the group increases. Also, if the group perceives the situation they are in as a crisis, there appears to be more desire for leadership. A factor that seems to decrease the need for leadership is if all members of the group perceive the group as consisting of people with similar levels of skills.

Group Tasks

Group meetings and interactions are more difficult in school than at work because each individual has their own class, work, sleep, and play schedule, and opportunities for regular group interaction on a daily basis are not very good. Also, group assignments may last only several weeks or one academic quarter. After that, each member goes their own way. In these circumstances, it is difficult to demand loyalty to what is only a fleeting alliance. Hence, effective group effort requires some discipline.

There is a tendency for newcomers to group projects to think that activities are engaged in by the entire group. You all go over to the library together to find the needed reference book or you all sit around a table performing the calculations together. This approach can be very inefficient and is susceptible to some group members, physically as well as intellectually, stepping on each other's toes. Give yourself some space. Only do as a group those activities and tasks that can best be done as a group. Many of the tasks can better be done by parceling them out to individual group members. With e-mail virtually universally available on college campuses, face-to-face group meetings don't have to occur as often as before.

Group project activities go in a cycle. The group meets, and tasks are assigned to individuals. The individuals go off to accomplish those tasks. At the next group meeting, each individual reports back to the group. After discussion and evaluation and execution of group-oriented tasks, new tasks are assigned to individuals and the cycle repeats itself. The balance between group-oriented tasks and individual-oriented tasks will vary from project to project. In some projects, most tasks can be allocated to individuals. In other projects, most of the tasks can be tackled as a group effort. Of course, the composition of the group may, to a certain extent, affect whether certain tasks are individual or group-oriented.

Over the course of a project, each individual in the group is expected to execute a set of tasks. However that assignment is made (whether by the group leader, by volunteers stepping forward, or by random lottery), the issue arises whether some personality types are better suited to carry out certain tasks. Some psychologists believe that a balanced team requires a good match between the individual members of the team and the key roles needed for the team to be successful.[7] One list of essential roles for a creative and innovative team is: generators, integrators, developers, and perfectors. Generators are people who have lots of ideas. Integrators are good at integrating other people's ideas into credible proposals. These proposals are advanced by the developers beyond the idea stage into a product or process. The perfector specializes in improving the product or process and reducing its cost.[8]

[7]Pearson and Gunz, pp. 139–163.
[8]Clutterbuck, p. 21.

The best advice is to recognize that design project group dynamics are fragile. Team members should always be alert to indications that the group is not functioning as effectively as possible. Individuals should feel free to bring these to the attention of the group leader. If the group does not have a leader, the issue of current effectiveness of the group should be a standing item on the agenda for each group meeting.

3.7. ORAL COMMUNICATIONS

Much technical information is exchanged verbally. This can occur in either an active or passive mode (see Sec. 3.2). Informal personal conversations represent the active mode of oral information exchange. These may be face-to-face with other individuals, telephone calls, or small group meetings that may occur spontaneously in the hallway (or almost spontaneously) when one member of the team gets an idea which they want to discuss with another individual or several members of the design team. The meeting may occur in someone's office, or in a small conference room if one is available. These interactions are covered by our group dynamics discussion in the previous section.

Formal Oral Presentations

Formal oral communications such as presentations tend to be more passive in nature, although many presentations have an active component in the form of a question and answer session at the end of the presentation. Some experienced presenters are willing to entertain questions at any time during their presentation. This is not a recommended format for less experienced engineers, because of the additional skills needed to handle frequent interruptions without being diverted from the main objective of the presentation. Nevertheless, oral presentations provide the opportunity to obtain instant feedback from the audience. You want to take advantage of that opportunity, and design and deliver the presentation in a way that elicits constructive feedback in a timely manner.

Never underestimate the importance of an oral presentation. Many company executives and clients must frequently make important decisions without the luxury of reading a detailed written analysis of all the ramifications. A decision of whether or not to approve your design proposal or to hire your design consulting firm instead of another may be based mostly, if not exclusively, on the impression you make when you deliver an oral presentation. Regardless of the specific content of the presentation, you want to leave your audience with the overall impression that you are a highly skilled professional design engineer and your proposals/results are technically sound.

Over the course of a design project, one or more of the design engineers most likely will be asked to provide oral briefings on the progress or accomplishments of the project. These presentations may be made to many different audiences—other members of the same design group, the leader of the group, company management which oversees your group and other design groups, your client or potential client, open-audience conference presentation to peers from other companies, or nontechnical groups such as government panels or neighborhood organizations. Presentations have to be tailored differently for each audience because each audience has a different level of familiarity with the particular project/technology and each audience will be interested in different kinds of information. Also, the level of candor of the presentation may change. Certain details which you would be encouraged to discuss during a presentation to an internal audience might have to be withheld during an open presentation in order to protect your company's competitive position.

Presentation Design

There is a well-known description of the three stages of a good oral presentation:

1. Tell them what you are going to say.
2. Say it.
3. Tell them what you just said.

Essentially this means that you provide an overview of the presentation before you get into the details, then summarize the presentation at the end. This list should be expanded because every presentation should begin with the following procedure:

> Give the title of the presentation. Identify yourself and other members of your group. Provide any other relevant information to put the upcoming presentation in the proper context.

When a group is asked to give a presentation, that doesn't necessarily mean that a group presentation is the most effective approach to take to get your message across. If one member of your group has the skills and the interest, the most effective approach may be for that individual to give the entire presentation. If you choose to go with a shared presentation, you have to deal with the added burden of assuring that there are smooth transitions between speakers. Another potentially awkward situation that arises in group presentations is how to provide smooth responses to questions from the audience. The last thing you want to happen is for different members of the presentation team to interrupt each other or contradict each other when responding to questions. One way to handle this is to designate one member of the team to field all questions and let that person answer the questions him/herself or call on other team members as appropriate.

Preparation

Engineers tend to underestimate the amount of time required to prepare an oral presentation. You should prepare for an upcoming formal oral presentation in the same way that you would prepare a formal written report. Start by developing an outline and expect the presentation to go through several revisions before you deliver it.

When satisfied with the content, prepare a set of visual aids (discussed below) that illustrate every point that you want your audience to focus on.

Some presenters feel that all they need is a good set of visual aids, and they can deliver their remarks solely from the cues contained in the visual aids. Others are more comfortable with a set of $3'' \times 5''$ index cards, one for each visual aid, to provide supplementary cues. Never, never plan to read directly from either the visual aids or the cue cards.

Plan to set aside enough time for several rehearsals, including in front of a mirror and in front of teammates, friends, and relatives. If English is not your native language, or it is but you have a strong regional accent, plan to put extra effort into your rehearsals and ask your rehearsal audience to alert you to pronunciation problems. Even if you think your pronunciation is fine, many novices tend to speak too softly, speak with their back to the audience, wander away from the microphone, let their sentences trail off without a sharp ending, and unconsciously resort to audible pauses such as "um, er," and "ah". These and many other practices can significantly detract from the effectiveness or your presentation, and you should use rehearsals to weed them out. If possible, rehearse in the same room in which the presentation is scheduled to be delivered, or one as close to it as possible in size and seating arrangement. Make at least one uninterrupted run in order to time the presentation. A quick way to make enemies is for your presentation to exceed the time allotted to it. Make sure to dress neatly and conservatively. You want your audience to focus on what you are saying and showing them, not on the gorgeous jacket you're wearing or the ketchup stain on your shirt.

Visual Aids

Good presentations make liberal use of visual aids. Several very sophisticated software packages are commercially available that can help you organize your presentation and prepare professional quality slides for use as visual aids. We discuss the design of visual aids in Section 3.9; here we concentrate on how to use them in your presentation.

There are two traditional formats for slides and the equipment for projecting them onto a screen. One is 35mm slides and a carousel-type projector; the other is transparencies and an overhead projector. The 35mm slides are photographic quality. The projector can feed them automatically and be controlled remotely. This gives the presenter freedom to move around during the presentation. If your presentation involves a display of photographs or other complicated images, this format is far superior to using transparencies.

On the other hand, transparencies are much simpler than 35mm slides to prepare—they normally can be produced on a standard office copy machine. The transparencies are standard $8\frac{1}{2}'' \times 11''$ note paper size so they can be conveniently stored and transported in three-ring binders. Also, an overhead projector is not as temperamental as a 35mm projector. We have all seen presentations interrupted because the projector jammed or the slides were inserted backwards or upside down, or both. Further, 35mm slides require the room to be darkened while transparencies show up well even in daylight. A darkened room is an invitation to some audience members (they're in every audience, it seems) to doze off.

Another advantage of overhead projectors is that they allow you to face the audience and look at the slide while you are explaining or discussing it. With the 35mm format, you have to turn away from the audience to look at the image on the screen. Since transparencies lie on top of the projector's horizontal surface in convenient reach of the presenter, the presenter can write on them with marker pens during the presentation to emphasize a point or to respond to a question. By stacking two or more transparencies, an overlay effect can be created to show a sequence of actions. One of the disadvantages of overhead projectors is that they are bulky and have to be placed fairly close to the screen, thereby potentially obstructing the view of some members of the audience.

A third format which has recently become increasingly popular is projecting the contents of a computer display terminal onto a large screen. This creates many opportunities for incorporating motion and simulations into presentations. It allows the presenter to make real-time changes in the visual aids by using a mouse or entering commands on a keyboard. The availability of low-cost video taping systems offers another option for enhancing the visual impact of presentations.

Regardless of the format used for the slides, a pointing device is an indispensable tool for calling the audience's attention to a particular portion of a slide. A pen or pencil will suffice for an overhead; use it to point to the slide as it sits on the projector surface, not to the image on the screen. In this manner, you don't have to turn your back to the audience. Do not use a light beam pointers; they are hard to control and the constant flickering can be a distraction. Likewise, don't play or fidget with the pointer. That's one disadvantage of a collapsible pointer, it is hard to resist fidgeting with it.

One other option is to distribute hard copies of the visual aids to each member of the audience at the beginning of the presentation. This frees the audience from having to try to retain the contents of each slide as the presentation progresses, and it also gives them an opportunity to jot down explanatory notes or formulate questions to ask later. However, these handouts can be a distraction. Some of your audience will stop listening to you and either scan the handouts ahead of you, or focus on an early handout long after you have already moved on to a later part of the presentation. Some of them may never get back in sync with you. Distributing hard copies of your slides is a real double-edged sword. A possible compromise strategy is to announce ahead of time that you have hard copies with you and that you will give them to individuals who request them at the end of the presentation.

Slides are not the only form of visual aids. You can significantly increase the effectiveness of your presentation by displaying an appropriate physical object, provided the object is large enough for each member of the audience to see the key features as you explain them. It is not a good idea to pass objects around for the audience to inspect. People get distracted, and by the time those in the back of the room receive the object, you have moved on to other stages of your presentation. One approach is to save a few minutes of your allotted time so that interested people can come up after the presentation to inspect the objects. Then they can ask you questions on the spot.

Physical Arrangements

Before giving an oral presentation, it is important that you review the physical arrangements and make whatever adjustments are necessary to eliminate obstacles to an effective presentation. Is the lighting dim enough so that your visual aids are sharp, but light enough for you to see your notes and the members of the audience? Many rooms designed for oral presentations have several localized lighting sources that can be adjusted independently. If time permits, practice projecting one slide to verify that the projector is properly focused.

Before beginning your presentation, carefully listen for background noises from air conditioners and especially from external sources such as adjacent rooms and hallways and nearby streets. For meetings in hotels some additional sources are kitchen noise associated with meal preparation and clean-up and music from a social event in the room next door. Close doors and windows and take other measures under your control to reduce such distractions to a tolerable level. This may include talking to the hotel/conference management staff and asking for their assistance.

Scope out the geometry of the room, particularly with respect to where you plan to stand and where the projector is located. You want to minimize the chances that your body and the projector will obstruct views of the screen. Use books or whatever is handy to raise or tilt the projector so that the image is high enough to clear all obstacles. Because of room configuration and seating arrangements, it may not always be possible to eliminate all obstructions.

Delivery

Never start a presentation until you have all your materials in order, you are familiar with the relevant physical arrangements (location of light switches, etc.), you are composed, and your audience is ready for you. If you are not introduced by another person, you can catch the audience's attention by stepping up to the podium, clearing your throat, and asking the members in the back of the room whether they can hear you.

Speak clearly and slowly and avoid use of specialized jargon that your audience may not be familiar with. Take care to distinctly pronounce all words, and never chew gum during a presentation. Vary the tone and volume of your voice during the presentation to keep things lively. Use a microphone if needed, but be careful that you don't have it set so high that you get feedback, or make audience members uncomfortable, or disturb the speaker in the next room. A microphone that clips onto your jacket is preferable because it lets you have your hands free and you don't have to worry about keeping it at the proper distance.

Maintain eye-contact with the audience, but do not fixate on one particular segment of the audience. Also, shift body stance to different parts of the audience. Avoid distracting mannerisms such as fidgeting with pen, keys, jewelry, or rings. Especially avoid any noise-generating distractions such as jingling coins, tapping fingers, and clicking ball-point pens. Audience members will focus on the smallest distraction, even repetitive shifting of feet, and you will have lost their attention. That doesn't mean you should be motionless. A relaxed comfortable presence is what you should be shooting for. Motion is fine, if it is related to what you are saying. Shift your stance, use hand gestures, and change your facial expressions in appropriate ways to punctuate your key points and enhance the audience's interest.

With regard to visual aids, allow about one minute per slide and remove each slide as soon as you have covered the relevant points. Leave the screen blank until you are ready for the next slide. If there is a long period between slides, shut off the projector. But be careful, frequent on-off cycling of the projector during a presentation could cause the bulb to burn out.

Responding to Questions

If you are entertaining questions, be prepared for some blunt criticism of your design efforts and some rude remarks from an unpleasant audience member. If it is not obvious that everyone in the audience knows each other, ask each questioner to identify themselves and their affiliation. Repeat the question, even if you think everyone heard it. This gives you a chance to confirm whether or not you correctly interpreted the inquiry. It also gives you a few seconds to organize your answer. Never lose your cool, and respond politely even to what you consider to be "off the wall" or malicious questions.

Some questions are well-intended, but will indicate that the questioner really didn't understand even the basic concepts. Rather than getting tied up in a lengthy response to such a question, offer to respond directly to that individual in detail after the presentation is officially concluded. That way you don't lose the attention of the rest of your audience, and the questioner will be impressed by your generous offer of individualized response.

A questioner may detect a flaw or weak spot in your design or supporting analysis. If so, don't be reluctant to admit it. Thank the questioner for their insight and indicate how you hope to follow-up on it.

3.8. WRITTEN COMMUNICATIONS

Many engineering students take a course in technical writing as part of their undergraduate curriculum. For those of you who have not taken that course yet, this section outlines some elements of effective written communications in a very abbreviated form.

While there are generally accepted principles of effective written technical communications, there is considerable variation in style from one company to another. Some companies have rigid rules regarding the style and format of all company documents. We will present one set of style guidelines that may be useful. They are intended to provide specific information on important elements of effective technical writing, while still allowing considerable flexibility to accommodate variations in style. To the extent these guidelines differ from specific practices you have been instructed to use by your boss, your client, or your professor, these guidelines should be deferred in favor of the more specific instructions.

Just as there are many modes of oral communications, so too are there a variety of written communication forms and formats. Probably the least formal is e-mail. Another informal, but extremely important form of written communication is a design notebook, sometimes referred to as a design journal.

Design Journal

A design journal is a *permanently bound* (no three-ring or spiral) notebook or "diary" which contains *dated entries* of *all* your notes, sketches, calculations, doodles, and any other record of your thoughts and activities related to your design projects. Each entry may take many different forms including: a list of assumptions you are using; a free-body diagram you are constructing; a record of a phone conversation you recently had; or a draft of a section of your written report. A design journal serves several purposes:

- It is a central record of all project activity and information. By including everything in the journal, you minimize the chance that important material will get misplaced on scraps of loose paper.
- The journal serves as a chronological record that allows you to recall information, decisions, or calculations from earlier phases of the project.
- It helps you keep track of your work effort for budgeting and billing purposes.
- In case of a future legal dispute over patent rights or professional liability, the journal could be submitted as evidence to a court. However, in order to be accepted as evidence, the journal must be permanently bound with all entries dated.
- The journal serves as a basis for communicating with other engineers, your supervisor, and your client.
- The ISO 9000 series of international standards discussed in Chapter 4 require documentation of design activity to be certified for conducting business in countries that use the standards.

Memos

Memos are short communications, usually internal, either requesting or directing that action be taken, or reporting the status of an activity. In many situations, they serve as internal status reports. The key feature of memos is that they are short and can be read almost immediately upon receipt. Some organizations have a rule that memos cannot take up more than one page. A typical format for a memo heading is:

Date:
To:
From:
cc:
Subject:

It is a good idea to make the subject entry the hook that catches the attention of the intended recipient(s) and encourages them to read the memo.

Technical Reports

During the course of a design project, several different kinds of formal technical report may be required. These include proposals, progress reports, and final reports. Since the content and format of proposals are usually specified by the client, we will not discuss them in this section.

Progress reports. A progress report is written to answer common questions which concern your client or boss, usually on a periodic basis. These include various questions: "Will the project be completed on schedule?", "What progress has been made so far?", "Have you encountered any unforeseen difficulties?", and "What are your expectations for the next reporting period?" Some clients may specify that the progress report be in letter or memo format; others want a more formal report appearance. The format may also depend on the size of the project, its duration, and the frequency of progress reports.

Final report. The final written report is the most significant product of many design projects. In many cases it serves as the only permanent record of the project. Some material which may have been included in the progress reports may be appropriate for inclusion in the final report.

Format for Technical Reports

Written technical reports should be clear and concise, using diagrams, tables, and graphs where appropriate. As an example of conciseness, compare the previous sentence with this wordier version: Progress and final technical reports should be written in a clear and concise style, making use of diagrams, tables, and graphs where appropriate. The first version contains fifteen words versus twenty-three words for the second version. This 35% reduction in words with no loss in content is just one example of many opportunities that are usually available to increase the readability of a report. As you are finalizing any technical report, examine each sentence to see if similar economies can be achieved.

Reports should be carefully proofread before submission to minimize typographical errors and to ensure that appropriate margins, pagination, and other requirements for professional appearance are met. With the availability of modern word processing software complete with spelling and grammar checkers, there is no excuse for not submitting reports that are professional in appearance and readability. Some format guidelines are presented in the remainder of this subsection.

Letter of transmittal.   A letter of transmittal is addressed to the customer and summarizes in a nontechnical fashion the contents of the report. The letter should be no more than one page, and it should be signed by someone with authority in the organization. The letter accompanies, but is not bound with the report. This letter is the official notice that the report was submitted and usually is filed away by the recipient in a correspondence file. The reports themselves tend to end up on someone's desk or bookshelf.

Report cover.   The cover serves primarily to protect the report. If the cover is opaque, it should contain the report title and author identification. If it has a window or is fully transparent, the relevant information on the title page should be visible.

Title page.   The title page should include the title, name, and address of the authors, the name and address of the client, and the report date. Other elements such as the contract number, or report number, may also be included.

Abstract.   The abstract is on a page by itself and should be labeled as such. Be brief (less than one hundred words) but highlight the objective, methodologies, and significant findings and designs resulting from the work described in the report. The abstract should help the reader decide whether to read the complete report.

Table of contents.   The table of contents should list the headings and major subheadings and the pages on which they appear. Do not include anything which appears before the table of contents or the table itself. List the titles and page numbers of all appendices. Include a separate list of figures and list of tables.

Pagination and captions.   All pages should be numbered except for the title page. All figures and tables must be assigned a number and caption. Only include figures and tables that are referred to in the body of the report. Figure numbers and captions appear at the bottom of the figure. Table numbers and captions appear at the top of the table.

Introduction.*[9]   Introduce the subject of the report by providing background for the reader. Make sure you explain why the project was undertaken and how this work builds upon previous studies or designs. Explain how you came to work on the project and list your objectives. Refer to previously published reports, books, and articles to help the reader understand why the project was approached in a certain way and what constraints existed.

Results and discussion.*   This is the heart of the report. Discuss in detail the design process or solution technique. Explain how you arrived at all major decisions

---

[9]There are many ways to organize the body of the report. The following asterisked sections are illustrative, not prescriptive. You can replace them by another set of sections more appropriate to the specific topic and audience.

and justify them with economic, mathematical, or experimental evidence as appropriate. Include a full description of hardware and software developed and the results from experiments and numerical simulations. Diagrams, graphs, and tables should be used along with the narrative to enhance the discussion. These should be appropriately numbered and given descriptive titles. They may be placed within the body of the report, as soon after they are first mentioned in the narrative, or grouped together at the end of the report. Detailed mathematical developments, long lists of data, computer program listings, and complete engineering drawings should be referred to in the text but should be placed in appendices.

Conclusions.* Conclusions are brief statements of the important inferences that can be drawn based on the study. They should naturally follow from the section that details results.

Recommendations.* Recommendations advocate actions which are based on conclusions. They should be specific and directed to the readers of the report.

References. The reference section contains a list of all reports, articles, and books that were alluded to in the main body of the report or in the appendices. Do not include material which has not been explicitly referred to, even if it was used somehow in your project. If it is important, then make sure you refer to it in the text. Indicate the reference in the body of the report by placing a number in square brackets immediately after the phrase or fact. If an entire paragraph is based upon the same reference, place the number at the end. Number the references sequentially and list all references in numerical order at the back of the report. In that listing, include the author's name, the document title and identification details. Always give complete, unambiguous citations. Many books are regularly revised and updated, so give the edition and year of publication. If your source is a book, give the page number. In general you want to cite your references so that the reader can go directly to the exact source. This happens frequently when there is a new development, or things change, and the client wants to know whether the information is still valid. Consider the following examples:

1. Robert C. Juvinall, *Fundamentals of Machine Component Design*, John Wiley and Sons, New York, NY, 1983, p. 316.
2. B. W. Webb and R. Viskanta, "Radiation-Induced Buoyancy-Driven Flow in Rectangular Enclosures: Experiment and Analysis," *Journal of Heat Transfer*, Vol. 109, May 1987, pp. 427–433.

Sometimes information is received in an unpublished or verbal form. This can be referred to by stating the source's name, the date on which it was transmitted to you, and the words "personal communication."

Appendices. This is the place for very detailed or cumbersome material which would be disruptive to the reader if it were placed in the main body of the report.

Only include materials that have been referred to in the body of the report. You can include neatly handwritten analytical work, specification sheets, product literature, and even previously published material (if it would be difficult for the reader to obtain otherwise). All appendices should have titles, be paginated, and be listed in the table of contents.

3.9. GRAPHICAL COMMUNICATIONS

Graphical forms of communication are an integral part of almost every phase of engineering design. It has been argued that a major difference between craftspersons and engineers is that craftspersons transform their mental images directly into physical artifacts, while engineers use graphical techniques to show others how to create the artifact.[10] In this section we examine the roles of graphic communication in engineering design, the techniques (including software) for generating graphic images, and the principles of good graphic design. Specific rules and techniques of traditional engineering graphics (orthographic projection, dimensioning, isometric views, descriptive geometry) are not covered here; they are the subject of entire books and are taught in separate courses in many engineering curricula.

Functions of Graphical Communication

Design engineers use graphical techniques for a variety of purposes. The key functions of graphics in the design process are briefly described below.

Problem formulation aids.   We saw in Chapter 2 how the House of Quality chart can be very useful in helping to formulate a design problem. We also saw in Chapter 2 that the objectives tree graphical depiction is a very effective tool for clarifying the relationships among design objectives. In both cases, we were able to use pictorial representations to clarify complicated relationships.

Conceptual design aids.   Many design concepts first take form as free-hand sketches. As a design engineer, you will frequently sketch shapes on paper or on a chalkboard (or on the back of the proverbial envelope) to help clarify design ideas in your own mind. Frequently these sketches are accompanied by equations, numerical calculations, and written notes indicating size, cost, materials, and functions. All these elements contribute to an enhanced understanding of the design concept.[11] You may do this when you are by yourself, with other engineers, or with your client as you try to clarify various aspects of the design. In Chapter 6 we will examine morphological boxes and morphological charts as graphical aids to concept generation.

[10]Ferguson, pp. 3–5.
[11]Dym, p. 83.

Analytical aids.   One of the distinguishing features of engineering is the development and use of graphical tools as analytical aids. This includes using graphs to record and interpret experimental results. Many engineers find the notion of a free body diagram indispensable to correctly identify and determine the forces acting on an object. Chemical engineers rely on process flow diagrams to depict the key steps in chemical transformation of raw materials into a finished product. Electrical engineers use network diagrams (see Fig. 3–2) to assist in clarifying the flow of electricity in a circuit, and block diagrams for understanding the feedback loops in control systems. Engineers concerned with fluid flows (civil, chemical, and mechanical) use a graphical device called a control volume to help achieve the necessary mass and momentum balances. Computer software engineers will rely on logic flow diagrams to reveal relationships not otherwise apparent in the instructions needed to accomplish certain tasks. Thermodynamic cycle diagrams are an indispensable graphic technique for identifying the major thermodynamic characteristics of engines, boilers, and other devices in which liquids and gases undergo sequences of temperature and pressure changes.

Solution techniques.   Several graphical techniques are not just helpful in analytical work, but are substitutes for mathematical analysis as a means of solving a problem. Descriptive geometry is one such powerful tool for determining and representing spatial relationships. A classical problem amenable to solution by descriptive geometry techniques is the location and length of the shortest connector between two non-coplanar pipelines. Civil engineers may use force polygons to determine the forces in individual members of truss structures; mechanical engineers will frequently use graphical methods to determine the velocities and accelerations of different parts of a mechanism.

Decision-making aids.   In Chapter 9 we will introduce decision trees and objective trees as graphical tools to help crystallize the nature of complicated decisions faced by design engineers when trying to select from among several alternative design concepts. These tools are powerful techniques for pictorially arranging and organizing the elements of the alternatives to assist in formulating the choices, facilitating the selection process, and documenting the choices that are made.

Explaining design to others.   Graphical forms of communication are part of a universal language, helping the design engineer explain to someone else key aspects of their design. The forms of graphics used for this purpose range from preliminary sketches to final detailed drawings and technical reports, to visual aids at meetings and conferences, to pictorial representations for purposes of display, assembly, and maintenance.

Engineering detail and assembly drawings serve as self-sufficient plans for implementing a design, for example, constructing or manufacturing the object or system depicted in the drawings. These drawings give unambiguous information regarding dimensions and tolerances, bill of materials, fabrication and assembly instructions, relationships to other components, and procurement information. They also identify the persons responsible for the design and the dates and nature of any modifications since the original design was released.

Graphics Software

An abundance of commercial software is available for presenting technical information in a graphical format. The major word processing and spreadsheet software packages have built-in graphing capabilities that are adequate for many simple graphs in the form of two-dimensional and three-dimensional bar charts, pie charts, line graphs, and scatter plots. More complicated graphs with nonlinear or multiple scales, and with special presentation requirements, may require a dedicated statistical or technical graphing software package.

Another category of graphics software includes packages that have built-in templates to facilitate construction of simple technical diagrams and flowcharts. These representations are characterized by straight line and standard geometrical and technical shapes. Some of the charting software packages have built-in templates dedicated to standard symbols and shapes used in civil, mechanical, and electrical engineering technical diagrams and drawings. The electric circuit depicted in Figure 3–2 was drawn using such a package.

A third category of software products are those that allow you to draw and paint virtually any picture. These are useful for pictorial representations that do not require the precision and accurate dimensional renditions of traditional engineering drawings.

The final category of graphics software of interest to engineers are a wide range of computer-aided drafting, design and manufacturing (CAD, CADD, and CAD-CAM) packages. Included in this category are small systems capable of producing a fully dimensioned two-dimensional engineering drawing of a small component. There also are very sophisticated packages that can produce detailed drawings for complicated systems and accurate three-dimensional multicolor pictorial representations including shading. Some CAD software allows these images to be rotated on the screen to show how an individual part looks from any angle or how parts of an assembly relate to each other, including identification of interferences. These more powerful graphics packages have permitted designs to go into manufacturing without having the usual mock-up constructed to identify how the components fit together. Some of them automatically calculate volume and weight and inertia properties of drawn shapes, and are linked to analysis packages or have built-in analytical capability, so that the designs are automatically modified to reflect the design changes dictated by the results of the analyses. This includes coupling the graphical representations with optimization routines.

Graphics Design Issues

Two books by Edward Tufte treat graphical and pictorial displays of information as art forms and design problems themselves. One of Tufte's principles of graphical excellence is, "Graphical excellence is that which gives to the viewer the greatest num-

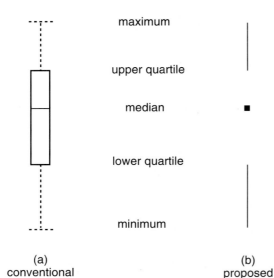

maximum

upper quartile

median

lower quartile

minimum

(a)
conventional

(b)
proposed

Figure 3–5. Two Versions of a Box-Whisker Plot *(Adapted from Tufte, 1983, p. 125, with permission from Edward Tufte.)*

ber of ideas in the shortest time with the least ink in the smallest space."[12] In this way, he extends to graphical communication of technical information some of the principles of good written technical communications. He adapts this general principle to the more restricted case of graphical display of data. In this context, it becomes, "Maximize the data-ink ratio, within reason,"[13] where the data-ink ratio is defined as the proportion of a graphic's ink devoted to the nonredundant display of data-information. A dramatic example of the application of this principle is Tufte's modification of a traditional box-whisker plot that displays the medians, upper and lower quartiles, and extreme values of data sets. Figure 3–5(a) shows the conventional configuration of a box-whisker plot and Figure 3–5(b) shows Tufte's proposed modification. The traditional version requires nine lines to construct, while the modified version requires only two lines and a filled circle. We don't lose any information in the transformation from version (a) to (b), but we substantially reduce the effort to construct the graph.

One of Tufte's arguments for his bare-bones approach to presenting technical information is muted by the availability of charting software that allows us to construct standard geometrical shapes with one click of the mouse, or with one keystroke. Also, the conventions used in Figure 3–5(a) may be so universally adopted that the revision proposed in Figure 3–5(b) is too radical to achieve widespread acceptance. Nevertheless, using Tufte's principles as general guidelines can be very useful in improving the effectiveness of graphic representations of engineering information. At the very least, design engineers should give as much thought to the effectiveness of their graphic representations as they do (or should do!) to their written communication forms.

[12]Tufte, 1983, p. 51.
[13]Ibid, p. 90.

Case Study of Graphic Representation of Trends in Steel Casting Technologies

Figure 3–6 shows the same data regarding technological trends in steel casting as Figure 3–4. Both were drawn with the same graphics software package, so the major differences are due to style and formatting choices. They both took about the same amount of effort, yet the version shown here is much more complicated than the earlier one and is not as effective in communicating key information. Let us analyze the reasons for this.

First, we look at the choice of the type of graph. In Figure 3–6 we plotted two curves, each showing the amount of steel produced using each casting technology. It is clear from the graph that continuous casting technology has increased substantially over the past decade while traditional ingot production has decreased. What isn't immediately apparent (though not that difficult to discern if we really try) is what has happened during that period to overall steel production. In Figure 3–4 we stacked the two graphs on top of each other so that the relative stability in total production was immediately obvious and an important part of the message.

Now let us examine several of the graphic elements in these two charts, keeping Tufte's principles in mind. Both the data points and the grid add a lot of clutter to Figure 3–6. Since our main message is overall trends, not precise numerical values, no useful information is lost by completely eliminating these elements, as was done in Figure 3–4. The short tick marks along the axes in Figure 3–4 locate the years and production levels as well as does the complete grid.

Comparing the vertical axes of these two graphs, we see that we used two techniques to reduce the clutter without sacrificing useful information. First, we chose units of million tons instead of thousand tons. Second, instead of placing tick marks

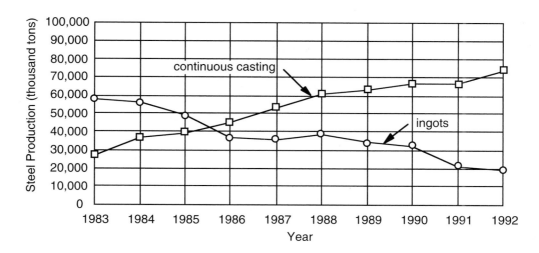

Figure 3–6.   Scatter Plot of Trends in Steel Casting

and labels every 10 million tons, Figure 3–4 only displays those elements at intervals of 20 million tons.

We used arrows in Figure 3–6 to associate labels with each of the two curves. One of the advantages of the stacked format used in Figure 3–4 is that we were able to eliminate the arrows and add shading to highlight the production trends.

Resisting the three-dimensional lure.   The commercial charting software packages give you the capability to present information in much more sophisticated ways than was previously possible. Users tend to become enamored with some of these features like three-dimensional effects, special fonts, and color treatments. However, a more elaborate format is not always as effective in delivering the message. It is easy to get carried away and overuse some of these features in situations where they really are not necessary. A good example of this is the three-dimensional version shown in Figure 3–7. In fact, the three-dimensional version obscures a lot of the message regarding the overall trends. Of course, when a third parameter or variable is an important part of a message, a three-dimensional chart can be very helpful.

Visual Aids Design

In Section 3.7 we discussed the importance of slides as part of an effective oral presentation. Readers have the luxury of examining text and graphs in a written document as long as necessary to absorb the needed information, and returning if necessary to the document to refresh his/her memory or focus more intensely on specific parts of the

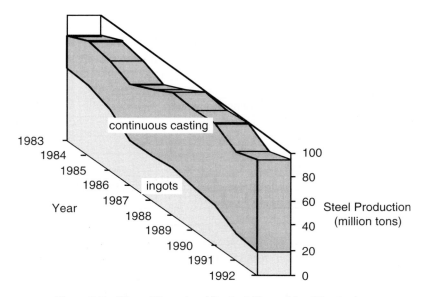

Figure 3–7.   Three-Dimensional Stacked Chart of Steel Production

text or graphic. On the other hand, a member of the audience at an oral presentation has only one shot, and a brief one at that, to digest the information displayed in a slide. Thus the emphasis in designing visual aids is on simplicity and clarity.

Text in visual aids.   Many slides used in oral presentations will be entirely text, but the choice of font and layout are so important to their effectiveness that it helps to think of slide design as a graphics issue. A crucial design issue is to make sure the font size is large enough so that it can be clearly seen from the back of the room. That usually means using boldface fonts of at least 20 points in size. If you commit yourself to this rule when you start designing the slides, you will automatically find that you are limited in the amount of material that you can include on each slide. That is good, because the amount of text on any slide should be limited. It is better to have more slides with less text on each one than fewer slides with more text on each one. There are three good rules of thumb to remember for displaying text on slides:

1. Each slide should contain only one idea.
2. Limit the number of words on each line to six.
3. Include no more than six lines of text on any slide.

Also, use mixed upper case and lower case rather than all upper case letters to improve readability.

We know that engineers love to deal with equations. But the observations above regarding text on slides hold doubly true for equations. The best policy is to place each equation on its own slide and never display an equation containing more than six terms or symbols.

Much engineering data can be effectively presented in a tabular form as part of a written document. But tables are notoriously ineffective formats for slides in an oral presentation. If you apply the above three text display guidelines to tables and count each number included in the table as a word, you'll find that it is very difficult for any but the most trivial table to satisfy these guidelines. In almost all cases, graphs are much more effective than tables as formats for displaying data in an oral presentation. However, if you must use tables, limit the number of rows and columns to three or four.

Charts and graphs.   Let's return to Figure 3–4, our model of good graphic design. As part of a written document such as this book, the graph and its accompanying caption serves its purpose. Readers have the luxury of examining a graphic in a written document as long as necessary to absorb the needed information, and returning if necessary to the document to refresh his/her memory or focus more intensely on specific part of the graphic. But some adjustments are needed if we want to use the graph as a visual aid in an oral presentation. A member of the audience at an oral presentation has only one shot, and a brief one at that, to digest the information. Thus the emphasis in designing a visual aid is on simplicity and clarity. The version of Figure 3–4 for use as a slide in an oral presentation is given in Figure 3–8.

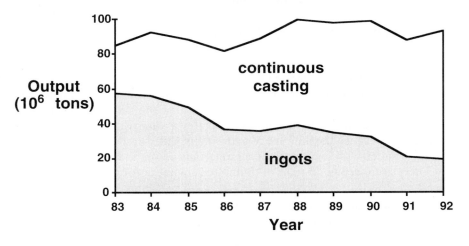

Figure 3–8.   Visual Aid Version of Steel Production Chart

A key modification we made was to increase the thickness of all lines and to convert all text to bold font. These changes are necessary to make the slide readable in a large room, but they make the chart appear more cluttered. To compensate for them, we dropped the first two digits in the labels for each year along the horizontal axis. We also rotated the caption for the vertical axis by 90° and wrote it in a more compact form. Finally we added a title and citation for the source of the data so that the slide is essentially stand-alone.

We show Figure 3–8 in black and white with gray shading added to enhance readability. Virtually all modern desktop computers have color monitors, and multi-color laser printers are becoming commonplace. However, unless you are really convinced that colors will enhance the effectiveness of the message, be careful of succumbing to the same temptation discussed above with regard to three-dimensional effects. Just because you have the capability to display color, don't assume that a multicolor slide is automatically more effective than a black and white slide. These principles of good design of visual aids also apply to flow charts and schematics.

3.10. CLOSURE

This chapter discussed the information collection and dissemination aspects of design. We treated both activities as special cases of a more general activity—the transfer of information. Though we discussed information collection in the first part of this chapter and information dissemination at the end of the chapter, a transition discussion in the middle of the chapter illustrated ways that information collection and dissemination overlap in interactive information exchange activities.

The chapter started with a brief treatment of different ways to organize our thinking regarding the information needs of design engineers. We then focused on different forms of technical information and the organizations that produce it. In the transition sections of the chapter we examined the simultaneous roles of engineers as both consumers of information and providers of information. That discussion included the topics of personal networks, group dynamics, and oral communications. The concluding sections of the chapter covered guidelines for effective written and graphical communications.

3.11. REFERENCES

American Iron and Steel Institute. 1992. *Annual Statistical Report of the American Iron and Steel Institute.* Washington, D.C.: American Iron and Steel Institute.

BENNETT, BYRON. Dec. 1988. New Filter Synthesis Technique—The Hourglass. *IEEE Transactions on Circuits and Systems.* Vol. 35, No. 12, pp. 1469–1477. New York: Institute of Electrical and Electronic Engineers.

CHEN, WAI-FAH. 1995. *The Civil Engineering Handbook.* Boca Raton, FL: CRC Press.

CLUTTERBUCK, D. Feb. 1979. R&D Under Management's Microscope. *International Management.* Vol. 34, No. 2.

CORSINI, RAYMOND J. 1994. *Encyclopedia of Psychology, 2nd Edition.* Vol. 2. New York, NY: John Wiley and Sons.

DIETER, G. E. 1991. *Engineering Design: A Materials and Processing Approach, 2nd Edition.* New York: McGraw-Hill, Inc.

DYM, CLIVE L. 1994. *Engineering Design: A Synthesis of Views.* New York: Cambridge University Press.

FERGUSON, EUGENE S. 1992. *Engineering and the Minds Eye.* Cambridge, MA: The MIT Press.

FINK, DONALD G. and H. WAYNE BEATY, ed. 1987. *Standard Handbook for Electrical Engineers, 12th Edition.* New York: McGraw-Hill Book Company.

FITZGERALD, NANCY. Dec. 1994. Talking Points: Guidelines for Giving Effective Engineering Presentations. *ASEE Prism,* Vol. 4, No. 4, pp. 28–31. Washington, DC: American Society for Engineering Education.

FORSYTH, DONELSON R. 1983. *An Introduction to Group Dynamics.* Monterey, CA: Brooks/Cole Publishing Company.

HOLLAND, F. A., F. A. WATSON, and J. K. WILKINSON. 1984. Process Economics. In *Perry's Chemical Engineer's Handbook, 6th edition,* eds. Robert H. Perry and Don Green. New York, NY: McGraw-Hill Book Co.

Index and Directory of Industry Standards. 1993. Englewood, CO: Information Handling Services.

International Trade Administration. 1994. *1994 U.S. Industrial Outlook, 35th Edition.* Washington, D.C.: International Trade Administration, U.S. Department of Commerce.

JAY, FRANK. 1984. *IEEE Standard Dictionary of Electrical and Electronic Terms, 3rd Edition.* New York: Institute of Electrical and Electronic Engineers.

KILEY, MARTIN and WILLIAM MOSELLE, eds. 1993. *1994 National Construction Estimator, 42nd Edition.* Carlsbad, CA: Craftsman Book Co.

LAVIGNE, JOHN R. 1993. *Pulp and Paper Dictionary, Revised Edition.* San Francisco, CA: Miller Freeman Books.

Marks' Standard Handbook for Mechanical Engineers. Tenth Edition. 1996. New York: McGraw-Hill Book Co.

McGraw Hill Encyclopedia of Science and Technology, 6th Edition. Vol. 6. New York: McGraw-Hill Book Company.

MEYERS, R. A., ed. 1987. *Encyclopedia of Physical Science and Technology.* Orlando: Academic Press, Inc.

PEARSON, ALAN W. and HUGH P. GUNZ. 1981. Project Groups. In *Groups at Work,* eds. Roy Payne and Cary L. Cooper. Chichester: John Wiley & Sons.

R. S. MEANS CO., INC. *Means Facilities Cost Data 1991, 6th annual edition.* 1991. Kingston, MA: R. S. Means Co. Inc.

SHAW, MARVIN E. 1981. *Group Dynamics: The Psychology of Small Group Behavior, 3rd Edition.* New York, NY: McGraw-Hill Publishing Company.

TUFTE, EDWARD R. 1983. *The Visual Display of Quantitative Information.* Cheshire, CT: Graphics Press.

TUFTE, EDWARD R. 1990. *Envisioning Information.* Cheshire, CT: Graphics Press.

VINCENTI, WALTER G. 1990. *What Engineers Know and How They Know It: Analytical Studies from Aeronautical History.* Baltimore, MD: The Johns Hopkins University Press.

VON HIPPEL, ERIC. Feb./March 1988. Trading Trade Secrets. *Technology Review.* Vol. 91, No. 2.

YOUNG, W. C. 1989. *Roark's Formulas for Stress and Strain, 6th edition.* New York: McGraw-Hill Book Co.

3.12. EXERCISES

1. Use *The Engineering Index* or *Compendex®* to identify (by author, title, and place and date of publication) five articles published in the last five years on the subject of control systems for industrial robots. Attach a photocopy of the title page of any one of the five articles you identified.

2. Use *The Engineering Index* or *Compendex®* to identify (by author, title, and place and date of publication) five articles published in the last five years on the subject of vibration in bearings. Attach a photocopy of the title page of any one of the five articles you identified.

3. As a design engineer for a high-performance bicycle manufacturer you have developed an innovative design for the front fork (the piece that holds the front wheel and is attached to the frame). What requirements are imposed on your design by the federal government? Your answer must be complete and unambiguous. Your client should be able to go directly to the exact source to verify your response.

4. You are designing a ventilation system for a new 30,000 ft^2 supermarket to be in conformance with the Energy Code in your state. How much outdoor air (ft^3/min) must your ven-

tilation system be capable of delivering to the interior of the supermarket? Provide complete and unambiguous citations of the specific provision(s) of a specific document to justify your response.

5. Determine the equation used in Section VIII, Division I of the *ASME Boiler and Pressure Vessel Code* for the minimum wall thickness *t* of a spherical shell of radius *R* subject to an internal pressure *P*. Your answer must be complete and unambiguous. Your client should be able to go directly to the exact source to verify your responses.

6. Determine the equation used in Section IV of the *ASME Boiler and Pressure Vessel Code* for the minimum wall thickness *t* of a cylindrical heating boiler of radius *R* subject to an internal pressure *P*. Your answer must be complete and unambiguous. Your client should be able to go directly to the exact source to verify your responses.

7. What is the minimum separation distance between two swings in a playground swingset according to the appropriate ASTM standard? Your answer must be complete and unambiguous. Your client should be able to go directly to the source to verify your response.

8. Determine from the appropriate ASTM standard the allowable wall thickness tolerance for a magnesium-alloy extruded tube with a 2.0″ outside diameter and 0.30″ nominal wall thickness.

9. Determine from the appropriate ASTM standard the minimum tensile strength for cold-drawn steel mechanical spring wire of diameter 1.0 mm.

10. You work for a German automobile manufacturer who wishes to sell passenger vans in the U.S. An existing model made by your company and now sold in Germany would be the preferred design but you must determine if it meets U.S. fuel economy standards. Your van is a four-wheeled vehicle, weighs 5,000 lbs gross, has two wheel drive, and carries twelve passengers. If introduced for the upcoming model year, what fuel economy must it achieve? Your answer must be complete and unambiguous. Your client should be able to go directly to the exact source to verify your response.

11. You are designing the elevator system for a new high-rise office building. You have proposed a layout as noted in the sketch. One of the senior engineers asks you, "Given the regulations implementing the Americans with Disabilities Act, how long must the elevator doors remain open?" What requirement does the ADA regulations impose on door opening time?

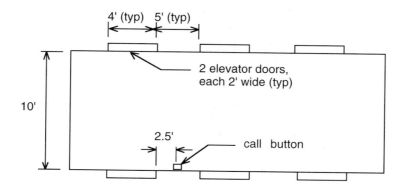

12. What is the ANSI/SAE standard recommended pressure test for snowmobile fuel tanks? Your response must be complete and unambiguous. Your client should be able to go directly to the exact source to verify your response.

13. The trade association for manufacturers of commercial and industrial air conditioning and refrigeration systems is the Air Conditioning and Refrigeration Institute (ARI). Use the ARI web site (www.ari.org) to summarize one of their current areas of research focus.

14. Starting with the American Petroleum Institute (API) web site (www.api.org) find the most recent API annual report on Petroleum Industry Environmental Performance. What are the latest statistics for release of toxic chemicals from petroleum refineries?

15. In 1970, an international treaty was concluded regarding the shipment of perishable foodstuffs between countries. According to that treaty, what is the maximum permissible rate of heat transfer through the walls of an insulated container that uses dry ice as the source of cold? Cite specific provisions of the treaty to justify your response.

16. Identify three technical resources in mechanical design available through the Canadian Technology Network (http://ctn.nrc.ca).

17. In order to economically ship natural gas across the ocean, it is cooled until liquefied, then transported in specially designed LNG (liquefied natural gas) tankers and regasified at the destination. You work for a firm which is designing a shore-based LNG facility for liquefaction and storage of the gas prior to loading up the tanker. Specifically, your current responsibility is to design all the penetrations in the 50,000 gallon LNG storage tanks. What requirements does the U.S. Department of Transportation impose on your design approach? Cite specific provisions of the *Code of Federal Regulations* to justify your response. Your answer must be complete and unambiguous. Your client should be able to go directly to the exact source to verify your response.

18. Required federal government test procedures for measuring the energy consumption of household appliances are published in Title 10 of the *Code of Federal Regulations*. In many cases, these test procedures refer to various existing voluntary consensus standards. Identify the ASHRAE standard cited in the CFR to be used for testing central air conditioners. Your answer must be complete and unambiguous. Your client should be able to go directly to the exact source to verify your response.

19. From the *ASHRAE Journal* or the ASHRAE web site (www.ashrae.org), describe a recent ASHRAE activity involving one of their standards.

20. ASME has issued a safety standard for the design of overhead guards on industrial forklift trucks. According to this standard, one of the design requirements is the ability of the guard to withstand repeated impacts of a 100 lb. weight dropped from a height of 5 ft. What is the maximum permissible permanent deformation in the guard as the result of this test? Provide complete and unambiguous citations of the specific provisions and documents to justify your response.

21. You are on your college crew team, and while rowing you developed an idea for making racing shell oars from fiber-reinforced composite materials. Before proceeding any further, you want to do a patent search to see if there may be a case of patent infringement. From the *Index to the U.S. Patent Classification System*, determine the class and subclass of an idea that may be similar to yours. Verify that the class and subclass numbers are close to what you need by comparing them with the *Manual of Classifications*. Use the CD-ROM system to search the class and subclass. How many titles are there in the class? Is there a patent that deals with the design of racing shell oars and the construction method? If there is such a patent, give patent number, issue number, and country code. Use the *U.S. Patent Office Gazette* to find the inventor's name, the date filed, the serial number, and a brief description of the patent. Provide the call number of the volume in which you located this information.

22. One day, while you were fishing, you thought of an idea for a fishing lure coupling apparatus. You know that this idea could make you millions from the various and sundry fishermen in the state. Envisioning millions, you apply for a patent but are rejected. Find the patent that you infringed, the creator, the year filed and the year approved, the serial number, the patent number, the state of application, and the call number of the corresponding gazette volume.

23. The Energy Information Administration of the U.S. Department of Energy regularly publishes a *Nonresidential Buildings Energy Consumption Survey*. According to the most recent survey, how many nonresidential buildings in the United States are equipped with boilers?

24. According to the *Manufacturing Energy Consumption Survey* published by the U.S. Department of Energy, how much energy in the form of wood chips and bark was consumed by paper mills in the most recent year for which data is available?

25. The U.S. Bureau of Census classifies all manufacturing industries according to a four-digit Standard Industrial Classification (SIC) number. Select a four-digit SIC designation whose second digit X (SIC _X_ _) matches the last digit Y of your phone number (_ _ _-_ _ _Y) and determine from the Annual Survey of Manufacturers the amount of electricity purchased by that industry in the most recent year for which data is available.

26. Using the National Technical Information Service (NTIS) database, identify a technical report written by an author whose name most closely matches your name. (Start by matching the first letter of the author's last name with the first letter of your last name. If there is more than one author which matches, then match the second letter, etc. Continue, using first and middle names if necessary, until you only have one match).

27. Use Chen, *The Civil Engineering Handbook*, to find the formula for determining the allowable design stress for a structural member made from wood.

28. Use *Marks' Standard Handbook for Mechanical Engineers* to find the equation for estimating the stopping distance of freight trains.

29. A thin steel circular arch is pinned at both ends and subject to an offset concentrated vertical load W as shown. The cross-section of the arch is circular of radius $r = 6''$. Determine the horizontal and vertical components of the reactions at A and B in terms of W using *Roark's Formulas for Stress and Strain*.

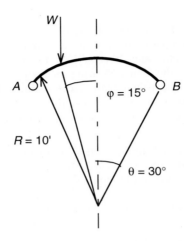

30. Search the Web site of a professional engineering society for the contents of the most recent issue of one of their specialty technical journals (not their general readership monthly magazine). Download the table of contents and submit a hard copy.

CHAPTER 4

Professional and Social Context of Design

4.1. OVERVIEW

The ABET definition of design presented in Chapter 1 refers to ". . . the benefit of mankind." This requires that environmental, legal, economic, and social considerations be an integral part of engineering design activities. This chapter describes that professional and social context in which engineering design decisions are embedded.

We begin with a discussion of the relationship between engineering, science, and innovation and their impact on society. We then offer a very brief review of historical developments. Next, the concept of engineering professionalism and the structure of the engineering profession is examined with a particular emphasis on the topic of engineering ethics.

As we discussed in Chapter 1, engineering design decisions inherently involve dealing with uncertainties. We address that aspect of design in Section 4.5 using the formalisms of risk analysis. The analysis of risk leads to a consideration of codes and standards and product and professional liability issues. We close with an examination of the interactions between engineering and public policy, with a special emphasis on the patent system.

4.2. ENGINEERING, TECHNOLOGY, AND SOCIETY

We live in a technological world. The very concept of civilization is inseparable from that of technology, and engineers are the only group dedicated to systematically applying technology for the benefit of society. In that sense, engineers are the stewards of social progress. This is a truly noble role. It is a role that engineers must take seriously. As engineers, we can never let ourselves be reduced to worker bees, focused only on the details of the immediate design task confronting us. We must always be sensitive to the wider implications of our work, assume personal responsibility for our professional decisions, and remain dedicated to using our talents for the benefit of humanity. Herbert Hoover, 31st President of the United States, said

> It is a great profession. There is a fascination of watching a figment of the imagination emerge through the aid of science to a plan on paper. Then it moves to realization in stone or metal or energy. Then it brings jobs and homes to men. Then it elevates the standards of living and adds to the comforts of life. That is the engineer's high privilege.
>
> The great liability of the engineer compared to men of other professions is that his works are out in the open where all can see them. His acts, step by step, are in hard substance. He cannot bury his mistakes in the grave like the doctors. He cannot argue them into thin air or blame the judge like the lawyers. He cannot, like the architects, cover his failures with trees or vines. He cannot, like the politicians, screen his shortcomings by blaming his opponents and hope that people will forget. The engineer simply cannot deny that he did it. If his works do not work, he is damned.
>
> On the other hand, unlike the doctor, his is not a life among the weak. Unlike the soldier, destruction is not his purpose. Unlike the lawyer, quarrels are not his daily bread. To the engineer falls the job of clothing the bare bones of science with life, comfort, and hope. No doubt as the years go by people forget which engineer did it, even if they ever knew. Or some politician puts his name on it. Or they credit it to some promoter who used other people's money. But the engineer himself looks back at the unending stream of goodness which flows from his successes with satisfaction that few professions may know. And the verdict of his fellow professionals is all the accolade he wants.

In this section we provide a brief overview of some of the philosophical and historical aspects of engineering design. Our hope is that the discussion will enhance your appreciation of the engineer's contributions to improving the human condition.

Science, Engineering, and Technology

An understanding of the relationships between science, engineering, and technology helps place the significance of engineering design activities in perspective. Think of technology as the product or result of sytematic efforts to modify (hopefully for the better) the human condition or natural environment. Engineering refers to a set of people, techniques, and processes that develop technology. However, not all technological change is the result of engineering; some is acheived by the efforts of nonengineering crafts and tradespersons. Science consists of the principles, and techniques for exploring those principles, that describe how the natural world functions. Engineers frequently use science, and many technologies are the outgrowth of scientific discoveries. However, there are some common misconceptions about these relationships.

The words 'science and engineering' are frequently used together; and usually in that order. This order of the words feeds the stereotype that engineering is a derivative of, or a subset of, science. Many in the engineering community have tried vigorously to argue that engineering is a lot more than applied science.[1]

---

[1]Florman, p. 59.

The words 'science and technology' are also frequently used together; and usually in that order. This order of the words implies that science precedes technology, and a popular contemporary model of technological development is that basic scientific research leads to applied research which leads to technological innovation. However, there is very powerful evidence that the relationship is much more complicated.

There are many cases in which successful technological innovations not only preceded an understanding of the associated scientific underpinnings, but served to stimulate the search for the scientific explanations. One example is James Watt's demonstration of a successful steam engine in 1765. This development preceded by thirty years the birth of Nicholas Carnot, one of the pioneers of the science of thermodynamics. Similarly, Robert Fulton's successful steamboat design in the early 1800s was based on empirical experiments of drag on ship hulls and inspired scientific research on fluid mechanics.[2] Vincenti describes the introduction of flush riveting in the aircraft industry in the 1930s as depending on changes in production methods, rather than as the result of a research program.[3] As these examples illustrate, new technologies are sometimes adopted before there is a full grasp of their scientific underpinnings; and their commercial success serves to stimulate the associated scientific research. Thus, engineering is much more than the practical application of science. In many ways it is one of the fundamental driving forces of society that feeds both social progress and scientific developments.

The Design Engineer's Heritage

It is one thing to talk about engineering in the abstract; it's another to identify with individual engineers who serve as role models; the giants on whose shoulders all contemporary engineers stand. America's debt to great engineers of the past is as great as its debt to political leaders like Thomas Jefferson and Abraham Lincoln; authors like Mark Twain and John Steinbeck; generals like George Washington and Douglas MacArthur; and composers like Aaron Copland and George Gershwin. Here are vignettes on a few influential engineers, who (like many other engineers) truly deserve credit for transforming the U.S. from an agrarian society to the world's foremost industrial power. Other countries have their own set of historically significant engineers.

We've picked one person to represent contributions in each of the four aspects of engineering (structures, machines, networks, and processes) that Billington claims are all necessary ingredients for an industrialized society.[4]

Structures.   John Roebling was the greatest structural engineer of the 19th century.[5] His masterpiece, the Brooklyn Bridge (Fig. 4–1), is one of the most familiar structures in America. Its role far surpasses its strictly utilitarian function. It has be-

---

[2]Billington, p. 53.
[3]Vincenti, p. 171.
[4]Billington, p. 19.
[5]Ibid, p. 207.

Figure 4–1.   The Brooklyn Bridge *(Chromosohm/Photo Researchers, Inc.)*

come a beloved national landmark and cultural treasure. But its place in history is not just because of its monumental grandeur. At the time of its construction (1869–1883) it was half again as long as any bridge previously built. One of Roebling's most important innovations was the technique for spinning the great wire suspension cables in-place, a method still used in suspension bridge construction.[6] John Roebling died in 1869 of injuries suffered during preliminary construction of the Brooklyn Bridge. He was succeeded as Chief Engineer on the project by his son, Washington Roebling.

Machines.   Igor Sikorsky was a student at Kiev Polytechnic Institute in the Ukraine in 1908 when he learned of the Wright Brother's pioneering accomplishments with powered heavier-than-air flight. He immediately dedicated himself to designing a flying machine that would rise "directly from the ground by the action of a lifting propeller" (see Fig. 4–2).[7] Thus started a career as an aeronautical engineer that lasted over fifty years. While much of his mid-career years were devoted to developing conventional aircraft, he returned in 1938 to his first love and thereby launched the modern era of helicopter design. The first test flight of his makeshift prototype lasted ten seconds in 1939. By the end of World War II, his company (Sikorsky Aviation) had delivered more than 400 helicopters to the military. The unique roles played

[6]Brooklyn Bridge. Spring 1997. *Grolier's Encyclopedia.* On line edition of Academic American Encyclopedia.

[7]Wohleber, 1993, pp. 26–39.

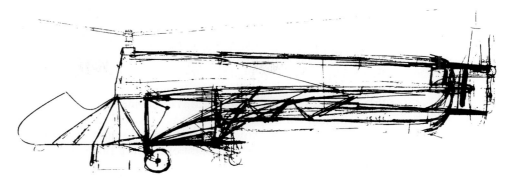

Figure 4–2. An Early Sikorsky Sketch of a Helicopter *(Reprinted with permission from Sikorsky Aircraft.)*

by modern helicopters in rescue, construction, surveillance, and military missions make them an indispensable part of any modern society.

Networks. George Westinghouse had over 400 patents and founded what is now one of the premier high-technology companies in the world. One of his first major engineering contributions was the design of the pneumatic braking system (still known as the Westinghouse air brake) for railroad trains. It greatly reduced the braking reaction time for long freight trains from 20 seconds to 2.5 seconds and vastly improved the safety record of railroads.

Westinghouse turned his attention in the 1880s to developing energy distribution networks. He started with designing and building the natural gas pipeline network for Pittsburgh. Then he became intrigued by the potential of distributing electricity. At the time, Thomas Edison's electricity distribution approach utilized direct current (DC). Its main drawback was the limited range of a DC network because of the high line losses associated with the low distribution voltage. Westinghouse contributed a major advance to transformer design that made alternating current (AC) networks cost-effective. He then found himself locked in a bitter public dispute with Edison for almost a decade over their rival electric power networks. The superiority of the Westinghouse system soon became apparent, and it has become virtually the exclusive form for distributing electricity.[8]

Processes. Frederick Taylor, a practicing mechanical engineer, invented the scientific management process. Among his many contributions were the concepts of time-study of tasks, techniques for division of labor, and the development of an incentive wage system for factory workers. Taylor's work had a huge impact on the way manufacturing firms were organized, in particular giving rise to the role of centralized

[8]Wohleber, 1997, pp. 28–42.

planning departments for organizing, scheduling, and coordinating labor, raw materials, and machinery.

One of Taylor's disciples was Henry Gantt, also a mechanical engineer. Gantt developed a system of charts for planning and keeping track of design and manufacturing activities.[9] The ubiquitous planning tool that still carries his name (the Gantt chart) is discussed in Chapter 7.

In addition to these and many other icons of engineering design, there are many engineers whose singular contributions are outside the mainstream of engineering design. Their engineering talents allowed them to make major contributions to other segments of society, and they are an important element of the heritage and continuing legacy of the engineering profession.

The engineer as statesman. President Hoover was a distinguished mining engineer, and served as national president of his professional engineering society (AIME) in 1919. His public service began during World War I while he was running his own mining engineering firm in London. American officials asked him to help resettle 120,000 Americans who were stranded in Europe by the war. He then organized food relief for Belgians after their country was overrun by Germany. When the U.S. entered the war in 1917, President Wilson appointed Hoover to be head of the U.S. Food Administration. He then served as Secretary of Commerce under both Presidents Harding and Coolidge before being elected president in 1928. After leaving office in 1932, he continued to devote more than thirty years to public service. Hoover wrote a series of books on American government and chaired two influential commissions from 1947–1955 on reorganizing the federal government.[10]

The engineer as artist. Kiyoyuki Kikutake received his mechanical engineering degree in 1968 from Chuo University in Japan. While working as an engineer he developed an interest in creating sculptures that help people connect with both technology and nature. After winning several competitions with works created in his spare time, he became a full-time sculptor in 1986. Today he is one of Japan's leading sculptors. Most of his works are large outdoor pieces that use sensors, gears, and motors to allow the sculptures to respond to their changing environment, particularly wind, temperature, sunlight, and people.[11]

The engineer as humanitarian. Some engineers' commitment to advancing the well-being of society reaches heroic proportions. Such is the case of Benjamin Linder, who received his BSME from the University of Washington in 1983.[12] Linder then served as a volunteer designing and constructing a small hydroelectric power-

---

[9]Layton, 1986.

[10]The World Book Encyclopedia, pp. 292–297.

[11]Normile, pp. 42–49.

[12]The material regarding Linder is drawn from Unger, pp. 44–48.

plant in an isolated and remote region of Nicaragua. Linder knew the risks he was taking, immersing himself in a country torn by armed conflict between the Contras and the Sandanista government.[13] Nevertheless, he felt a calling to use his talents to uplift people out of abject poverty.

He was able to secure United Nations funding to purchase a 100 kW generator for his project, but tools, raw materials, and skilled labor were in short supply. Linder's efforts on the project included teaching the local people the skills to build this power plant and additional ones on their own. The power plant went into operation in May 1986, providing electricity to a machine shop, a medical center, and a school.

Linder then turned his attention to collecting rainfall data in preparation for designing and building a second hydroelectric power plant in the same region. However, in April 1987 while on his way to collect data, Ben Linder and two others in his party were killed in an ambush by the Contras. In 1988, one branch of the main electrical engineering professional society posthumously recognized Benjamin Linder for his, "courageous and altruistic efforts to create human good by applying his technical abilities."

4.3. PROFESSIONALISM

The quote from Herbert Hoover in the prior section begins with the description of engineering as "... a great profession." In this section, we explore the concept of profession in more detail and examine the ways that engineering manifests itself as a profession, with a special focus on those aspects of professional conduct and responsibility that affect the behavior of design engineers and the design decisions they make.

Characteristics of a Profession

In the United States, employment in a profession bestows a status and prestige that is not associated with nonprofessional employment. It is no wonder that workers in many occupations seek to have their jobs accorded professional standing. To understand the distinction between a nonprofessional job and a professional position, consider the following four characteristics normally associated with a profession.

1. The work involves exercising skills, judgment, and discretion which is not entirely routine or subject to mechanization. This latter phrase has interesting implications for the many aspects of engineering that are computerized.
2. Preparation requires extensive formal training. This implies significant post-secondary formal training, although not necessarily in a university or college. Many vocational and technical institutes provide formal training to prepare peo-

---

[13]The Contras were the U.S. backed guerrillas who were fighting against the Nicaraguan Sandanista government during the 1980s. See Unger, p. 46.

ple for careers as beauticians, masseuses, surveyors, flight attendants, computer repair technicians, or automobile mechanics.

3. A specialized organization exists to set standards and codes of practice. These organizations define the extent and nature of training needed for someone to be accepted into the profession. In some professions, there are internships that must be served and standardized examinations that individuals must pass before they are granted professional status. Some professions also require evidence of subsequent activities to assure continuing professional competence. These organizations also maintain rules of practice that provide guidelines regarding professional behavior and relationships with peers and customers.

4. A professional makes a commitment to serve the public good. This concept of devotion to the public takes precedence over loyalty to the company, or concern over job security. This is especially important in professions where the health and safety of the public depends on uncompromising dedication by the practitioner.

Not all professions exhibit these characteristics to the same degree. For example, the medical profession requires substantially more formal training than the engineering profession, while engineers require more formal training than welders. In most states, the right to practice in some professions is based on passing a state-administered examination and being granted a license to practice by the state government.

However, there really is no standard measure of the degree of professionalism exhibited by any field. The final arbiter of professional status is public perception. If the public perceives your activities as professional, then that makes you a professional. Public perception, built on observation, understanding, and trust, is fragile. The same public that holds engineers in high esteem can quickly lose its confidence in engineers as a result of highly publicized incidents that cast engineers in bad light. Therefore, engineers and all others who aspire to the mantle of professionalism exhibit a strong obligation to build and maintain public trust.

Structure of the Engineering Profession

The third item in the above list of characteristics of a profession refers to the organizations that serve the needs of its members, promote the interest of the profession, and provide the mechanisms for its members to serve the public good. Most people have heard of the American Medical Association (AMA), the professional society of most medical doctors. For lawyers, there is the American Bar Association (ABA). There is no counterpart to the AMA or ABA within the engineering profession. In fact, the engineering profession is organized in a very decentralized, even fragmented, fashion. We describe the kinds of engineering societies in this subsection.

Technical engineering societies.   Five of the oldest professional engineering societies are known as the Founder Societies:

American Institute of Chemical Engineers (AIChE)
American Society of Civil Engineers (ASCE)
American Society of Mechanical Engineers (ASME)
Institute of Electrical and Electronic Engineers (IEEE)
American Institute of Mining, Metallurgical, and Petroleum Engineers (AIME)

These societies are comprehensive in terms of their coverage of technical interests within their discipline. Also, there are dozens of societies devoted to more specialized technical interests. These include the following:

American Institute of Aeronautics and Astronautics (AIAA)
Institute of Industrial Engineers (IIE)
Society of Automotive Engineers (SAE)
Society of Manufacturing Engineers (SME)
American Society of Heating, Refrigeration, and Air-conditioning Engineers (ASHRAE)
Institute of Traffic Engineers (ITE)

In addition to these societies, which are organized primarily on the basis of common technical interests of its members, other engineering societies have occupational or other professional activity or status as their organizing principle. Included in this category are the National Society of Professional Engineers (NSPE), American Consulting Engineers Council (ACEC), American Society for Engineering Education (ASEE), Society of Women Engineers (SWE), and the Society of Hispanic Professional Engineers (SHPE).

Table 4–1 lists these engineering societies along with their membership. These organizations are voluntary associations, in which individual members pay annual dues. Membership is in no way a requirement for practicing engineering. Many practicing engineers are not a member of any engineering society. On the other hand, many engineers belong to more than one. For example, consider a civil engineering professor who has a PE license (see the subsection later in this section entitled Registration) and who has his own consulting engineering practice. That individual might be a member of ASCE, which meets his need for keeping current on technical developments in civil engineering. The professor might also be a member of ASEE to keep abreast of engineering education issues. NSPE and ACEC may fill his professional and business needs.

The closest thing to an umbrella engineering organization is the American Association of Engineering Societies (AAES). AAES, formed in 1979, is a federation of twenty-two societies that voluntarily join together to pursue common interests and address profession-wide issues. AAES is the latest in a series of efforts to unify the engineering profession. However, it has never achieved the critical mass and consensus needed to approach the profession-wide support similar to that enjoyed by the AMA and ABA. It has undergone several reorganizations since its inception, most recently in 1996 when it responded to a severe financial crisis by deciding to concentrate its efforts on profession-wide public policy issues.

TABLE 4–1.   PARTIAL LISTING OF PROFESSIONAL
ENGINEERING SOCIETIES

Professional Society	Membership
IEEE	274,000
ASME	121,000
ASCE	110,000
AIME*	92,000
NSPE	75,000
SME	75,000
SAE	60,000
ASHRAE	55,000
AIAA	45,000
IIE	35,000
SWE	16,000
SHPE	6,000
ACEC	5,500

Source: Encyclopedia of Associations, 1995.
*A federation of four quasi-independent societies.

Coordination among the specialized engineering societies also can occur at the local level. For example, the Puget Sound Engineering Council (PSEC) in the Seattle metropolitan area is an umbrella organization involving local chapters of approximately twenty national engineering societies.

Honorary societies.   The engineering profession also includes many honorary societies and organizations. The most prestigious is the National Academy of Engineering (NAE). Spun off in 1964 from the congressionally chartered National Academy of Sciences (NAS), the NAE consists of approximately 1,200 members, elected on the basis of their outstanding accomplishments as engineers. Together with NAS and their other sister society, the Institute of Medicine, NAE operates the National Research Council (NRC) to conduct studies at the request of government entities on a variety of scientific, technological, and health issues. As an example of the relevance of the NAE to the subject of design, NRC published in 1991 a landmark report on engineering design practice, education, and research.[14]

Tau Beta Pi is the profession-wide academic honor society for engineering students. It is to engineering students what Phi Beta Kappa is to liberal arts majors. In addition, most engineering disciplines have their own specialized honor society, like Eta Kappa Nu for electrical engineers and Pi Tau Sigma for mechanical engineers.

[14]National Research Council, 1991.

Activities of Professional Engineering Societies

Most professional engineering societies, such as the ones listed in Table 4–1, have a wide range of technical, professional, educational, and public policy activities. Here we will focus on examining the technical and geographical dimensions of a typical professional engineering society.

Technical specialties.   As an example of the way a professional society organizes to deal with technical interests of its members, consider the organizational chart for ASME's Council on Engineering, displayed in Figure 4–3. The Council is headed by an ASME senior vice president (a volunteer member, just like you and me), and each of the eight groups is led by an ASME vice president (also a volunteer). The group of most interest to us is the Systems and Design Group because one of its divisions is the Design Engineering Division, although most of the other divisions also have relevance to design. Each division is led by a nationally elected group of officers (all volunteers) with support provided by a small professional staff. It is at the division level where most of the planning for national and international technical conferences is conducted, and where most of the information dissemination activities occur (research journals, conference proceedings, etc.). Each division publishes its own newsletter to keep its members up to date on divisional activities.

Geographical subdivisions.   Not all professional activities of the engineering societies occur at the national level. Typically there are regional and local units of societies. For example, ASCE is divided into four geographical zones, each headed up by a nationally elected vice president. Within these zones are 83 sections and 143 branches. Typically these units meet about once a month and offer a variety of technical, management, public policy, and career development programs. Participation at the local level offers a great way to build a professional network and keep abreast of local trends in professional employment.

Student membership.   Most professional societies encourage student membership and sponsor student sections at engineering colleges. The student sections of the professional societies elect their own officers, hold their own meetings, and participate in regional and national student conferences. Many awards and competitions, including design competitions, are sponsored by the societies specifically for their student members. Joining and becoming active in a student section of an engineering professional society is a great way to expand your horizons and broaden your education, especially in the area of group dynamics and communications skills which are so important to design activities. It's also a terrific opportunity to begin constructing your professional network.

Other activities.   Other major activities of professional societies include programs in developing codes and standards; education issues spanning K–12 through post-college continuing education; and involvement in public policy issues at the local, state, and federal levels. In addition, all the engineering societies have honors and awards programs for recognizing outstanding accomplishments of engineers.

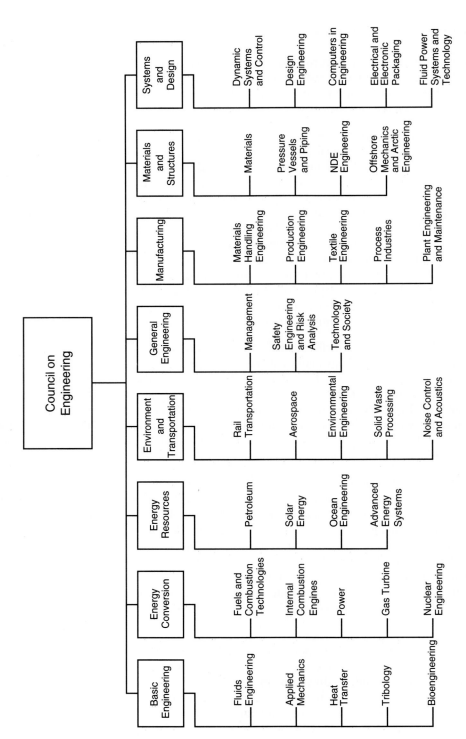

Figure 4-3. Organization Chart for the ASME Council on Engineering

The major engineering societies, realizing the importance of new information technologies, have created their own pages on the World Wide Web. Everything from joining a society to registering for conferences to obtaining copies of conference papers can be conducted electronically.

Registration

The National Council of Examiners for Engineering and Surveying (NCEES) represents the licensing boards of each state. NCEES has developed a model law for the registration of engineers. Although NCEES represents only the consensus of its individual members, every state has patterned their engineering registration law after this model. While there are some minor differences among the states, the key elements of the model law are common to all states.

An engineer who has been licensed to practice engineering in a state is called a Professional Engineer (PE). In most states, only those engineers with a PE license are legally allowed to perform consulting and certain other design functions. However, there is an industrial exemption for most engineers who work for companies that design and manufacture products. The process of becoming a Professional Engineer involves several stages of education, experience, and examination.

Undergraduate students are eligible to take the Fundamentals of Engineering Examination (FEE) several months prior to receiving an accredited four-year engineering degree. This is an eight hour examination, offered several times a year under the auspices of the State Board of Engineering Registration. Individuals who successfully pass the FEE are considered Engineers in Training (EIT).

After acquiring an additional four years of progressively more responsible experience, individuals are entitled to take another eight-hour examination, the Principles and Practice Examination (PPE). Registration as a PE is granted after successfully completing the PPE. Some states offer only a general PPE examination, while others offer separate PPE exams for each major discipline. Also, some states allow an individual registered in another state to practice in their state without having to take a separate examination.

Many universities, professional engineering organizations, and private firms offer review courses for those individuals preparing for the FEE or PPE. Resource materials are available, including reference manuals, sample examinations, and solutions manuals for both examinations, including specialized versions for civil, electrical, mechanical, and chemical PEE. In 1993, the format of the FEE was changed to allow each examinee to utilize a booklet with approximately sixty-five pages prepared by NCEES as a reference handbook during the examination.

Getting a PE license is no longer the last step towards achieving professional status. Recognizing that engineering knowledge and design skills have a relatively short half-life, many engineering leaders are advocating the need for regular evidence of continued professional competence. Quite a few states now have a requirement that annual renewal of a PE license requires evidence of Continuing Professional Competency (CPC). NCEES developed model rules for CPC, which have been adopted by ten states as of January 1997. Some engineers who are licensed in more than one state are concerned over having too many rules to satisfy.

4.4. ENGINEERING ETHICS

As described in the previous section, two characteristics of any profession are the establishment of rules of professional behavior and a profession-wide commitment to serve the public. The engineering profession incorporates both of these aspects of professionalism in a code of ethics. In this section, we provide some background on important ethical principles, discuss several engineering codes of ethical conduct, and examine strategies for implementing the codes.

An Interesting Airplane Trip

To get you psyched up for this discussion, pretend that you are traveling on an airplane to another city to present your company's approach to designing a retractable roof for a new sports stadium (see Fig. 4–4).

Your presentation, scheduled for tomorrow, will reveal your innovative approach to designing the roof structure, the mechanisms for opening and closing the roof, and the computerized system for controlling the motion of the roof panels. Your only concern over your ability to win the roof design contract is that the cost of building your design may be more than the prospective client will be willing to pay. You

Figure 4–4. A Model of the Retractable Roof Stadium in Seattle *(Photograph by Alan Berner, Seattle Times. Reprinted with permission from the Seattle Times.)*

had tried to keep the cost down, but the 30% increase in the price of structural steel announced two days ago by the steel supplier came too late for you to modify the design or find an alternate steel source. The person sitting next to you on the plane is absorbed in reading a technical report. As your eyes wander, drawn by the interesting diagrams in the document, you realize that your seat-mate is from a competing design engineering firm. The report he is reading is that firm's cost estimate for their version of the retractable roof. It sure would be nice if you knew what steel price they used in their cost estimate. If they used a value substantially below yours, that difference alone could swing the contract their way.

You could keep quiet, even pretend to be asleep, and hope you can catch a glimpse of some key cost data. Or you could strike up a conversation with your fellow passenger, with the intent of steering the conversation toward the topic of steel costs, hoping that he will reveal crucial information to you. Of course, you wouldn't volunteer your involvement in the stadium project.

On the other hand, you could introduce yourself and indicate the purpose of your trip without mentioning the extent of your interest in the specific topic of the price of steel. Then, any such information he chooses to disclose to you is fair game, isn't it?

Or consider a slightly different scenario. Suppose before any conversation took place between you, he got up to go to the restroom and left the report face up on the seat. Do you take advantage of the golden opportunity presented to you? What if he placed the report face down on the seat? Would you turn it over?

Here's another variation on the same theme. Instead of sitting next to you, he is sitting in the row ahead of you with another engineer from the same firm. You can't help overhearing their conversation regarding their strategy for tomorrow's meeting. You could continue to eavesdrop without being too obvious about it.

What would you do in each of these situations? Would your actions depend on the risk of getting caught? What do you think an engineering code of ethics would determine about each situation? If you knew the code of ethics guidelines, would you proceed to act contrary to it if you thought you could get away with it?

Let's add one final little twist to make this really interesting. Suppose you do decide to eavesdrop on the conversation. You overhear the two engineers congratulating each other on their firm's brilliant last-minute switch to another steel supplier to avoid the huge cost increase that burdens your proposal. You can hardly restrain yourself when you learn that their new steel supply company is a front for an individual who has previously been imprisoned for bribing inspectors and falsifying test reports on inferior quality materials. Your concern is no longer simply one of economics, but one of public safety. This raises a whole different set of questions about what is the right thing to do. Finally, what if your company is in a tenuous financial situation? It will go out of business unless it wins this contract. How should these circumstances affect your behavior?

Most people could answer these questions using common-sense knowledge of right and wrong. In turn, that knowledge probably has been heavily influenced by parental guidance and religious upbringing. Since those influences vary so greatly among individuals, it is often helpful to use a profession-wide framework to help analyze situations such as these. We devote the reminder of this section to examining the elements of such a framework, relying heavily on the work of Martin and Schinzinger.

The Study of Ethics

Engineering ethics is: (1) the study of the moral problems confronted by individuals and organizations involved in engineering; and (2) the study of related questions about moral conduct, character, ideals, and relationships of people and organizations involved in technological development.[15] This definition requires us to address the issue of what is meant by morality.

A moral problem is a problem in which one's actions involve consideration of others. Moral conduct is based on concern for other people; it is not reducible to self-interest, law, or religion.[16] Philosophers and theologians over the ages have proposed various theories regarding moral behavior, and they continue to study and debate the issues even today. However, a consensus appears to have developed within the community of philosophers that there are four general theories that can serve to provide a framework within which to examine moral conduct.

Moral Theories

Four main approaches to examining moral behavior are described in this section.[17] While each theory provides a distinctly different perspective on the issue, they are not mutually exclusive. In many real-world situations involving complex moral arguments, it may be extremely useful to analyze the situation from more than one perspective. At the very least, it is important to realize that part of the basis for rigorous disagreements over specific engineering activities may be because the protagonists are using different moral theories to justify their actions.

Utilitarianism. This theory proposes that the best course of action is the one that does the most good for the most people. In its most simplistic form, utilitarianism focuses on the consequence of the actions, and attempts to list each outcome in either the "good" column or the "bad" column. The moral choice is the option which has the best net "good" score.

Consider the choice between building a coal power plant or a new hydroelectric dam to bring electric power to a rural region in a developing country. Keep in mind that this is an oversimplified problem. In reality, there would probably be a lot more options than these. Both options provide the benefits of electricity. After construction, the coal plant might provide more jobs at the coal mine and in the power plant. On the other hand, the new dam will provide flood control benefits. On the downside,

[15]Martin and Schinzinger, pp. 4–5.
[16]Martin and Schinzinger, p. 33.
[17]Martin and Schinzinger, pp. 33–42.

the strip mining activities may produce sludge run-off that will contaminate the local supply of drinking water, and the air pollution from the coal plant may have a particularly adverse effect on people with asthma. The negative effects of the dam include the loss of the fertile farm land that will end up at the bottom of the reservoir. Also, the fish runs that are a significant source of food for the local population will be substantially reduced.

Clearly the job of quantifying the entries in the "good" and "bad" columns is very difficult. However, assuming that we've done the best job we could in that regard, our utilitarian perspective tells us to recognize that there will always be items in the "bad" column. We can never eliminate them completely. The best we can do is attempt to at least partially compensate for those losses, then grit our teeth and choose the option with the best net score. The next two theories, duty-based morality and rights-based morality, take issue with the fundamental assumption of utilitarianism that all that counts is the results. We now turn to examining each of them.

Duty-based morality.   A duty-based theory of morality emphasizes that any action you take should not violate the obligations you have to others. These obligations include your responsibility to be fair and honest in your dealings with other people. It means you should keep your promises and not betray another's trust. This approach emphasizes the relationship between people and the way they treat each other, in contrast to the utilitarian emphasis on results.

Rights-based morality.   A rights-based theory focuses on ensuring that the action doesn't violate another's rights. As with duty-based morality, the emphasis is on human relationships rather than on the results. Compared to the duty-based emphasis on the individual who is faced with a decision, the rights-based approach focuses on the person who will be affected by the actions. The focus is on that person's fundamental rights that you must respect such as life, liberty, pursuit of happiness, privacy, and property. According to this moral theory, any duty you might have towards another person arises out of that person's rights.

Virtue-based morality.   This moral concept focuses less on conduct or interactions with others but on identifying ideal character traits which people should aspire to. Moral character, according to this theory, is based on motives, attitudes, aspirations, and ideals.[18] Some examples of moral virtues are courage, generosity, integrity, loyalty, and discretion.

Responsibilities of Engineers

Four elements of morally responsible engineering behavior can be derived from the vision of engineering design as a social experiment:[19]

[18]Martin and Schinzinger, p. 51.
[19]Martin and Schinzinger, p. 72.

- A primary obligation to protect the safety and respect the right of consent of human subjects
- A constant awareness of the experimental nature of any project, imaginative forecasting of its possible side effects, and a reasonable effort to monitor them
- Autonomous, personal involvement in all steps of a project
- Accepting accountability for the results of a project

An important ingredient of each of these elements is personal awareness and responsibility. Engineers must not uncritically rely on directives from a supervisor, traditional ways of doing things, or even a written code of ethics, to guide and justify our actions. We must consciously and continuously analyze the moral implications of our design decisions and assume personal responsibility for their impacts.

Sources of Ethical Dilemmas

Ethical dilemmas are situations in which a person is confronted by conflicts among two or more moral considerations. One such conflict arises when you have conflicting loyalties to your employer, your employees, your customers, and the general public. As we will see later, most engineering codes of ethics place your responsibility to the public in a preeminent position. The problem is that it is not always clear what constitutes the public interest: not all members of the public are affected equally by a given action and not all members of the public view a given action from the same perspective. There are many ways in which the public can be organized into groups whose interests do not always coincide: old versus young, healthy versus sick, male versus female, rich versus poor, urban versus rural, tall versus short. Contemporary debates over issues such as abortion, physician assisted suicide, and funding for welfare illustrate that the "public welfare" is a very elusive concept.

Engineering Codes of Ethics and Rules of Conduct

To sort these issues out, many people will rely on their upbringing, sense of decency, and common sense. Still, it is helpful to have guidelines for decision making and a framework for critical analysis of options and behavior. Adoption of such a code of ethical behavior by an engineering professional organization serves as a unifying force and also provides a forum for group discussion of ethical issues confronted by its members.

ABET code.   There are several different versions of codes of ethics for engineers. The one adopted by the Accreditation Board for Engineering and Technology consists of four fundamental principles and seven fundamental canons.[20]

The Fundamental Principles
Engineers uphold and advance the integrity, honor, and dignity of the engineering profession by:

---

[20]Martin and Schinzinger, p. 342.

I. Using their knowledge and skill for the enhancement of human welfare;

II. Being honest and impartial, and serving with fidelity the public, their employers and clients;

III. Striving to increase the competence and prestige of the engineering profession; and

IV. Supporting the professional and technical societies of their disciplines.

The Fundamental Canons

1. Engineers shall hold paramount the safety, health, and welfare of the public in the performance of their professional duties.

2. Engineers shall perform services only in the areas of their competence.

3. Engineers shall issue public statements only in an objective and truthful manner.

4. Engineers shall act in professional matters for each employer or client as faithful agents or trustees, and shall avoid conflicts of interest.

5. Engineers shall build their professional reputation on the merits of their services and shall not compete unfairly with others.

6. Engineers shall act in such a manner as to uphold and enhance the honor, integrity, and dignity of the profession.

7. Engineers shall continue their professional development throughout their careers and shall provide opportunities for the professional and ethical development of those engineers under their supervision.

While the ABET Code is fairly brief, ABET also has issued a more detailed set of suggested guidelines for use with the fundamental canons of ethics.[21] Many engineering professional engineering societies have adopted their own code of ethics, some of which are very similar to the ABET code. They all have the common feature of emphasizing the engineer's responsibility to serve the public welfare.

It is important to note that such codes of ethics do not carry the force of law. They are adopted by organizations which engineers voluntarily join. While each of the adopting societies has its own internal mechanisms for investigating and rendering judgments on alleged violations of their code, they have virtually no enforcement powers. Typically their strongest punishment would be expelling the offender from the society.

NCEES model rules of professional conduct.   The National Council of Examiners for Engineering and Surveying (NCEES) has issued a set of model rules of professional conduct. These rules are essentially a code of ethics for engineers and are a lot more detailed than the ABET Code. The complete text of these rules is pro-

---

[21]Martin and Schinzinger, pp. 343–349.

vided in the appendix to this section. Like the ABET and similar codes, the NCEES rules represent the consensus of a voluntary organization and do not have the force of law. However, since NCEES designs the questions used by professional engineering license examinations in every state, their rules of conduct have special significance.

Starting in Autumn 1996, questions on engineering ethics are included in the Fundamentals of Engineering Examination (see Section 4.3). The NCEES reference handbook that exam takers can use during the examination includes the text of the NCEES rules together with some brief background information on engineering ethics.

Further, the NCEES rules are intended to serve as a model for engineering registration laws in the states. Some states adopt the NCEES language verbatim into their professional engineering registration laws; others modify the NCEES language slightly before adopting it. In either case, this leads to rules of ethical conduct that are legally binding on all professional engineers licensed to practice in those states. An individual who violates those rules has broken a law; penalties may include loss of PE license or fines as determined by the state law/regulation.

Implementation

Now we have a set of rules to serve as guidelines for ethical behavior and some additional analytical background regarding considerations of moral conduct. That puts us in a better position to decide how we should behave in situations such as those encountered during the airplane ride described at the beginning of this section. Of course, the existence of a code of ethics doesn't mean that a cookbook recipe exists for resolving every professional ethics dilemma. Quite the contrary, the burden is still on the (autonomous) individual to decide what is right. But the code provides a mechanism for organizing your thinking.

The situation you encountered on the airplane is not the only type of dilemma you are likely to encounter in your career. Another category involves circumstances in which you are asked by your boss, or client, or vendor, to act in a manner which may be contrary to the code. Still another type of dilemma arises when you witness another person acting in a manner contrary to the code, or you become aware of a situation in which an organization's activities pose a threat to public welfare.

Action alternatives.   It's one thing to struggle conceptually with the issues of right or wrong and morality or immorality. It's quite another thing to take action on the basis of your convictions. Disagreements regarding design decisions may be the result of legitimate differences in design philosophy or the assumptions used to arrive at design decisions. There can also be legitimate professional differences over the magnitude of the risk associated with a proposed activity or the cost-effectiveness of a proposed risk reduction measure. There may be a conflict over management's desire to meet a production deadline and the engineer's desire for additional testing to verify the effectiveness of a new safety feature. It is not always easy to distinguish between situations involving legitimate differences in professional judgment and unethical behavior which creates what you consider to be an unreasonable risk. The next big hurdle, once you are confident that you have made a reasoned judgment that unethical behavior is occurring, is to decide what action to take.

Suppose you are aware that an organization or individual has, or is about to, violate the code of ethics. You might be tempted to remain silent, on the grounds that the activity is outside the scope of your professional jurisdiction (it is somebody else's problem). Unfortunately, the concept of moral responsibility does not let you off the hook so easily. Having observed wrongful behavior, you have a moral obligation to become involved. A detailed discussion of strategies at this point is provided by Unger.[22] We provide just the highlights of his analysis here.

First, make sure you have done your homework. In particular, you need to be sure you have all the relevant technical information. If you are going to report an action or behavior you consider to be unethical, be sure that your position is technically credible and well-documented. There are several options available to you, depending on the nature and severity of the existing or pending violation. If the action has not yet occurred, nor reached a point of causing harm, one approach is to treat the incident as a private matter between you and the offending party. Your approach should be direct, matter-of-fact, and focused on the action or behavior and its consequences without resorting to personal attacks on the individuals involved. You can inform them of your intent to report their actions if they persist.

If the situation cannot be solved informally, you can bring it to the attention of the appropriate first-line supervisor in the organization. In many situations, the manager will take the appropriate corrective action and the issue will be quietly and quickly resolved. The chances of this happening increase if you bring the matter to management's attention in a timely manner so that a relatively small problem can be corrected before it escalates out of control.

But what if the manager refuses to act? There may be many reasons for ignoring or rejecting your concerns. The manager may see the situation as one of differing professional judgments, rather than one of unethical behavior. The manager may feel that you are interfering in a matter which is outside the scope of your responsibilities. The manager may want to avoid incurring the extra costs or project delays associated with taking corrective action. If the situation occurred within their area of responsibility, they may be concerned that it will make them look bad to upper management. He/she may feel that corrective action would draw public attention to a situation that will prove embarrassing for the organization. This sort of engineering-management conflict characterizes many engineering ethics situations. It is a classic struggle between the engineer's concern for the public welfare and management's loyalty to the bureaucracy.

Many larger corporations and government agencies have formal internal appeal/grievance procedures that allow you to take your concerns to a higher level

---

[22]Unger, pp. 227–237.

within the organization. This approach is not without its perils, since if formal procedures are instituted, you could be the target of charges of incompetence or trouble-making by the managers whose decision you are appealing. Getting support for your position from engineering colleagues within the organization can enhance your own credibility and reduce your vulnerability to counter-attack.

Appeals within the "system" can have varying degrees of success. One form of resolution involves not rectifying the specific incident that triggered your concerns, but modifying practices so that it doesn't happen again. Don't lose sight of the fact that your goal is not necessarily to win, but to modify unethical behavior. Whether such an arrangement meets your ethical standards depends on the specific details.

Of course, there is always the possibility that internal appeals will fall on deaf ears. Not all executives are willing to "do the right thing" if it means exposing the company to unfavorable publicity, law suits, or significantly reduced profits. They believe that their responsibility is to "keep the lid on" and their strategy may include ignoring or intimidating you. If the situation has reached this point, your job and career may be in jeopardy. However, if the consequences of abandoning your crusade are as great as those that caused you to get involved in the first place, you need to press on.

This usually means going outside the organization. There are several paths available at this point. You could resign, thereby disassociating yourself from the unethical practice. However, while disassociation may separate you from the source of the problem, it may not do anything to solve the problem. It may be necessary to share the information with outsiders. One way to do this is to consult privately and in confidence with trusted colleagues from your professional engineering society. Their perspectives and advice can give you moral support and provide you with ideas for pursuing the issue.

If the matter involves the violation or potential violation of a regulation, you could bring it to the attention of the appropriate regulatory body. This could be done anonymously, if you are concerned about being publicly identified as the source of the complaint. But if no regulation has yet been violated, the regulatory agency may be powerless to intervene, or the agency may be understaffed and unable to respond in a timely manner to prevent harm from occurring.

In that case, your last resort may be to contact a local politician or investigative journalist. If you can get them interested, they may take up your cause and pursue it with vigor and resources unavailable to you. The mere indication that such an outsider is pursuing the issue may be enough to convince top management to act before public disclosure creates an embarrassing, if not potentially liable situation.

Appendix 4A: NCEES Model Rules

We print on p. 137 the complete text of the NCEES Model Rules of Professional Conduct.[23]

---

[23]NCEES, pp. 80–81.

Preamble

To comply with the purpose of the (*identify state, registration statute*) which is to safeguard life, health, and property, to promote the public welfare, and to maintain a high standard of integrity and practice, the (*identify state board, registration statute*) has developed the following "Rules of Professional Conduct." These rules shall be binding on every person holding a certificate of registration to offer or perform engineering or land surveying services in this state. All persons registered under (*identify state registration statute*) are required to be familiar with the registration statute and these rules. The "Rules of Professional Conduct" delineate specific obligations the registrant must meet. In addition, each registrant is charged with the responsibility of adhering to standards of highest ethical and moral conduct in all aspects of the practice of professional engineering and land surveying.

The practice of professional engineering and land surveying is a privilege, as opposed to a right. All registrants shall exercise their privilege of practicing by performing services only in the area of their competence according to current standards of technical competence.

Registrants shall recognize their responsibility to the public and shall represent themselves before the public only in an objective and truthful manner.

They shall avoid conflict of interest and faithfully serve the legitimate interests of their employers, clients, and customers within the limits defined by these rules. Their professional reputation shall be built on the merit of their services and they shall not compete unfairly with others.

The "Rules of Professional Conduct" as promulgated herein are enforced under the powers and vested by (*identify state enforcing agency*). In these rules, the word "registrant" shall mean any person holding a license or a certificate issued by (*identify state registration agency*).

I. Obligation to Society

a. Registrants, in the performance of their services for clients, employers, and customers, shall be cognizant that their first and foremost responsibility is to the public welfare.

b. Registrants shall approve and seal only those documents and surveys that conform to accepted engineering and land surveying standards and safeguard the life, health, property, and welfare of the public.

c. Registrants shall notify their employer or client and such other authority as may be appropriate when their professional judgment is overruled under circumstances where the life, health, property, or welfare of the public is endangered.

d. Registrants shall be objective and truthful in professional reports, statements, or testimony. They shall include all relevant and pertinent information in such reports, statements, or testimony.

e. Registrants shall express a professional opinion publicly only when it is founded upon an adequate knowledge of the facts and a competent evaluation of the subject matter.

f. Registrants shall issue no statements, criticisms, or arguments on technical matters which are inspired or paid for by interested parties, unless they explicitly identify the interested parties on whose behalf they are speaking, and reveal any interest they have in the matters.

g. Registrants shall not permit the use of their name or firm name by, nor associate in the business ventures with, any person or firm which is engaging in fraudulent or dishonest business or professional practice.

h. Registrants having knowledge of possible violations of any of these "Rules of Professional Conduct" shall provide the state board information and assistance necessary to the final determination of such violation.

II. Obligation to Employer and Client

a. Registrants shall undertake assignments only when qualified by education or experience in the specific technical fields of engineering or land surveying involved.

b. Registrants shall not affix their signature or seals to any plans or documents dealing with subject matter in which they lack competence, nor to any such plan or document not prepared under their direct control and personal supervision.

c. Registrants may accept assignments for coordination of an entire project, provided that each design segment is signed and sealed by the registrant responsible for preparation of that design segment.

d. Registrants shall not reveal facts, data, or information obtained in a professional capacity without the prior consent of the client or employer as authorized or required by law.

e. Registrants shall not solicit or accept financial or other valuable consideration, directly or indirectly, from contractors, their agents, or other parties in connection with work for employers or clients.

f. Registrants shall make full prior disclosures to their employers or clients of potential conflicts of interest or other circumstances which could influence or appear to influence their judgment or the quality of their service.

g. Registrants shall not accept compensation, financial or otherwise, from more than one party for services pertaining to the same project, unless the circumstances are fully disclosed and agreed to by all interested parties.

h. Registrants shall not solicit or accept a professional contract from a governmental body on which a principal or officer of their organization serves as a member. Conversely, registrants serving as members, advisors, or employees of a government body or department, who are the principals or employees of a private concern, shall not participate in decisions with respect to professional services offered or provided by said concern to the governmental body which they serve.

III. Obligation to Other Registrants

a. Registrants shall not falsify or permit misrepresentation of their, or their associates', academic or professional qualification. They shall not misrepresent or exaggerate their degree of responsibility in prior assignments nor the complexity of said assignments. Presentations incident to the solicitation of employment or business shall not misrepresent pertinent facts concerning employers, employees, associates, joint ventures, or past accomplishments.

b. Registrants shall not offer, give, solicit, or receive, either directly or indirectly, any commission or gift, or other valuable consideration in order to secure work, and shall not make any political contribution with the intent to influence the award of a contract by public authority.

c. Registrants shall not attempt to injure, maliciously or falsely, directly or indirectly, the professional reputation, prospects, practice or employment of other registrants, nor indiscriminately criticize other registrants' work.

4.5. RISK ANALYSIS

Failures occur all around us. Pencils break, light bulbs burn out, sidewalks crack, trains derail, and computer disk drives crash. These are real problems that impact design engineers—we continuously face the possibility that our designs can produce negative effects. Since no one can predict with absolute certainty the future environment under which their design will operate, the possibility of failure is an integral part of an engineer's life. The notion of failure and its central role in the engineering design process has been explored in elegant detail by Henry Petroski (see reference at end of chapter).

Serious negative effects of engineered systems include air pollution from industrial facilities and toxic chemicals leaking into municipal water supply systems. We also observe many other unpleasant events that are less directly associated with engineering failures: babies fall down abandoned water wells, people contract AIDS, earthquakes topple buildings, and teenagers suffer hearing losses from listening to loud rock music. When we think about it, there is no activity of society that is risk free. Yet society is demanding greater protection from risks, and the burden for meeting this need frequently falls upon design engineers. We engineers are increasingly finding ourselves in the midst of controversies surrounding the real or perceived negative effects of our design activities. We need to understand the important concepts of risk analysis if we are to live up to society's expectations and produce designs that merit the public's trust. Many engineers have taken on this challenge and specialize in safety engineering (Fig. 4–3 lists Safety Engineering and Risk Analysis as one of ASME's thirty-six technical divisions). Several other professional fields, such as public health, psychology, and environmental studies, have also contributed to the systematic analysis of risk. Interest in the study of risk has grown to the point where it has spawned its own professional society and scholarly research journal. As with many other topics in this book, all we can do here is expose you to some of the basic concepts and principles.

In this section, we examine the concept of risk using a framework widely adopted within the risk analysis professional community. We begin by defining risk as the potential for something of value (people, works of art, an animal species, an eco-

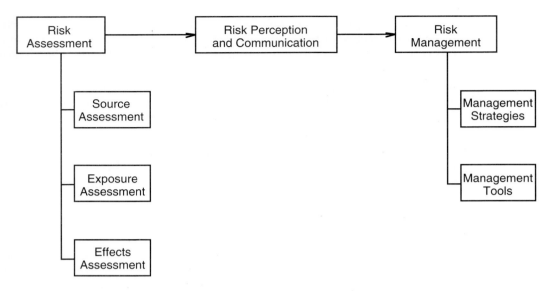

Figure 4–5.   Components of Risk Analysis

logical habitat) to be adversely impacted by an event. We will divide our discussion of risk into three components: risk assessment, risk perception and communication, and risk management. Within each of these components, we further subdivide the analysis as appropriate. Figure 4–5 provides an overview of our treatment of risk analysis in this section.

Risk Assessment

Risk assessment (sometimes referred to as risk characterization) deals with describing the nature and magnitude of a risk. This can be done at several levels of detail, depending on the nature of the risk, the amount of data available, the cost of doing the assessment, and the anticipated use of the information.

   At a coarse level, the assessment can consist simply of a numerical measure of the risk expressed as the consequence per unit of time. For example, the risk of an individual being killed by lightning in any year is one in 3,000,000. Using this approach, it generally is useful to express the risk as the product of a magnitude term and a frequency term as indicated in Equ. (4–1), where the magnitude term is an indication of the seriousness of an event while the frequency term is a measure of how likely the event is to occur.

$$\text{Risk}\left(\frac{\text{consequence}}{\text{unit time}}\right) = \text{magnitude}\left(\frac{\text{consequence}}{\text{event}}\right) \times \text{frequency}\left(\frac{\text{event}}{\text{unit time}}\right) \quad (4\text{–}1)$$

For the example of lightning, this factoring takes the form

$$\frac{\text{risk of being killed by}}{\text{lightning in a year}} = \frac{\text{risk of dying after being}}{\text{struck by lightning}} \times \frac{\text{chance of being struck}}{\text{by lightning in a year}}$$

with the numerical values[24]

$$\left(\frac{1}{3,000,000}\right) = \left(\frac{1}{4}\right) \times \left(\frac{1}{750,000}\right)$$

The context of any particular risk analysis and the purposes for which it is being conducted will give us guidance regarding which "consequences" are most appropriate for that analysis. Consider the risks associated with automobile travel. We can express the "consequences" as injuries, deaths, or amount of property damage. Similarly, there are several possible choices for the "time" parameter. For many risks, a useful "unit of time" is the calendar year. Hours of exposure might be a more appropriate time unit for analyzing risks associated with occupational activities. On the other hand, the risk of dying from lung cancer might conveniently be discussed in terms of life span as the time parameter.

The choice of the "event" in Equ. (4–1) is even more subtle. It could be the specific event that potentially triggers the consequence: the automobile crash. Or it could be the general activity that only rarely has adverse consequences: the number of miles traveled. Another consideration is that activities with large frequencies can lead to large risks, even though the magnitude of each consequence can be very small. For example, compare the risks of injury from walking up/down stairs with that of rappelling or climbing up cliff sides. Many more people are injured each year falling down stairs than from falling off a cliff. That's because many more people engage in the activity of walking up and down stairs than engage in cliff climbing, and they do it much more frequently. Does that mean that walking up and down stairs is riskier than climbing cliffs?

One way to deal with this issue is to normalize the quantitative measure of risk by the number of people exposed, and/or the amount or frequency of their exposure or engagement in the activity. Some alternative normalization schemes for automobile travel risks are as follows:

$$\left(\frac{\text{fatalities}}{\text{year}}\right) \text{ or } \left(\frac{\text{fatalities}}{\text{person–year}}\right) \text{ or } \left(\frac{\text{fatalities}}{\text{passenger–year}}\right) \text{ or } \left(\frac{\text{fatalities}}{\text{passenger–mile–year}}\right)$$

The choice of the normalization scheme can make a big difference in comparative risk assessment. For example, consider the relative risks of an individual traveling by airplane or automobile between New York City and Boston. The distances between the two locations are approximately the same for the two modes of travel. However, the time spent in traveling between those locations is substantially different for the two modes. A comparison of the risks of the two travel modes based on a "time of expo-

---

[24]*Discover*, pp. 82–83.

sure" normalization will give substantially different results than one based on distance traveled or on the number of trips. In addition, many risks involve more than one consequence, such as fatalities, injuries, and property damage.

These issues require careful thought. Many risk analyses are comparative in nature: Does design option A pose a greater risk than design option B? It would be easy to prejudice such an analysis during the process of selecting the consequence, event, and time parameters. There may not be any purely objective approach to risk analysis. However, by involving both advocates of A and B in the process along with observers who do not have a direct stake in either A or B (for example, insurance companies, medical doctors), we can increase the prospects for a successful risk analysis.

Data problems. Decisions regarding the selection of parameters for a risk analysis are sometimes made by default because the data is available only in a certain form. Sources of data for risks such as automobile travel include hospitals, police, fire departments, emergency rescue units, and the insurance industry. Sometimes the data is incomplete, inconclusive, or otherwise inadequate and there is controversy over the extent to which we should rely upon it. For example, some crimes such as rape and some diseases such as AIDS may be under-reported. Data on risk of some injuries, like whiplash, may be inflated as a result of fraudulent efforts to collect from insurance companies. An important, but often overlooked aspect of risk assessment is an estimate of the level of uncertainty associated with any quantification of risk.

A frequent problem with risk data involving public health issues arises because some adverse health effects can be associated with minute levels of very toxic chemicals. We may not be able to directly measure such small levels of concentration because of limitations of existing measurement techniques. This forces us to extrapolate from available data on high levels of concentration in order to estimate risks at smaller levels of concentration. There are inevitable controversies over the nature of the extrapolation. Consider the situation depicted in Figure 4–6 where the data points justify constructing the straight line segment AB to depict the relationship between risk and the level of chemical concentration (note that both axes in Fig. 4–6 have logarithmic scales). Given that no data is available for lower levels of chemical concentration, the issue to address is how to extrapolate the segment AB. Figure 4–6 shows how different extrapolation assumptions can lead to differences in risk of several orders of magnitude at very low levels of concentration.

Another well-known data difficulty is associated with estimating cumulative risks to humans due to long-term exposure to suspected carcinogenic substances. Since cancer may not develop in humans until many years after initial exposure, it is very difficult to collect data that isolates the effect of the suspected carcinogen from that of other effects. The standard procedure is to conduct controlled experiments on animals (usually mice). Controversy arises when we use conclusions regarding the risks to mice to draw conclusions regarding the risk to humans.

Sometimes the risk assessment approach just described is inadequate since it doesn't help us understand the reasons for the risks and what opportunities may be available to reduce the risks. We can conduct a more refined analysis that recognizes

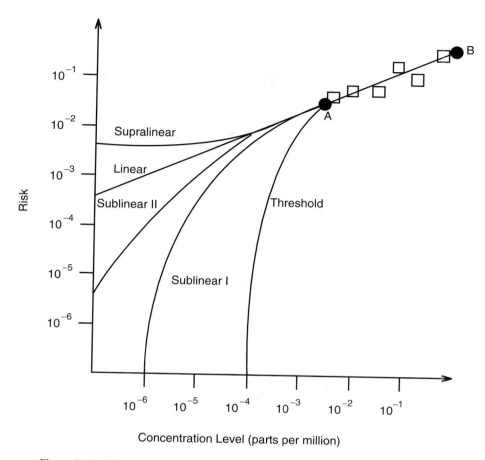

Figure 4–6. Alternative Extrapolations of Risk Data at Low Levels *(Adapted from Faustman and Omenn, p. 80, with permission from The McGraw-Hill Companies.)*

several dimensions to risk assessment: identify the source of the risk; understand the nature of the exposure; and evaluate the consequences or effects of that exposure. We will discuss each of these dimensions separately. Figure 4–5 depicts the role these issues play in the overall risk analysis framework.

Source assessment. This phase of risk assessment identifies the source and magnitude of the risk. It sometimes is referred to as hazards analysis. In the case of, say a structural failure due to a material defect, this phase of the analysis might attempt to quantify the probability that the defect will cause the structure to fail (techniques for doing this are discussed in Chapter 5). For another type of risk, say the particulates emitted into the air when coal is burned, source assessment might focus on

TABLE 4–2. SOURCE OF HAZARDS AND THEIR EFFECTS

Agent	Effects	Description
Natural or unknown agent largely beyond human control	Gradual, chronic	Infectious and degenerative diseases (flu, arthritis)
	Sudden, catastrophic	Natural disasters (floods, hurricanes, etc.)
Unintentional design or manufacturing defect, operational negligence, misuse, "human error"	Sudden, dispersed	Discrete small-scale accidents
	Sudden, concentrated	Failures of large technological system (hotel fire, plane crash, etc.)
	Gradual, chronic	Low-level, delayed effect (asbestos, PCB, global warming, etc.)
Anti-social behavior		Sociopolitical disruption (street crime, war, etc.)

Source: Lowrance, pp. 8–10.

the concentration of the emissions. We organize six different classes of hazards in Table 4–2 according to their causal agents and nature of their effects.

Note that Table 4–2 describes "human error" as one of the agents responsible for a class of hazards. The notion of "human error" is an interesting one; a legitimate question to ask is, "What *isn't* 'human error' in a human-engineered system?"

Exposure assessment. This dimension of risk assessment starts with the result of the source assessment just described and analyzes who (or what) is exposed to the risk, and the nature and amount of that exposure. In some cases, this aspect can be quite complicated. For example, if we are examining the risks of cigarette smoking, the source assessment would lead to quantitative estimates of levels of tar and nicotine (and other substances) in the tobacco. The exposure assessment would focus on how much of those substances reach the lungs or are absorbed into the blood stream of the smoker and/or those in the immediate vicinity exposed to second-hand smoke.

Exposure assessments can have several stages, especially for risks associated with impacts along a portion of the food chain. Consider, for example, the situation in which a source assessment determines the concentration of dioxin released from an industrial facility into a nearby river. The first stage of an exposure assessment might involve determining how much of the dioxin is taken up by recreational fish species that inhabit the river. The exposure mechanism could be directly to the fish and indirectly through plants and insects consumed by the fish. You can imagine the difficulty of conducting such an assessment. Just conducting the engineering analysis of the spatial and temporal distribution of the dioxin throughout the river would be an extremely formidable challenge. Let's assume we are successful in combining that work with the similarly complicated biological analysis of absorption of the dioxin by the native fish (several species?). That completes the first stage. The next stage of the exposure assessment might be an estimate of dioxin intake by humans who consume the dioxin-contaminated fish. An example of such an assessment conducted by the

U.S. Environmental Protection Agency (EPA) for the lifetime average daily dose ($LADD$) is provided in Equ. (4–2).[25]

$$LADD = \frac{\dfrac{3 \cdot 10^{-9}\text{mg dioxin}}{\text{g fish}} \times \dfrac{150\text{ g fish}}{\text{meal}} \times \dfrac{3\text{ meals}}{\text{year}}}{70\text{ kg body weight} \times \dfrac{70\text{ years}}{\text{lifetime}} \times \dfrac{365\text{ days}}{\text{year}}} = 5.3 \cdot 10^{-11}\text{mg/kg/day} \quad (4\text{–}2)$$

The potential for uncertainty in each term in Equ. (4–2) is obvious. How many meals of recreationally caught fish do residents along the river consume annually? In addition to the $LADD$ estimate, EPA uses a different set of numerical values in Equ. (4–2) to arrive at a high-end exposure estimate ($HEEE$) of $1.2 \cdot 10^{-9}$ mg/kg/day. The $HEEE$ is, "designed to represent a 'plausible estimate' of the exposure of the individuals in the upper 90th percentile of the exposure distribution."[26]

Effects assessment.   Once we know who (or what) is exposed to the hazard and the level and duration of that exposure, we can turn to examining the effects or consequences of that exposure. The two key steps in this process are:[27] (1) establishing the causal relationship between the exposure and the effect; and (2) quantifying the magnitude of the effect. This brings us back to the magnitude term in Equ. (4–1). Of course, now that we've conducted the source and exposure assessments, we're in a much better position than we were before to estimate the magnitude of the consequences. However we are still faced with the problems associated with inadequate data discussed earlier in this section. In the context of effects assessment, curves such as shown in Figure 4–6 are called dose-response curves. For the specific example of consuming dioxin-contaminated fish, the dose, measured in mg/kg/day [see Equ. (4–2)] would be plotted along the horizontal axis. The response, which could be the probability of dying from that level of exposure, would be plotted along the vertical axis.

As complicated as it might be to conduct a risk assessment according to the three stages described here, we sometimes have to do it in reverse. That is, we observe an adverse effect and try to chase it back to a source. This frequently is the situation we find ourselves in when investigating accidents and many other kinds of failures of engineered systems. An important engineering tool for identifying and exploring the connections between the failures of complex engineering systems and the potential sources of those failures is fault tree analysis.

Fault tree analysis.   Fault tree analysis (FTA) is a graphical technique for displaying the relationships between component failure (source assessment) and system behavior (effects assessment) for complex engineering systems. We use a standard set of symbols to display those relationships in a fault tree format. Figure 4–7 shows a

---

[25]Reprinted from Faustman and Omenn, p. 84, with permission from The McGraw-Hill Companies.
[26]Faustman and Omenn, p. 83.
[27]AAES, p. 10.

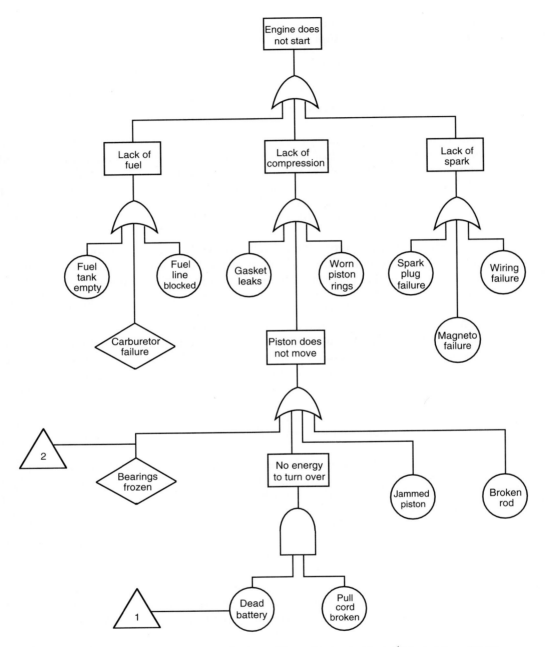

Figure 4–7.   Fault Tree for Failure of a Lawn Mower Engine to Start. (*Adapted from Dieter, p. 557, with permission from The McGraw-Hill Companies.*)

fault tree for the failure of a lawn mower engine to start. We construct the tree from the top down, starting with the effect assessment and tracing it back to possible sources. The notation is defined in Table 4–3. We can combine fault tree analysis with the probabilistic methods described in Chapter 5 to develop quantitative estimates for the probabilities of various events occurring.

Comparative risk assessment.   There are other important dimensions to risk assessment that we need to touch upon before moving on to the other phases of risk analysis. First, is the fact that most activities and systems that pose a risk also produce a benefit. Assessment of the risk should be accompanied by a similar assessment of the associated benefits. Second, most activities and systems for which we conduct a risk analysis are activities and systems for which alternatives exist. Risk assessment of a single activity or system should be placed in the context of the risks associated with the available alternatives. This notion of relative or comparative risks is central to risk assessment. Some argue that the best approach to quantifying comparable risks is to express all alternatives in terms of a common unit of measure, specifically the reduction in life expectancy.[28] Others argue that most risks are multidimensional, and no single measure can capture all their ramifications.[29] This latter approach leads directly to the issues of risk perception and communication, the second phase of risk analysis (see Fig. 4–5).

TABLE 4–3.   DEFINITIONS OF FTA SYMBOLS

Symbol	Name	Meaning
▭	Output Event	The resulting event due to prior events occurring lower in the tree
◁	OR gate	Output event immediately above OR gate occurs when at least one of the input events immediately below occurs
◯	Independent Event	An event that has no preceding events
◇	Undeveloped Event	An event that is not pursued in greater detail
⌂	AND gate	Output event immediately above AND gate occurs only when all the input events immediately below occur
△	Transfer Symbol	Used to make connections between different branches of the fault tree
⌂	Normal Event	An event that is expected to occur during normal operation

Source: Adapted from Ertas and Jones, p. 318, with permission from John Wiley & Sons.

[28]AAES, p. 19.
[29]Fischoff, Watson, and Hope.

Risk Perception and Communication

Risk perception is the interpretation and evaluation of the significance of a risk and the determination of whether a given level of risk is acceptable. Research studies have documented that people's perceptions of risk are frequently at-odds with the quantitative estimates of risk obtained from the risk assessment methods discussed above. Many engineering and science experts on risk attribute this discrepancy to the public's inadequate understanding of the complexities of risk assessment and the underlying scientific principles governing the systems being analyzed. If the problem was inadequate public understanding of science, the technical community could bridge the gap between public attitudes toward risk and the risk assessment facts by better explaining risk to lay audiences. This presumption gave rise to the concept of risk communication as a tool for educating the public on risk issues. However, further studies show that lack of technical knowledge only partially explains the disparity between public perceptions and the "technical facts" as revealed by risk assessments. Understanding these more subtle influences on risk perception is helping to lead to more sophisticated approaches to risk analysis that embody, rather than filter out, considerations of risk perception.[30] Dealing with people's perceptions of risk is a much more subjective phase of risk analysis than risk assessment. However, we can still gain important insights into key risk perception issues that will help us better understand individual and collective attitudes towards risk. This appreciation can go a long way toward enhancing our ability to design systems that meet the real needs of our clients and the public.

To illustrate key risk perception issues, we examine a situation where risk A is quantitatively much higher than risk B, but their relative acceptability appears to be inconsistent with the risk assessment results. Consider a comparison of the risks of airplane and automobile travel, conducted according to the framework introduced in Equ. (4–2) and summarized by the hypothetical numbers in Table 4–4.

We would probably tolerate the level of risk $R = 30,000$ for cars as the price we are willing to pay for the advantages of automobile travel. However, an annual rate of thirty

TABLE 4–4. HYPOTHETICAL COMPARATIVE RISKS FOR TRAVEL MODES

Travel Mode	Risk (fatalities per year)	Magnitude (fatalities/ accident)	Frequency (accidents/year)
Automobile	30,000	0.5	60,000
Airplane	3,000	100	30

Note: These numbers are hypothetical and were selected solely to facilitate the comparisons.

[30]Lamarre, pp. 20–29.

commercial aviation crashes averaging 100 fatalities per crash leading to a value of $R = 3,000$ for airplanes would be totally unacceptable. Why would we consider $R = 3,000$ to be unacceptable while we tolerate $R = 30,000$? Clearly there are alot more factors affecting people's attitudes toward risk than just the numbers contained in Table 4–4.

Some of the factors that individuals and society use, unconsciously or consciously, in their perception of the relative risks of automobile travel and airline travel are listed here:

- Those of us that drive automobiles feel that we have control over our car. Upon hearing about an accident, we feel that such an accident couldn't happen to us because we are better drivers (we don't drink and drive, we don't speed, we drive defensively). That may partially explain why many of us don't wear seat belts, which are known to substantially reduce the risk of injury. We do not feel a similar sense of control over the airplane in which we are flying—we have to rely on the pilot. This issue of control has an important effect on attitudes and perceptions of risk. We are more willing to tolerate risks in situations where we feel (whether or not that feeling is justified) that we have control over the activity.

- Car accidents are widely dispersed in time and location. Each automobile accident by itself is not a major catastrophe (except for the victims and their immediate family and friends); and does not warrant the focused attention of the general public. On the other hand, every major plane crash receives extensive media coverage that graphically portrays and describes the horrible details. (If 95% of all automobile accidents were to occur in Kansas during the same weekend, the story would also make national headlines.) Because of these differences in geographical and temporal dispersion, people tend to perceive airplane travel as riskier than automobile travel. Here is another example of how a small number of well-publicized catastrophic events foster perceptions of risk that are at variance with the facts. Most people are very surprised to learn that the combined annual deaths in the U.S. from tornadoes, earthquakes, hurricanes, and floods are less than the annual deaths due to appendicitis.

More generally, there are many other factors that influence our perceptions of risk:

- Most people are more willing to tolerate a given level of risk if the source of the risk is from a familiar agent than if it originates from an unfamiliar agent. This may partially explain why many people are more tolerant of the risks associated with a coal-burning electric power plant than they are with the risks associated with a nuclear power plant.

- The amount of risk people will accept depends on whether their exposure to the risk is voluntary or involuntary. This explains why some heavy cigarette smokers will object to air pollution from a nearby industrial facility. It also explains why many people will gladly spend much more money to play state-sponsored lotteries than they grudgingly pay in state taxes.

- The timing, permanence, and certainty of the consequences will influence people's perceptions. For example, some teenage smokers rationalize their activity because they perceive that the serious medical consequences are far into the future. Some forms of environmental degradation, such as oil spills, may be more tolerable than others because the adverse impacts are not irreversible. And issues for which there is significant uncertainty regarding the consequences, such as global warming, are perceived differently from those for which the consequences are more predictable.

- Attitudes toward risk also depend on who is affected. Risks that disproportionately adversely affect sensitive populations (elderly, infants, people with asthma) are less likely to be tolerated than those whose consequences are evenly distributed across the population. Also, some people are more willing to voluntarily accept a risk than others, especially when they personally are the immediate beneficiaries of the risk-taking behavior. People with risk-prone personalities, who willingly engage in thrill-seeking activities such as sky-diving approach risk differently than people with risk-averse personalities.

- Trust in the institution that is responsible for managing the risk is another important shaper of risk perception. An unwillingness to assume a risk (such as the risk of oil spills from supertankers) may be a reflection of an underlying mistrust of the responsible organization (in this case the oil companies).

Figure 4–8 consolidates various risk perception issues into a two-dimensional risk "space" within which risks are located according to their degree of controllability and observability. Risks located in the first quadrant of Figure 4–8 are most likely to trigger public demands for governmental regulation, even though they may not be considered major from a traditional risk assessment perspective.

Risk Management

The third phase of risk analysis is risk management, which refers to any action taken to eliminate or reduce a risk to an acceptable level. The three-stage model of risk assessment discussed earlier (source, exposure, and effects assessment; see Fig. 4–5) suggest strategies for risk management.

Risk management strategies.   Here we offer a set of general strategies for managing risk. We illustrate each in the context of managing risk from radiation releases at nuclear power plants.[31]

- Modify the Source—We can redesign the power plant and/or operating procedures to reduce the amount of radiation released. This can include measures such as thicker containment domes and additional filters for ventilation systems.

- Reduce the Exposure—To reduce the public exposure to radiation releases, the power plant can be located in remote locations downwind from major population centers. Another measure in this category is to adopt emergency commu-

---

[31]This is a modified version of the strategies described in Morgan, 1990, p. 18.

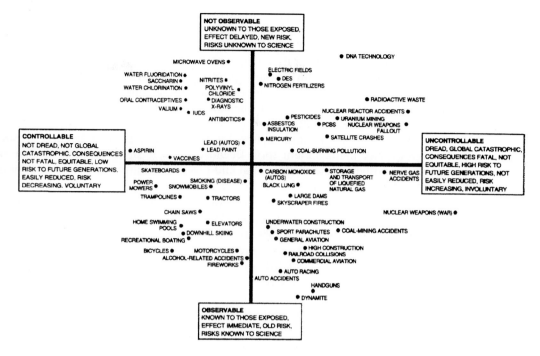

Figure 4–8.   Rating Risks According to Their Observability and Controllability *(Reprinted from Morgan, 1993, pp. 32–41. Copyright © 1993 by Scientific American, Inc. All rights reserved.)*

nity evacuation plans that would be implemented when the threat of an imminent accidental release of large amounts of radiation occurs.

- Reduce the Effects—Some of the effects of exposure to radiation releases can be reduced through ingestion of iodine tablets.

- Mitigate or Compensate for the Effects—Prompt medical attention can mitigate some of the effects. Exposure victims can be compensated monetarily for ill-health effects caused by the exposure.

- Modify the Environment—Energy conservation measures can reduce the need for electricity, hence avoiding construction of the power plant in the first place. Or the electricity can be generated from a different source (coal, natural gas, wind turbines, hydroelectric, photovoltaics).

   Risk management tools.   We can use a variety of useful tools to implement the risk management strategies discussed above. We discuss four important tools,[32] within the context of managing the risk of radioactive releases from nuclear power plants.

[32]Morgan, 1990, p. 23.

- Tort and Other Common Law—Many existing laws related to negligence, liability, nuisance, and trespass can be used as risk management tools. Known as torts, these have evolved and continue to be invoked as the result of civil suits filed in court. It is difficult to anticipate which way the jury or judge will rule in a given case and the amount of damages that will be awarded if the suit is resolved in favor of the plaintiff. The federal government, through the Price-Andersen Act, has capped the liability of nuclear power plant owners in case of a catastrophic release of radiation.

  Many individuals and groups don't have the resources needed to pursue torts as a risk management tool. Because of their significance to design engineers, we will examine professional and product liability issues in more detail in Section 4.7.

- Insurance—Both power plant operators, customers, and residents of nearby communities can provide protection against the costs associated with radioactive releases by purchasing insurance to cover such eventualities.

- Voluntary Standards—Many industry trade associations and professional societies develop safety standards governing the design, construction, inspection, operation, and maintenance of engineered systems. In particular, a major section of ASME's Boiler and Pressure Vessel Code is devoted to rules governing design of mechanical equipment for nuclear power plants.[33] See Section 4.6 for more detailed discussion of voluntary codes and standards.

- Government Regulation—Municipal, state, and federal governments frequently adopt regulations designed to protect public safety. This includes forming government agencies whose primary, if not sole, mission is risk management. An example relevant to this discussion is the federal Nuclear Regulatory Commission. We examine the interaction between government and design engineers more closely in Section 4.8.

Risk-benefit trade-offs.   Whatever risk management technique is adopted, it will probably cost money. The dilemma then arises of how much money we are willing to spend to reduce a given risk to a certain level. Risk-benefit and cost-benefit analysis techniques can help us keep track of the costs of risk reduction and the value of the consequent benefits, even though we recognize that many benefits (environmental, aesthetic, human) are difficult to quantify. Further, risk-benefit and cost-benefit analyses are techniques which rest on a foundation of a utilitarian moral philosophy.[34] As we have already discussed in Section 4.4, there are many valid arguments for rejecting utilitarianism in favor of alternative moral theories.

In many cases, this question of the cost of reducing risk boils down to asking how much a human life is worth. While many people object to the concept of placing a dollar value on human life, there are two arenas in which that decision is being made.

---

[33]Section III: Rules for Construction of Nuclear Power Plant Components, ASME Boiler and Pressure Vessel Code.

[34]Kelman, pp. 129–137.

First, juries struggle with this question in civil suits when they are asked to award damages to the survivors of victims of wrongful death. Second, some government regulatory agencies have to justify their expenditures in terms of the numbers of lives saved by their safety regulations.

One consideration in compensating survivors in wrongful death suits is the estimated forgone earning power of the deceased. Compensatory damages awarded by juries in wrongful death suits vary widely among states and on the basis of the age, gender, current income, and number of dependents left behind by the deceased. The average compensatory damage awards by juries in 1992 ranged from $150,000 for single women in the age range of 70–79 with no surviving minor children to $2,100,000 for married men in the age range of 30–39 with two surviving minor children.[35]

The amount of money government agencies spend for each life saved by their regulations also varies widely. One estimate is that the Federal Aviation Administration spends approximately $2,500,000 per life saved while the U.S. Environmental Protection Agency spends $7,000,000 for each life saved.[36] However, another study of cancer prevention regulations concludes that EPA spends between $15,000,000 and $45,000,000 for each case of cancer avoided.[37]

Another dimension to the risk-benefit trade-off is distributional in the sense that safety regulations may actually increase the risk to certain segments of the population. This was true in the case of passenger-side air bags in automobiles. While credited with saving 1,600 lives during the 1986–1996 time frame, they have also been associated with fifty-one fatalities, including thirty-one children. The public and governmental reaction to this situation demonstrated clearly that this was an unacceptable trade-off. The automobile safety regulatory agency (the National Highway Traffic Safety Administration, NHTSA) moved quickly to respond to these revelations. NHTSA allowed existing passenger-side airbags to be disconnected, and called for a refined design in which the bag deployment speed is a function of the weight of the occupant in the front passenger seat.

4.6. CODES AND STANDARDS

Our discussion of codes and standards in Chapter 3 was limited to their role as sources of information for design decisions. Here we want to examine the nature of codes and standards, the reasons why we have them, and the processes by which they are developed and enforced. Design engineers should understand these aspects of codes and standards because they form an important part of the context within which design decisions are made.

In Chapter 3, we described standards as necessary in order for objects to fit together (staples in staplers, three-ring notebook paper in three-ring binders, etc.). There are several other justifications for standards.

[35]Ostrom, 1993.

[36]Ibid.

[37]Van Houtven, pp. 6–10.

One large group of standards are those that ensure conformity among products, other than the dimensional conformity previously discussed. For example, there are industry-wide standards in the lumber industry regarding moisture content, strength, and appearance for lumber. Pharmaceutical companies have agreed upon standard dosage levels for common over-the-counter medicines such as aspirin.

Some standards are developed primarily or exclusively to protect public safety. In the late 1800s in the beginning of the steam powered industrial revolution, approximately 50,000 people died and two million were injured each year in steam boiler explosions. The adoption of a boiler and pressure vessel code by the American Society of Mechanical Engineers (ASME) dramatically altered the situation. Now the ASME Boiler and Pressure Vessel Code is used worldwide. Other examples of public safety standards are codes for design, construction, maintenance, and inspection of elevators and escalators. Fire protection and structural codes for buildings also exist. Federal safety standards exist for automobile headlights, taillights, turn signals, windshield glass, seat-belts, and airbags. Some of the many other areas covered by federal safety standards include flammable fabrics, dead-man switches on power lawnmowers, and children's toys.

Another category of standards advance the general health and welfare. Emission standards for cars and wood stoves to reduce air pollution protect public health, even though emissions may not be an imminent safety issue. Energy efficient standards for buildings and appliances also deal with public interest and welfare, with indirect connection to public health and safety.

Standards have also been adopted for terminology. For example, the IEEE *Standard Dictionary of Electrical and Electronic Terms* is itself a standard (ANSI/IEEE Std 100-1984). Such standards serve to facilitate communications between designers, customers, vendors, and the public.

Another class of standards are those that, rather than addressing the characteristics of an engineered system, establish procedures and documentation requirements for designing and testing those systems. The objective of these standards is to assure the quality of the design process itself. In turn, that gives customers increased confidence in the products and systems. These kind of quality assurance standards are increasingly being used as strategic tools to streamline processes and reduce costs in order to remain competitive.[38] Later in this section we will examine in detail a set of such standards (ISO 9000) that has enormous implications for engineering design activities.

Performance Versus Prescriptive Standards

Design standards and performance standards are the two main kinds of standards. Performance standards look at the bottom line; they address the issue of how the system performs and functions. An example is the energy efficiency portion of a local building code that says a new building can be of any design and use any construction

---

[38]American National Standards Institute, 1997.

materials, as long as the heat loss through the building's walls doesn't exceed a specified amount of BTU/hr-ft^2 under given weather conditions. On the other hand, a design or prescriptive-oriented standard dealing with building energy efficiency might specify the level of insulation, may require a certain amount of weather stripping, and could limit the amount or type of glass windows installed.

The advantage of a performance standard is that it gives the designer maximum flexibility in achieving the desired performance level. However, it isn't until the system is built and tested that you, the designer, can be sure that the standard has been satisfied. A sophisticated computer simulation can give you confidence in the validity of the design, and a simulation performed by a certified user with an approved computer simulation software package may be acceptable to the regulatory group as evidence that the standard has been satisfied. Nevertheless, performance standards can be difficult and expensive to satisfy and enforce.

A design standard is simpler, less expensive, and involves less uncertainty. It's easy for the designer, builder, and the inspector to know when the correct level of insulation has been installed. No complex analysis or sophisticated engineering design judgments are necessary to know whether or not the standard has been satisfied. Design standards represent more of a cookbook approach than performance standards. On the other hand, that is also their biggest drawback. Design standards may lock design engineers into specified practices and prevent them from applying innovative approaches. In some cases, new technological approaches do not become commercially successful because they don't satisfy the existing design code, and the political obstacles to modify the code may be too great to overcome. Some companies that are successfully marketing their products under the existing code may feel that their market share is threatened if the new technology is allowed to enter the market under a modified code; hence, they oppose changing the code.

Voluntary Versus Mandatory Standards

Another important way to categorize standards is according to whether they are voluntary or mandatory. Voluntary (frequently also called consensus) standards do not have the force of law. They are simply agreements among manufacturers and other private sector organizations such as insurance companies that are involved in a particular industry or product. Consensus standards are developed by various users, suppliers and other interested parties who arrive at a consensus as to the content of a standard. In many cases, the consensus is strong enough so that every company in the industry adopts the standard voluntarily. Occasionally, it is not possible to arrive at a consensus standard. Good examples are the unsuccessful efforts to develop standards for audio tapes (eight-track cartridges versus cassettes), video tapes (Beta versus VHS), and operating systems for personal computers (Apple, DOS, OS/2, Windows 95).

Mandatory standards are those adopted by a unit of government. They can be imbedded within a law or be issued by a duly constituted regulatory body, which is charged by law to develop a detailed regulation to implement a more general directive embodied in the law. Any such regulation has the full force of the law which authorized it, and violation of the regulation is punishable in accordance with the enabling legislation. At the federal level, some of the safety-oriented regulatory bodies are the Nuclear Regulatory Commission (NRC), Consumer Product Safety Commission (CPSC), Federal Aviation Administration (FAA), National Highway Traffic Safety Administration (NHTSA), and the Occupational Safety and Health Administration (OSHA). Other federal regulatory agencies manage public resources (for example, the Federal Communications Commission and the Environmental Protection Agency), and protect the free market system from abuse (for example, the Federal Trade Commission).

Mandatory standards frequently take advantage of an existing voluntary standard. A regulatory agency may adopt an existing voluntary standard intact; just by citing the voluntary standard by reference in its regulation. The regulating agency may also modify an existing voluntary standard before adopting it. Or it can ignore an existing voluntary standard (unless directed otherwise by law) and write its own from scratch. Frequently, voluntary standards get adopted into law and become mandatory by reference. Consider for example, the federal safety standards covering bicycle helmets discussed in Chapter 3. The law specifies that the Consumer Product Safety Commission (CPSC) use the ANSI Z90.4-1984 and the ASTM (American Society for Testing and Materials) F1447 standards, which were developed as consensus standards. The proposed CPSC regulation uses the ASTM standard as required by the law but combines it with the ANSI provisions regarding required head coverage. The regulation also refers to SAE test instrumentation and ISO headforms.

Standards Making Organizations and Their Procedures

There are many private sector organizations that issue consensus standards. Generally, they fall into two categories: trade associations and professional engineering societies. Examples of the former are the National Electrical Manufacturer's Association (NEMA) and the Association of Home Appliance Manufacturers (AHAM).

Professional engineering societies that issue consensus standards include the American Society of Mechanical Engineers (ASME), Society of Automotive Engineers (SAE), and the Institute of Electrical and Electronic Engineers (IEEE). As an example of the scope of these activities, ASME issues approximately 1,300 different codes and standards. Approximately 4,000 volunteer ASME members participate on ASME standards committees, supported by a full-time staff of about eighty professionals. The composition of the committees and the process they follow are spelled out in considerable detail and are designed to bring the best technical expertise to bear on the issue, while ensuring that the views of designers, manufacturers, vendors, customers, and the public are adequately represented. Figure 4–9 depicts the procedure used in ASME's standards activities.

The process by which a mandatory federal regulation is developed in the U.S. is spelled out in most cases by the Administrative Procedures Act. This is designed to ensure the agency acts in an open and fair way, and provides interested parties ample opportunity to comment on proposed rules. This involves publishing a Notice of Pro-

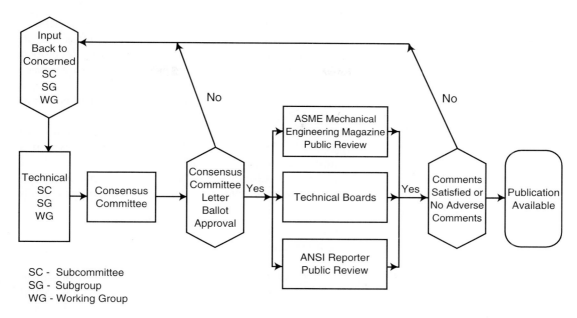

Figure 4–9. Process for Creating or Revising an ASME Standard *(ASME Manual MM-2, Fall 1996, p. 35.)*

posed Rulemaking (NOPR) and sometimes an Advance Notice of Proposed Rule-making (ANOPR) in the *Federal Register*. The regulatory body may also hold public hearings in different parts of the country to solicit feedback on its plans. Usually, the agency is required to summarize the input it receives and to respond to specific comments. Any regulation that is adopted must be explicitly justified by the adopting agency. Failure to do so places the agency in jeopardy of having its actions overturned in court.

Government regulation can have a major effect on entire industries or classes of technologies. For example, the Price-Anderson Act passed in the 1950s gave a jump-start to the civilian nuclear power industry by limiting the liability of utility companies in the event of an accident at a nuclear power plant. Emissions regulation has had a big effect on air quality while developing new environmental technologies. Government regulation can have the opposite effect also. Consider the fate of the domestic single-engine light aircraft industry in this country. Annual production levels fell from approximately 16,000 in the late 1970s to 500 in 1991. One reason for this decline is that light aircraft have been required to meet the same federal safety standards as large commercial airliners.

State and Local Government

Many regulatory activities in the U.S. occur at the state government and local government levels. For example, codes and standards for residential and commercial buildings with respect to structural integrity, fire safety, electrical, plumbing, water supply, and energy efficiency are regulated at the state level. Also, standards for other aspects of the community infrastructure, such as road systems, traffic control, mass transit, waste collection, disposal and recycling, and sewer treatment systems are developed through a regulatory process. Each state has the authority to establish its own rules and regulations in areas such as these, unless they have been preempted by federal legislation. But clearly, it would be very inefficient if each state worked in isolation from the other states in such matters. Several coordinating mechanisms have been established such as the National Governors Association (NGA) and the National Conference of State Legislators (NCSL). Sometimes a federal agency will develop a model code which it offers to any state that wants to adopt it. Also, states and local governments will frequently make use of voluntary standards. For example, many energy efficiency state codes are adoptions or adaptations of a voluntary standard developed by ASHRAE.

American National Standards Institute

The American National Standards Institute (ANSI) is a private, nonprofit organization that coordinates most voluntary standards operations in the United States. ANSI serves as an umbrella for the standards activities of its 1,400 member organizations, including professional societies, trade associations, manufacturing and service companies, consumer and labor organizations, and government agencies. ANSI has three key roles:

- ANSI certifies standards making procedures used by other organizations.
- ANSI initiates new standards making activities, either by inviting one of its member organizations to develop a standard, or developing it on their own. Many standards developed by ANSI member organizations carry identifying designations of both organizations, for example, ANSI/ASME 32.3.
- ANSI represents the U.S. on the International Organization for Standardization (ISO).

International Standards

As international trade barriers crumble and the world economy becomes more globally integrated, international codes and standards become increasingly important. The two dominant organizations dealing with international standards are the International Electrotechnical Commission (IEC) and the International Organization for Standardization (ISO). IEC has responsibility for electrical and electronic engineering standards. Standards in all other fields are handled by ISO, and we will limit our further discussion of international standards to ISO.

International organization for standardization (ISO). ISO is a nongovernmental, voluntary federation of national standards organizations. It consists of one member from each of approximately one hundred countries.[39] Each member is the national body most representative of standardization in its country (ANSI represents the United States). ISO has produced almost 1,000 standards, many of which affect familiar objects and consumer products. For example, the size of credit cards and the symbols used on automobile dashboards were established by ISO standards.

ISO's standards activities involve 2,700 technical committees, subcommittees, and working groups. Each standard is adopted by a consensus process, requiring approval by two thirds of the members actively involved in developing that standard and three fourths of all votes cast.

ISO 9000.   Spurred by the establishment of the European Community integrated market in the early 1990s, ISO developed a series of quality assurance standards for products and services, including the engineering design process and related activities. This series, known as ISO 9000, has been adopted in more than eighty countries by companies engaged in international trade. In the United States, ISO 9000 guidelines have also been incorporated into contractual requirements of government agencies such as the Department of Defense, National Aeronautics and Space Administration, and the Federal Aviation Administration.[40] ISO 9000 consists of the following set of five standards:

ISO 9000—Quality Management and Quality Assurance Standards: Guidelines for Selection and Use

ISO 9001—Quality Systems: Model for Quality Assurance in Design/Development, Production, Installation, and Servicing

ISO 9002—Quality Systems: Model for Quality Assurance in Production, Installation

ISO 9003—Quality Systems: Model for Quality Assurance in Final Inspection and Test

ISO 9004—Quality Management and Quality System Elements: Guidelines

The ISO 9000 standards establish procedures, practices, and documentation requirements that an organization must adopt in order to demonstrate that their activities are at an acceptable level of quality. Most recently revised in 1994, the portion of this series that is most directly relevant to design is ISO 9001. More specifically, of the twenty requirements of ISO 9001, Section 4.4 (Design Control) explicitly addresses the design process. This language is reproduced in the appendix to this section. As you can see, it requires extensive documentation of virtually every planning activity

---

[39]The descriptive material regarding ISO was obtained from http://www.iso.ch/infoe/intro.html, Jan. 17, 1997.

[40]Torodov, pp. 2–3.

and design decision. As this and related requirements become the norm, it will have a major impact on the way engineering design activities are conducted and documented. One indication of the significance of these developments is that books are being published that help businesses implement the standards.[41]

Appendix 4B: Section 4.4 of ISO 9001 (1994 version)

4.4.1 General

The supplier shall establish and maintain documented procedures to control and verify the design of the product in order to ensure that the specified requirements are met.

4.4.2 Design and development planning

The supplier shall prepare plans for each design and development activity. The plans shall describe or reference these activities, and define responsibility for their implementation. The design and development activities shall be assigned to qualified personnel equipped with adequate resources. The plans shall be updated, as the design evolves.

4.4.3 Organizational and technical interfaces

Organizational and technical interfaces between different groups which input into the design process shall be defined and the necessary information documented, transmitted, and regularly reviewed.

4.4.4 Design input

Design-input requirements relating to the product, including applicable statutory and regulatory requirements, shall be identified, documented, and their selection reviewed by the supplier for adequacy. Incomplete, ambiguous, or conflicting requirements shall be resolved with those responsible for imposing these requirements.

Design input shall take into consideration the results of any contract-review activities.

4.4.5 Design output

Design output shall be documented and expressed in terms that can be verified against design-input requirements and validated (see 4.4.8).

Design output shall:

---

[41]Novack, 1995.

a) meet the design-input requirements;

b) contain or make reference to acceptance criteria;

c) identify those characteristics of the design that are crucial to the safe and proper functioning of the product (e.g., operating, storage, handling, maintenance, and disposal requirements).

Design-output documents shall be reviewed before release.

4.4.6 Design review

At appropriate stages of design, formal documented reviews of the design results shall be planned and conducted. Participants at each design review shall include representatives of all functions concerned with the design stage being reviewed, as well as other specialist personnel, as required. Records of such reviews shall be maintained (see 4.16).

4.4.7 Design verification

At appropriate stages of design, design verification shall be performed to ensure that the design-stage output meets the design-stage input requirements. The design-verification measures shall be recorded (see 4.16).

4.4.8 Design validation

Design validation shall be performed to ensure that product conforms to defined user needs and/or requirements.

4.4.9 Design changes

All design changes and modifications shall be identified, documented, reviewed, and approved by authorized personnel before their implementation.

4.7. PROFESSIONAL AND PRODUCT LIABILITY

> If a builder has built a house for a man and has not made strong his work, and the house he built has fallen, and he has caused the death of the owner of the house, that builder shall be put to death. If he has caused the son of the owner to die, one shall put to death the son of the builder.

Though the form of punishment has changed since the Babylon King Hammurabi issued this rule 4,000 years ago,[42] we have retained the principle that engineers are legally responsible for the integrity of their designs.

[42]Streeter, p. 3.

The Hyatt Walkways

The worst accidental structural failure in a building in the United States was the 1981 collapse of the suspended walkways spanning the lobby of the Hyatt Regency Hotel in Kansas City. Over one hundred people were killed in the disaster. The subsequent well-publicized investigation and lawsuits, combined with the magnitude of the tragedy and the elementary character of the engineering misjudgment, transformed the incident into one of the most widely cited illustrations of engineering liability issues.

     The several elevated walkways crossing the Hyatt lobby were suspended by rods from the ceiling of the hotel atrium (see Fig. 4–10). Walkway support beams were located at the appropriate distances along the span of the walkways. A change in the de-

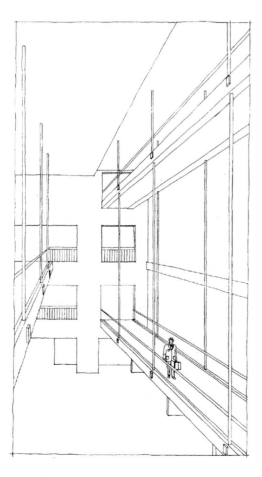

Figure 4–10. Architect's Rendering of Hyatt Walkways *(Marshall, p. 21.)*

tails of how the walkway support beams were attached to the suspension rods transformed an unusual, but marginally safe, design into a disaster waiting to happen. Failure of the support beam on the fourth-level walkway (located at the upper right of Fig. 4–10) triggered the collapse.

Examining the situation from an elementary structural perspective emphasizes the importance of validating the integrity of any design or design modification, regardless of how trivial it may appear at first glance. The Hyatt walkway story also vividly illustrates the importance of the implementation phase of design (Step 9 in the design process model discussed in Chapter 1).

Figure 4–11(a) shows a cut-away view of the original connection between each fourth-level walkway support beam and one of the threaded suspension rods. The bottom flange of the support beam rested on a washer kept in place by a nut threaded onto the suspension rod. The suspension rod continued on down and connected to the support beam for the second-level walkway in the same manner.

Because of the difficulty of threading the long rods, delivering them to the site, and attaching the walkway support beams to the rods, the design was subsequently

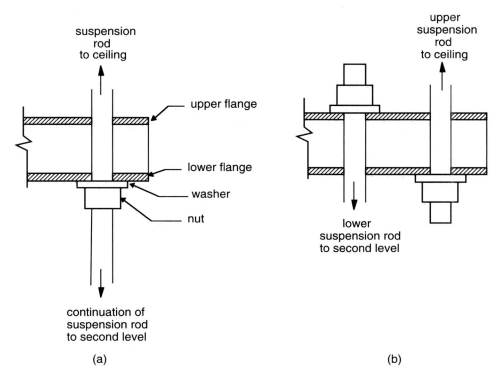

(a)   (b)

Figure 4–11. Original and Modified Support Configurations of Fourth-Level Walkway

modified. In the modified version, shown in Figure 4–11(b), the sets of long vertical suspension rods were replaced by two sets of shorter rods. The upper set was used to hang the fourth-level walkway from the ceiling; the lower set was used to hang the second-level walkway from the fourth-level walkway. This new design greatly simplified the fabrication of the rods (which now only had to be threaded at their ends) and assembly of the walkway system. However, it complicated (fatally) the details of the connection between the fourth-level walkway support beam and the suspension rods. The upper set of rods were in contact with the bottom flange of the support beams for the fourth-level walkway in the same manner as in the original configuration. The lower set of rods were supported by the upper flange of the fourth-level walkway beams. Let's compare the load imposed on the lower flange of the support beam in configurations (a) and (b).

Let P_2 represent the load transferred from the second-level walkway to the suspension rod and let P_4 be the load transferred from the fourth-level walkway to the suspension rod. Figure 4–12(a) shows that in the original configuration, the bottom flange of the fourth-level support beam transfers the load P_4 to the suspension rod. With the modified design, the fourth-level support beam also has to transfer the load P_2 from the lower suspension rod to the upper suspension rod. As displayed in Figure 4–12(b), the bottom flange has to transfer $P_2 + P_4$ to the suspension rod. The walkway failure was precipitated by this overloading of the bottom flange.

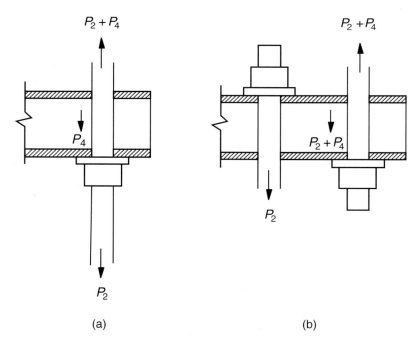

Figure 4–12. Load Transfer Between Fourth-Level Support Beam and Suspension Rod

The lengthy complicated legal developments following the collapse has been described in detail elsewhere. For the purposes of our discussion, we note that both the chief engineer and his boss were found guilty of gross negligence and had their licenses revoked.

Legal Concepts

As demonstrated by the above story, the issue of legal liability is crucial to all design engineers. We explore its various aspects and ramifications in the remainder of this section. In order to provide the appropriate background for that discussion, we begin with a short overview of relevant legal concepts.

Liability law in the United States is, for the most part, part of our system of civil (as opposed to criminal) law. There are several important distinctions. In criminal law, the concepts of innocence and guilt exist. The accused is presumed innocent and the prosecution has the burden of proving guilt beyond a reasonable doubt. Unanimity is required in order for a jury to reach a decision in a criminal case. Individuals convicted in criminal cases may suffer punishments which deprive them of their freedom and possibly of their life, if convicted of a capital crime.

In civil law, the issue is not one of innocence or guilt; it is a question of who is at fault in a dispute, or who violated an agreement, or who failed to fulfill obligations. Neither the plaintiff (the one who files the complaint) nor the defendant has a built-in advantage. A jury does not require unanimity to find in favor of either party in a civil case.[43] The test for ruling in favor of one of the parties is that of "the preponderance of the evidence." This is a much easier test than the "beyond a reasonable doubt" test required in criminal cases. Depending on the nature of the dispute, winners in civil cases are usually awarded the disputed property, monetary damages, or custody rights (as in divorce cases).

Most criminal law in the United States is based on specific statutes enacted by state or federal government (statutory law). On the other hand, much of our civil law is common law. Common law is a collection of tradition and prior court decisions that serve as precedents to guide every new case. While some statutes have been passed to clarify or modify some elements of common law, the bulk of common law has no specific statutory language that clarifies ahead of time who the likely winners and losers will be in any specific civil suit.

Liability law belongs to that branch of civil law known as torts. Tort law does not include disputes over contracts and warranties. Much of the current debate regarding overhaul of the liability system is carried under the heading of tort reform.

One of the similarities between criminal and civil law is that while there is some activity at the federal level, most of the legal action occurs in state courts. Court decisions on liability issues in one state may contradict decisions on the same or similar issue in another state. This lack of consistency among the states presents a difficult

---

[43]In Washington State, civil cases may be heard by juries with either six or twelve members. Five members are needed to reach a verdict for the smaller body, and ten for the larger group.

legal landscape in the liability arena. Design engineers and manufacturers who are designing and producing products for a national market face many uncertainties regarding the legal ramifications of their actions.

Liability Legal Issues

Now we turn to the legal issues that are specific to product and professional liability. First, we need to clarify the differences between product liability and professional liability as they affect design engineers. Product liability claims by a plaintiff address the characteristics of a product. These characteristics may be the result of design decisions, in which case the engineer(s) who designed the product will clearly be targeted in the lawsuit, either directly or indirectly as an employee of the company that made the product. Product liability claims may also be the result of the way the product was manufactured, serviced, or repaired. Claims may also be based on the manuals that provide assembly, maintenance, or use instructions for the product. Claims based on these other characteristics of the product would not involve design engineers if these characteristics were not their responsibility.

Not all design engineering decisions are product-oriented. Engineers design processes and provide consulting advice and other services. Liability suits filed against engineers as a result of these kinds of activities come under the heading of professional liability. Even in a case involving product design, the term 'product liability' may be used in the context of claims made against the manufacturer and the term 'professional liability' may be used when referring to the engineers involved. Sometimes the distinction is hard to make because it may not be clear whether the problem arose as the result of a design decision or a manufacturing activity. In fact, as our design process model in Chapter 1 emphasizes, design responsibilities extend into an implementation phase. Aside from the issue of who is the responsible party, the other key liability issue is the degree of responsibility. This is also a complicated matter and the law in this area is still evolving.

Privity and Negligence.   Until the early part of the 20th century, the prevailing liability doctrine was that engineers and manufacturers were responsible for harm done by their product or design only if they had an explicit contract with their customer. This is known as the privity doctrine. Beginning with a famous case in 1916 involving injuries sustained when a wheel fell off an automobile, the privity standard was replaced by the doctrine of negligence. This meant, that even without an explicit contract, the manufacturer (or engineer) had the responsibility to act in a careful and prudent manner. Negligence can be based on any of the following conditions:

- The design created a concealed danger
- Failure to incorporate appropriate safety devices
- The product was made from inadequate materials
- Failure to warn user of the danger

Differences in the amount of care missing from the action and the extent to which the lack of care is willful gives rise to many different forms of negligence. *Black's Law Dictionary* actually defines nineteen kinds of negligence! Three forms interest us most:[44]

- Simple negligence—failure to exercise for protection of others that degree of care and caution that would, under prevailing circumstances, be exercised by an ordinarily prudent person
- Gross negligence—the intentional failure to perform a manifest duty in reckless disregard of the consequences as affecting the life or property of another
- Criminal negligence—such a flagrant and reckless disregard of the safety of others, or willful indifference to the injury liable to follow, as to convert an act otherwise lawful into a crime when it results in personal injury or death.

Strict liability.   Beginning in the 1960s the concept of negligence as the basis for product liability gave way to the doctrine of strict liability, which puts greater burden on the design engineer and manufacturer. Under strict liability, the plaintiff does not have to prove negligence. Instead, the manufacturer is liable if:

- The product was defective and unreasonably dangerous.
- The defect existed at the time the product left the defendant's control.
- The defect caused the harm.
- The harm is appropriately assignable to the defect.

The important distinction between negligence and strict liability is that negligence addresses the behavior of the defendant; strict liability considers the behavior irrelevant, and all that counts is the condition of the product. Strict liability is the prevailing doctrine for product liability in most states, while negligence, for the most part, is still the standard for professional liability when no product is involved.[45] The concepts of "defective" and "unreasonably dangerous" within the strict liability doctrine deserve additional elaboration.[46]

Defects.   Defects can be the result of either a production error or a design error. A production error occurs when the product is not manufactured as intended due to a substandard manufacturing process. Usually a production error gives rise to the issue of product liability for which the manufacturer may be held responsible. Whether the design engineer is also responsible depends on the role of the engineer in the manufacturing phase. Design defects are errors in engineering judgment for which the design engineer bears the responsibility. As mentioned earlier, whether a claim on

---

[44]Definitions taken from *Black's Law Dictionary.*
[45]Streeter, p. 10.
[46]Ashley, pp. 46–47.

the basis of a design defect is a case of product liability or professional liability depends on the relationship between the engineer and the manufacturer.

Unreasonable danger. We can best examine the concept of unreasonable danger using the risk-benefit approach introduced in Section 4.5. That is, the test of unreasonableness has to be made in the context of the benefits associated with the product. Seven considerations have been proposed for use in balancing the risks against the benefits:[47]

- The product's usefulness
- The availability of safer products to meet the same need
- The likelihood and probable seriousness of injury
- The obviousness of the danger
- The public expectation of the danger
- The avoidability of injury by care in the use of the product, including the effect of instructions and warnings
- The manufacturer's or seller's ability to eliminate the danger of the product without making it useless or unduly expensive

Attempting to achieve this balance is no easy task for a design engineer. You can imagine how difficult this balancing act is for a jury of nonengineers who are asked to render a verdict in a liability law suit. For example, juries in the state of Washington are instructed to decide liability on the basis of a product being not reasonably safe as designed, if:[48]

> . . . at the time of manufacture, the likelihood that the product would cause injury or damage similar to that claimed by the claimant, and the seriousness of such injury or damage outweighed the burden on the manufacturer to design a product that would have prevented the injury or damage, and outweighed the adverse effect that an alternate design that was practical and feasible would have on the usefulness of the product; or

> the product is unsafe to an extent beyond that which would be contemplated by an ordinary user. In determining what an ordinary user would reasonably expect, you should consider the relative cost of the product, the seriousness of the potential harm from the claimed defect, the cost and feasibility of eliminating or minimizing the risk, and such other factors as the nature of the product and the claimed defect indicate are appropriate.

Compensation. Once a judgment has been made in favor of the plaintiff in a liability case, a monetary award is made. This award is established at a level designed to compensate the plaintiff for the damages they suffered (compensatory damages).

---

[47]Vargo, p. 46.
[48]Washington . . . , p. 534.

In some cases punitive damages can also be awarded when the level of compensatory damages is not deemed to sufficiently spur the defendant to modify their behavior.

Implications for Design Decisions

Our discussion paints a picture of a design swamp filled with submerged mines primed to go off at the first ripple of design activity. How can any engineer approach design in an environment so full of uncertainty at-best and booby-traps at-worst? Some authors have written about the stifling effect the liability threat has had on creative engineering design and technological innovation. Others have argued that the threat of liability law suits have actually spurred design engineers and manufacturers to be more sensitive to safety issues and to address them in more creative and innovative ways.[49]

More directly, engineering design strategies that reduce exposure to liability have been suggested. One key concept is for design engineers to recognize that products will be used in careless or abusive ways, or for purposes not intended by the designer. Let's face it, every once in a while consumers (including you and me) do some pretty stupid things! Courts have made it clear that careless use is not grounds for absolving the manufacturer or engineer of liability.[50] Anticipating such circumstances and incorporating features into products to make them more robust in such situations is a prudent design strategy. Specific procedural steps include using fault tree analysis (see Section 4.6), conducting extensive tests, fully documenting all design and manufacturing details, carefully preparing instruction and maintenance manuals, and obtaining legal advice throughout the design process.[51]

4.8. INTELLECTUAL PROPERTY

We have already seen how government regulatory actions can have a big effect on the activities of design engineers. In this section we discuss the federal government's role in protecting intellectual property. The most important of these activities for design engineers is the patent system.

Patents

One major influence of government on design engineers is the patent system under which the government protects new technological developments. In Chapter 3 we focused on how to search through the patent literature to determine whether one of your design concepts has already been patented by someone else. Here we want to

---

[49]Both sides of this argument are presented in Huber and Litan, 1991.

[50]Cortes-Comerer, pp. 40–42.

[51]For a detailed compilation of twenty guidelines for reducing product liability risks, see Middendorf, 1990, pp. 36–40.

concentrate on how to use the patent system to protect your designs so that others cannot exploit them without your permission and without compensating you. This is important for you because one of the designs you are working on now, or will soon be working on, may be patentable.

The patent process.   In Section 3.3, we described a patent search for a multiple-speed drive for manual wheelchairs. That search was conducted in January 1994 as part of a senior design project that I and four of my students were working on with Mr. Ben Jeffries, a Seattle-area engineer who first proposed the project to me. Since our patent search failed to reveal an existing patent that satisfied our needs, we proceeded to spend the Winter 1994 academic quarter designing a two-speed, hub-mounted drive with a quick-disconnect mechanism for easy wheel disengagement. We will concentrate our attention here on the process we went through to obtain the patent, not on the specifics of the design.[52] We provide the following brief description of the design as background information.

The location of the shifting mechanism on the wheelchair is shown in Figure 4–13. In conventional wheelchairs, the handwheel (item 32 in Fig. 4–13) by which the occupant propels the chair is rigidly attached to the wheel assembly (item 14). In this

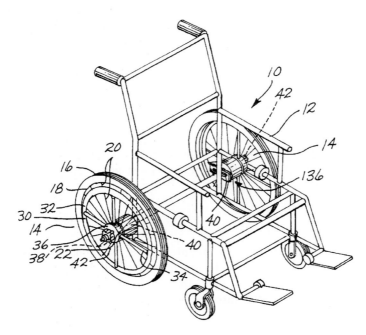

Figure 4–13.   Hub-Mounted, Quick-Release, Wheelchair Shifting Mechanism *(U.S. Patent 5,482,305)*

[52]For additional drawings and explanation of the design, see U.S. Patent 5,482,305.

system, the handwheel is connected to the wheel assembly only via the handwheel hub (36) and the wheel assembly hub (22). In the conventional mode, the handwheel and wheel assembly rotate at the same speed. When the chair occupant encounters a ramp or other slope that requires additional torque to traverse, they rotate the knob on the handwheel hub to engage a gear mechanism mounted inside the wheel assembly hub (22). That allows the handwheel to rotate at about twice the speed as the wheel assembly and transfer about twice the torque to the wheel.

An assembly drawing of the shifting mechanism is given in Figure 4–14. Another important feature of this design is that the wheel can be attached to the frame of the wheelchair by depressing the button (item 116 in Fig. 4–14) in the center of the handwheel knob. This allows the wheels to be easily retrofitted onto existing wheelchairs. Similarly, depressing the button disconnects the wheel from the frame, thereby allowing for easy stowage of the chair in the back seat or trunk of a car. Both Figure 4–13 and Figure 4–14 are reproduced from the patent document.

Most design engineers that develop patentable ideas do so within the context of their professional employment. In many cases, that means that the patent rights belong to the employer. However, many employers share income derived from the patent with the inventors. In our case, since the work was done by the students as part of their course of study at the University of Washington, and my contribution was made in my capacity as a faculty member, the university has the rights to any resulting patent.

After completing the design in March 1994, we contacted the university's Office of Technology Transfer (OTT) to ascertain whether they were interested in pursuing the patentability and commercial potential of the mechanism. After a preliminary evaluation by OTT of the prospects for the device to be commercially successful, each

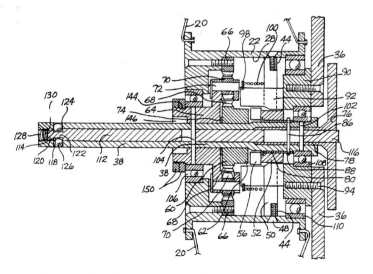

Figure 4–14.   Assembly Drawing of Wheelchair Shifting Mechanism *(U.S. Patent 5,482,305)*

of the students, Mr. Jeffries, and myself signed a formal disclosure notice in June 1994. The significance of the formal disclosure is that it essentially starts the clock for all patent-related matters. Specifically it requires the patent application to be filed within a year of the disclosure.

Work then proceeded simultaneously on four fronts. First, each of my students and I signed an agreement with the university that transferred all rights to the invention to the university. Second, because Mr. Jeffries was willing to invest some money towards securing a patent and in further developing the device, he negotiated a separate co-ownership agreement with the university. Both agreements included a compensation formula under which income derived from sales of the product would be distributed among the university and the individual inventors. The students and I were pleased with the agreement since none of us had either the funds, time, or expertise to pursue a patent and commercial opportunities. (The attorney's fees for this project exceeded $15,000.) Third, the university hired a patent attorney (who, as required for patent attorneys, had a technical degree) to conduct a thorough search of existing patents on multi-speed wheelchairs. When the search revealed that our concept did not infringe on any existing patents, the university authorized the attorney to file a patent application. Fourth, the university began to seek out entrepreneurs interested in manufacturing and marketing the mechanism. That search included a market analysis.

The patent application was filed in January 1995. A market analysis was completed in April 1995 and in the same month the university signed an option agreement with a firm that was considering commercializing the invention. The patent was awarded in January 1996 and the university signed its first licensing agreement for this mechanism in November 1996.

The patent system.   The patent system is based on Article 1, Section 8 of the U.S. Constitution which gives Congress the power, ". . . to promote the progress of science and useful arts by securing for a limited time to authors and inventors the exclusive right to their respective writings and discoveries." The patent laws that have been enacted since then are embodied in Title 35 of the United States Code (35 USC). These laws are administered by the U.S. Patent and Trademark Office. Significant changes in U.S. patent law occurred in 1993 and 1994 when the North American Free Trade Agreement (NAFTA) and the General Agreement on Tariffs and Trade (GATT) were enacted. These changes were made to bring patent laws of individual countries into conformity with each other.

There are three types of patents. Utility patents protect new, useful processes, machines, manufactures, or chemical compositions. This type of patent is available to cover new engineering designs; they are valid for twenty years.[53] Design patents protect new, original and ornamental designs (how something looks). They are granted for three and one half, seven, or fourteen years depending on the fee the inventor chooses to pay. The third type are plant patents, which protect new, distinct, asexually reproduced biological plants. Plant patents are granted for seventeen years.

---

[53]The twenty-year period from the date of filing was one of the changes associated with GATT. Previously, the term of utility patents was seventeen years from the date the patent was issued.

More than 5.3 million U.S. patents have been awarded and approximately 90,000 new ones are issued each year. With an acceptance rate of about 50%, that means approximately 180,000 patent applications are filed each year. The system has served us well and lived up to the expectations of the founding fathers.

To be eligible for a patent, an invention must be novel, must perform the task claimed for it, and must be nonobvious to a person skilled in the field. Obtaining a patent means that the patent office has found that these conditions have been satisfied. It does not mean that the device is likely to be commercially successful. Many patents never are transformed into commercially successful products; some of these have more value as sources of amusement. Figure 4–15 depicts a device cited as the Wacky Patent of the Month for March 1997(U.S. Patent 4,583,939).[54]

Sometimes more than one person makes a claim for the same invention. In that case, the one who had the idea first is recognized by the courts as the inventor. This emphasizes the importance of keeping a complete, dated, design journal (see Section 3.8). It may be crucial to resolving a patent dispute.

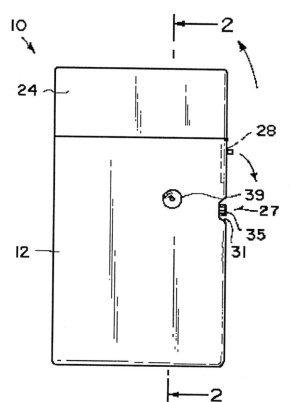

Figure 4–15. A Combination Cigarette Lighter and Perfume Dispenser *(U.S. Patent 4,583,939)*

[54]http://colitz.com/site/wacky.htm.

Once the patent has been filed in the Patent Office, it gains the status of patent pending. You have probably seen that marked or displayed on many products and their descriptions. This status does not give the inventor any protection until and unless the patent is granted. During that time, someone who wants to use (steal?) the idea and rush to market with a product, cannot be restrained from doing so by the inventor. Patent pending is a notice to any potential usurpers that, if the patent is eventually awarded, the inventor can sue the usurper not only for actual damages (the value of the income that the inventor was deprived of) but for triple damages (three times the actual damages). Thus, the patent pending status does serve to discourage potential infringers.

Not all developers of innovative designs seek the protection offered by the patent system. Inventors have to balance the competing pressures between disclosure, marketing an invention, and securing a patent. Some fear that the disclosure required by filing a patent removes more protection than is obtained by securing the patent. In areas involving rapid changes in technology, some feel that the best protection against competitors is offered, not by the patent system, but by aggressively and ceaselessly innovating. This strategy is based on the expectation that by the time a competitor copies one of your designs and brings the product to market, you will have advanced the technology to a higher level of innovation.

If a patent search conducted early in the design process reveals existing patents that are similar to your design concept, there are several approaches you can take. One approach is to pay a royalty or licensing fee to the patent holder in exchange for using the patented device as part of your system. Under certain circumstances, this may be preferable to designing a new device from scratch. If you decide to proceed with your own design, the patent documentation alerts you to design strategies to avoid infringing on the existing patent.

Other Intellectual Property

Patents are one of four types of intellectual property. The other three types are copyrights, trademarks, and trade secrets. We discuss each of them briefly here.

Copyrights.   Most copyrights apply to published material, but copyright protection also extends to models, reports, and engineering drawings. The copyright applies only to the written/pictorial depiction of an engineered object or idea, not the object or idea itself. Copyright law protects the work of design engineers by preventing others from copying and distributing plans and other documents without permission. Sometimes the copyright holder does grant permission but imposes a copyright fee as a form of compensation.

U.S. copyright law changed significantly in 1978. Works created after January 1, 1978 are protected for the author's life plus fifty years. Works that were federally copyrighted prior to 1978 are protected for twenty-eight years and may be renewed for another forty-seven years. Unpublished pre-1978 work that had not been copyrighted is treated as post-1978 work.

Trademarks. Trademarks are protection granted to brand names, slogans, and symbols that distinctly identify a product or a company. Trademark protection is granted for twenty-eight years and can be renewed in perpetuity as long as it continues to be used by the trademark holder.

Trade secrets. Trade secrets consist of information that an individual or company considers would be harmful to their competitive position if disclosed to an unauthorized person. If the company hired you as a consulting design engineer on a project for which that information is relevant, they can protect the proprietary information by asking you to sign a nondisclosure agreement before hiring you. While such confidentiality agreements are between two private parties, they are legally binding contracts that can be enforced by the courts.

4.9. GOVERNMENT AND ENGINEERING DESIGN

In this section we discuss several other ways in which government decisions can impact engineering designers.

Research and Development

About half of all the research conducted in the U.S. is funded by the federal government. The level of federal research and development (R&D) and the distribution of those activities among federal agencies for fiscal year 1997 is shown in Figure 4–16. Government research and development funding can stimulate interest and activities in new technological arenas. Frequently, momentous national decisions involve the commitment by the federal government to take the lead role in promoting and/or developing entire classes of technologies. Nuclear energy, the Internet, air bags and other automobile safety features, the Apollo and Space Shuttle programs, the Strategic Defense Initiative, and environmental control technologies, are all examples of technological changes brought about by government action. Engineering students should be sensitive to the implications that future government decisions could have on their career opportunities. More specifically, insight into how key decisions on government R&D programs are made will broaden your understanding of how intertwined our technological progress is with social and political forces.

Other Government Policies

There are a myriad of other ways that government policies affect technological change and thus affect the type of engineering design activities in which we engage. Probably one of the most significant effects on the nature and role of design education in a typical engineering curriculum during the 1990s has been the increased funding by the National Science Foundation (NSF) for design education curriculum reform.

Government tax credits for research expenditures and export licenses for high-tech equipment are two other tools used by government to stimulate technological change. These and other forms of governmental action have historically been used for these pur-

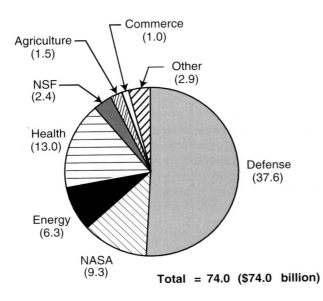

Agriculture (1.5)
Commerce (1.0)
Other (2.9)
NSF (2.4)
Health (13.0)
Defense (37.6)
Energy (6.3)
NASA (9.3)

Total = 74.0 ($74.0 billion)

Figure 4–16. Fiscal Year 1997 Research and Development Budget for Federal Agencies *(Issues in Science and Technology, Winter 1996–97, pp. 21–22.)*

poses. For example, a key factor in the success of Robert Fulton's steamboat operations on the Hudson River in the early 1800s was the twenty-year monopoly granted to his operations by the New York State Legislature.[55] More recently, during the energy crisis years of the mid-1970s–1980s, federal tax breaks and grants were used to stimulate investments in a wide range of energy conservation and alternative energy technologies.

Engineering Input to Government Decisions

Over the past several decades, the engineering profession has realized how important it is for engineers to actively participate in the government decision making process. During that time, many of the professional engineering societies have developed sophisticated government relations programs that empower their members to be more effective contributors to public policies that have a substantial engineering component. This includes activities such as providing rapid response to requests for technical expertise by members of Congress; convening task-forces to analyze R&D budget proposals of federal agencies; issuing public statements that represent the consensus of engineers on specific issues; and nominating prominent engineers for presidential appointment to key administrative and regulatory agencies.

4.10. CLOSURE

We have tried in this chapter to provide a broad context within which to examine the various dimensions of engineering design. A major theme of this examination is the constant reminder of the many ways in which our design activities can contribute to

---

[55]Billington, p. 42, 54.

"... the benefit of mankind." A corollary that we've tried to emphasize is how fragile that relationship between engineering and the benefit of mankind is. There are many pressures that tend to distract us from keeping that linkage strong and healthy. It requires constant vigilance to ensure that individual design engineers, and the engineering profession in general, live up to our obligation and potential to use their talents to improve the human condition. To put a face on these principles, we provided some brief anecdotes regarding a few engineers who dedicated themselves to these lofty ideals.

We also examined the relationship between engineering, science, innovation and their impact on society. Next, we introduced the concept of engineering professionalism and the structure of the engineering profession, with a particular emphasis on the topic of engineering ethics.

Engineering design decisions inherently involve dealing with uncertainties. Many of those uncertainties introduce an element of risk into design decisions. We discussed the risk analysis framework as a tool for addressing those aspects of design. The analysis of risk led to a consideration of codes and standards and product and professional liability issues. We closed with an examination of the interactions between engineering and public policy, with a special emphasis on the patent system.

4.11. REFERENCES

American Association of Engineering Societies, (AAES). 1996. *Risk Analysis: The Process and Its Application.* A Statement of the Engineers' Public Policy Council. Washington, DC: AAES.

American National Standards Institute. Jan. 17, 1997. Standardization: A Management Tool for Building Success. American National Standards Institute, Jan. 17.

ASHLEY, STEVEN. March 1991. Confronting Product Liability. *Mechanical Engineering.*

Associations Yellow Book: Who's Who at the Leading U.S. Trade and Professional Associations. New York, NY: Leadership Directories, Inc.

BILLINGTON, DAVID P. 1996. *The Innovators: The Engineering Pioneers Who Made America Modern.* New York: John Wiley & Sons.

Black's Law Dictionary, Abridged Fifth Edition. 1983. St. Paul, MN: West Publishing Co.

CORTES-COMERER, NHORA. August 1988. Defensive Designing: On Guard Against the Bizarre. *Mechanical Engineering.*

DIETER, G. E. 1991. *Engineering Design: A Materials and Processing Approach, 2nd Edition.* New York: McGraw-Hill, Inc.

Discover. May 1996. A Fistful of Risks.

ERTAS, A. and J. C. JONES. 1993. *The Engineering Design Process.* New York: John Wiley & Sons.

FAUSTMAN, E. M. and G. S. OMENN. 1996. *Risk Assessment.* In Klaassen, C. D. ed. *Casarett and Doull's Toxicology: The Basic Science of Poisons 5th ed.* New York: McGraw-Hill Book Co.

FISCHOFF, B., S. R. WATSON, and C. HOPE. *Defining Risk.* In Glickman, T. S. and M. Gough, eds. 1990. *Readings in Risk.* Washington, D.C.: Resources for the Future.

FLORMAN, SAMUEL. 1987. *The Civilized Engineer.* New York: St. Martins Press.

FLORMAN, SAMUEL. July 1997. Subsumed by Science? *Technology Review.* Vol. 100, No. 5.

Grolier's Encyclopedia. Spring 1997. On-line edition of Academic American Encyclopedia.

HUBER, PETER W. and ROBERT E. LITAN, 1991. *The Liability Maze: The Impact of Liability Law on Safety and Innovation.* Washington, DC: The Brookings Institution.

KELMAN, STEVEN. Cost-Benefit Analysis: An Ethical Critique. In Glickman Theodore S. and Michael Gough, eds. 1990. *Readings in Risk.* Washington DC: Resources for the Future.

KRANZBERG, M. February 1988. Technology and the U.S. Constitution. *Engineering Education.*

LAMARRE, LESLIE. Oct/Nov 1992. What Are You Afraid Of? *EPRI Journal.* Palo Alto, CA: Electric Power Research Institute.

LAYTON, EDWIN T. 1986. *The Revolt of the Engineers: Social Responsibility and the American Engineering Profession.* Baltimore, MD: The Johns Hopkins University Press.

LOWRANCE, W. The Nature of Risk. In SCHWING, R. C, and W. A. ALBERS JR, eds. 1980. *Societal Risk Assessment: How Safe is Safe Enough?* New York: Plenum Press.

MARSHALL, R. D. et al. 1982. *Investigation of the Kansas City Hyatt Regency Walkways Collapse.* NBS Building Science Series No. 143. National Bureau of Standards.

MARTIN, M. W. and R. SCHINZINGER. 1989. *Ethics in Engineering, 2nd Edition.* New York: McGraw-Hill Book Co.

MIDDENDORF, WILLIAM H. 1990. *Design of Devices and Systems, 2nd Ed.* New York: Marcel Dekker, Inc.

MORGAN, GRANGER M. Choosing and Managing Technology-Induced Risk. In Glickman Theodore S. and Michael Gough, eds. 1990. *Readings in Risk.* Washington D.C.: Resources for the Future.

MORGAN, GRANGER M. July 1993. Risk Analysis and Management. *Scientific American.*

National Council of Examiners for Engineering and Surveying (NCEES). January 1996. *Fundamentals of Engineering (FE) Discipline Specific Reference Handbook.* Clemson, SC: NCEES.

National Research Council. 1991. *Improving Engineering Design.* Washington, DC: National Academy Press.

NORMILE, DENNIS. August/September 1996. Responsive Sculpture. *Technology Review.*

NOVACK, JANET L. 1995. *The ISO 9000 Documentation Toolkit: 1994 Revised 9001 Standard.* Englewood Cliffs, NJ: Prentice Hall.

OSTROM, CAROL. 1993. How Much for Human Life? *Seattle Times.* February 26.

PETROSKI, HENRY. 1985. *To Engineer is Human: The Role of Failure in Successful Design.* New York: St. Martin's Press.

STREETER, HARRISON. 1988. *Professional Liability of Architects and Engineers.* New York: John Wiley & Sons.

TORODOV, BRANIMIR. 1996. *ISO 9000 Required: Your Worldwide Passport to Customer Confidence.* Portland, OR: Productivity Press.

UNGER, STEPHEN H. 1994. *Controlling Technology: Ethics and the Responsible Engineer, 2nd edition.* New York: John Wiley & Sons.

VAN HOUTVEN, GEORGE L. and MAUREEN L. CROPPER. Winter 1994. When is a Life Too Costly to Save? The Evidence from Environmental Regulations. *Resources.* Washington, DC: Resources for the Future.

VARGO, JOHN. October 1995. Professionally Speaking: Understanding Product Liability. *Mechanical Engineering.*

VINCENTI, WALTER G. 1990. *What Engineers Know and How They Know It.* Baltimore, MD: The Johns Hopkins University Press.

Washington Supreme Court Committee on Jury Instructions. 1989. *Washington Practice, Vol. 6. Washington Pattern Jury Instructions—Civil, 3rd Ed.* St. Paul, MN: West Publishing Co.

WOHLEBER, CURT. Winter 1993. Straight Up. *American Heritage of Invention and Technology.* Vol. 8, No. 3.

WOHLEBER, CURT. Winter 1997. 'St. George' Westinghouse. *American Heritage of Invention and Technology.* Vol. 12, No. 3.

World Book Encyclopedia. 1979. Chicago, IL: World Book-Childcraft International, Inc., Volume 9.

4.12. EXERCISES

1. An important tool for a successful career in engineering design is familiarity with the professional engineering societies, trade associations, and government agencies which are major sources of data, standards, regulations, and other technical information. Of course, most such organizations are frequently known by their acronyms rather than by their full names. Complete the "Acronymania" crossword puzzle depicted below. The answer to each puzzle clue is the acronym of a well-known organization that is part of the "engineering enterprise."

Engineering Acronymania Puzzle

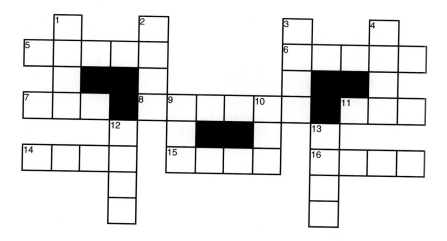

Across

5. The arm of the U.S. Dept. of Transportation which sets safety standards for automobiles.
6. Many Boeing engineers are members of this organization.
7. An environmental advisory body within the Executive Office of the President.
8. Many HVAC engineers belong to this professional society.
11. A professional society of ceramic engineers and scientists.
14. A federation of societies representing mining, metallurgical, and petroleum engineers.
15. The research and development arm of the electric utility industry.
16. This professional society focuses on aerospace engineering.

Down

1. Organization of American engineers with Hispanic heritage.
2. Government agency which is responsible for aeronautics and aerospace programs.
3. The major professional society for most mechanical engineers.
4. This federal regulatory body oversees many activities of electric and gas utility companies.
9. Professional society of engineers interested in design of transportation vehicles.
10. Major trade association of oil companies.
12. Largest professional engineering society in the world.
13. Trade association of manufacturers of home appliances powered by natural gas (ovens, clothes dryers, etc.).

2. Explain which professional engineering society comes closest to meeting your needs at this stage of your career. Include excerpts from that society's literature to support your explanation.

3. Should all engineering professors be required to have a PE license? Explain.

4. The first assignment for a recent engineering graduate hired by a large manufacturing firm was to redesign a plastic crash helmet. She learned that one of the major customers for this helmet was a foreign country that issued the helmets to its military pilots. That country was currently at war and was expecting the redesigned helmet to significantly improve the performance of its pilots as well as provide them added protection. Extensive media coverage of the war made it clear that the country purchasing the helmets was committing atrocities against its enemy's civilian population as well as restricting the civil rights of its own citizens. What guidance does the ABET Code of Ethics provide this engineer? Cite specific provisions of the Code in your response.

5. Compare the language in the NCEEE Model Code with the Rules of Conduct required of all registered professional engineers in your state.

6. Apply the risk analysis framework discussed in Section 4.5 to two risks associated with an electric clothes dryer. For each risk, describe the nature of the hazard, how you would conduct the risk assessment, what you think the key risk perception issue is, and your recommendation for managing the risk.

7. Apply Vargo's seven "balancing" criteria to conclude whether a power lawnmower without a "deadman switch" is an unreasonably dangerous product.

8. Identify three agencies of the federal government whose primary responsibility is safety. Briefly describe the mission of each.

9. The following incident allegedly occurred in the aircraft industry. A mechanical engineer from Company A had conducted tests of a certain aircraft tail assembly configuration in his company's wind tunnel and knew that devastating vibrations could occur with that configuration under certain circumstances, leading to destruction of the aircraft. Later, at a professional meeting, Company A's engineer heard an engineer from Company B, a competitor, describe a tail assembly configuration for one of Company B's new aircraft which would run the risk of producing just the kind of destructive vibrations which Company A's engineer had discovered in his tests. The engineer from Company A has an obligation, both as a matter of morals and of law, to maintain company confidentiality regarding Company A's proprietary knowledge. On the other hand, engineers are supposed to bear a responsibility

for public safety and welfare. If the engineer from Company A remains silent, Company B may not discover the possibility of the existence of destructive vibrations until a dreadful crash occurs, killing many people. What would you do if you were the engineer from Company A? What follow-up actions would you take if your initial efforts were unsuccessful? What are the limits (if any) beyond which you would not go? Justify your reasoning by reference to specific provisions of the ABET Code of Ethics.

10. In his article, "Reclaiming the High Ground: An Engineering Ethic for the New Age of Engineering," published in the April 1991 issue of *Engineering Education*, Brian Stimpson argues that a new engineering ethic is needed. Briefly summarize (in less than one hundred words) the major similarities/differences between the approach to engineering ethics taken by Stimpson and that taken by Sam Florman in Chapter X of his book, *The Civilized Engineer*.

11. An article on the social role of engineers by Professor P. Vesilind of Duke University was published in the April 1991 issue of *Engineering Education*. In the article, the author labels engineers as "utilitarians" and "positivists" while most members of the general public are labeled "ethical egoists" and "idealists." The article notes the inherent conflict between these personality types and finally queries: should engineers at times "practice their profession as applied social scientists instead of applied physical scientists?" Discuss (in 200–400 words) whether Professor Vesilind's analysis and approach is encouraged or discouraged by the ABET Code of Ethics, and how or whether you would incorporate this approach into your engineering activities.

12. An op-ed article on medical ethics was published on July 3, 1988 in the *New York Times*. The author, a medical student, argues for an increased emphasis on ethics as part of the medical school curriculum. Prepare a follow-up article of approximately the same length which addresses the related issue of engineering ethics and its role in the engineering curriculum.

13. Discuss (400–600 words) under what circumstances you think engineers are justified in participating in the design and manufacture of products with built-in obsolescence (products which wear out rapidly and cannot be repaired). What guidance does the ABET Code of Ethics provide for this situation?

14. Conduct a telephone interview with an engineer who works as a full-time staff member in the codes and standards area for one of the professional engineering societies. Prepare a written report (suggested length of 500 words) describing this individual's educational and professional background, their current job responsibilities, and the key issues they are currently dealing with.

15. Describe and analyze (in approximately 1,000 words) in detail the contents of a recent issue of the nationally distributed monthly magazine published by a professional engineering society. Address these questions: Was the magazine too technical for you? Which articles were most useful to you? What about the advertisements? What were your impressions of the letters to the editor? Were there particular features which you found most appealing?

16. Attend a regularly scheduled monthly meeting of the local professional (not student) chapter of a society. Prepare a report (suggested length of 500 words) describing the activities which took place at the meeting and your reactions to those activities. Which aspects of the meeting impressed you? Which ones disappointed you? Interview two professional (not student) members who also attended that meeting. Find out about their educational back-

ground, their jobs, their most recent design experience, and the nature of their involvement in the society. Append to your report a brief biographical sketch of the two engineers who you interviewed, and attach their business cards.

17. Select a newspaper article published within the last five years dealing with a public safety issue. Supplement that article with at least two government reports written on the same subject within the past five years. Using the risk analysis framework discussed in Section 4.4, discuss (200–400 words) the nature of the hazard, and the role of the engineering profession in the assessment, perception, and management of the risk described in the article and the accompanying reports. Attach to your response a copy of the newspaper article and the title pages from your additional references. For the purposes of this problem, a government report is any document listed within the past five years in either the Government Reports Index, Congressional Information Service Index, NTIS, or the Government Printing Office Catalog.

18. Required federal government test procedures for measuring the energy consumption of household appliances are published in Title 10 of the Code of Federal Regulations. In many cases, these test procedures refer to various existing voluntary consensus standards. Identify the ANSI standard that the CFR states should be used to measure the interior volume of refrigerator-freezers for purposes of testing their energy consumption.

19. A plastic knob on a floor gearshift lever deteriorated due to sunlight exposure and developed hairline cracks. A collision occurred and a passenger was thrown against the gearshift. The knob broke and the lever penetrated the passenger's chest. Of the million cars already on the market with this knob, this is the only fatality. Your company manufactures 100,000 cars/year with this knob. A sun-resistant knob would cost your company an extra $10/car. What approach would you take, as the chief design engineer for the automobile company, regarding the shift lever knob? Discuss your reasoning from both a professional ethics and risk management perspective. Does your position change if the extra cost of the new knob is 25¢?

◆ *Source:* Middendorf, p. 62.

20. Do you plan to take the Fundamentals of Engineering Exam prior to graduation? Explain why or why not.

CHAPTER 5

Probabilistic Considerations in Design

5.1. OVERVIEW

In Chapter 4 we explored the various facets of risk analysis, including the concept of risk management as a way to reduce risk. Inherent in that discussion was the issue of dealing with uncertainty. This chapter presents a particular approach to dealing with several types of uncertainties; others are dealt with in Chapters 7, 8, and 9.

A traditional risk management tool used in engineering design is the safety factor. The safety factor is used to account for the uncertainties in material properties, fabrication procedures, and operating environment. Many times such an approach leads to a design that is too big, too heavy, or too expensive. In addition, there are circumstances for which the safety factor approach does not adequately guard against failure. In this chapter, we introduce a more sophisticated approach to engineering design that deals explicitly and rigorously with the probabilities of failure.

The context for the first half of this chapter is the design of a slender cylindrical rod under tension. This is perhaps the simplest possible structural design problem. We examine how to address the problem by explicitly considering probabilistic aspects of the applied tension, the material properties, and dimensions of the rod. We then introduce the concept of reliability to deal with failures as a function of time. We conclude by looking at the key relationships between the reliability of a system and the reliabilities of the components within the system.

5.2. YIELD STRENGTH AS A RANDOM VARIABLE

Consider a structure fabricated from slender cylindrical rods made from a material with a yield strength in tension y. Suppose the supplier has a large inventory of rods in stock. Because of variations in metallurgical processes and fabrication methods, we can expect random variations in yield strength among the rods. In this and the several

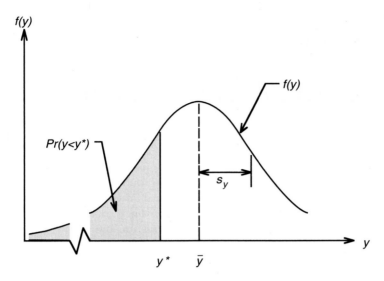

Figure 5–1. Normal Probability Distribution

following sections, we'll assume that the tensile yield strength y can be modeled as a normally distributed random variable with a mean value $\bar{y}$ and a standard deviation s_y.[1] We begin our discussion by examining the basic properties of the normal distribution.

The Normal Distribution

The normal distribution is the familiar bell-shaped curve shown in Figure 5–1. The equation that defines the curve is known as the probability density function for the normal distribution and is given as

$$f(y) = \frac{1}{s_y\sqrt{2\pi}}\, e^{-[(y-\bar{y})^2/2s_y^2]} \tag{5–1}$$

where $\bar{y}$ is the mean value of y and s_y is the standard deviation. As shown in Figure 5–1, the probability density function $f(y)$ for a normal distribution is symmetrical with respect to the mean value $\bar{y}$, and $f(y)$ is positive for all values of y.

Changing the numerical value of $\bar{y}$ in Equ. (5–1) has the effect of shifting the entire $f(y)$ curve along the y-axis. Also, the standard deviation s_y is depicted in Figure

---

[1] Later in this chapter, we will discuss probabilistic values of stress. This raises a notation problem since the Greek symbol σ is universally used by structural engineers to designate stress. The same symbol is used throughout the probability theory literature to designate standard deviation. To avoid confusion, we will adopt σ for stress and use s_x for standard deviation, where the subscript on s refers to the random variable x. So for example, we will designate the standard deviation of stress as s_σ.

5–1 as the distance from the mean to the inflection point on the $f(y)$ curve. Therefore, the standard deviation is a measure of the spread of the y distribution.[2]

It can be rigorously shown that the area under the $f(y)$ curve is unity:

$$\frac{1}{s_y\sqrt{2\pi}}\int_{-\infty}^{\infty} e^{-[(y-\bar{y})^2/2s_y^2]}\, dy = 1$$

The property that

$$\int_{-\infty}^{\infty} f(y)\, dy = 1$$

that is, the area under the curve is unity, is a requirement for all probability density functions $f(y)$, not just for the normal distribution. This means that the probability of encountering a value which is within the population is 100%.

The probability that a bar selected at random from the stock will have a yield strength less than a specified value $y*$ is

$$Pr(y < y*) = \frac{1}{s_y\sqrt{2\pi}}\int_{-\infty}^{y*} e^{-[(y-\bar{y})^2/2s_y^2]}\, dy \tag{5–2}$$

As depicted in Figure 5–1, $Pr(y < y*)$ is the area under the probability density function to the left of the $y*$ line. Since the total area under the curve is unity,

$$Pr(y > y*) = 1 - Pr(y < y*) \tag{5–3}$$

Since the integrand in Equ. (5–2) is positive for all values of y, $Pr(y < y*)$ and $Pr(y > y*) > 0$ for all values of $y*$. This is a characteristic of the mathematical definition of the normal distribution that may be at odds with the physical meaning of the random variable y modeled by Equ. (5–2).

For example, here the random variable y represents the tensile yield strength of the material. Consider a particular example where the numerical values of the mean and standard deviation are known to be

$$\bar{y} = 50{,}000 \text{ psi}, \; s_y = 6{,}000 \text{ psi} \tag{5–4}$$

Suppose we want to find the probability that a bar selected at random from this stock will have a tensile yield strength less than $y* = 0$. Now physically it doesn't make any sense to have a negative tensile yield strength. But Equ. (5–2) will provide a positive (although a very small) number; the mathematical model tells us that $Pr(y < 0) > 0$. What this means is that the normal distribution model of the random nature of the yield strength is subject to error for values of the yield strength that are very low relative to the mean value. We will show in the next section that the magnitude of this error is very small.

---

[2]There are many probability density functions other than the normal distribution described here. We select the normal distribution because it is the most commonly used and broadly applicable one. For other distributions, even those that are not symmetric, the mean value locates the distribution along the y-axis and the standard deviation is a measure of the spread.

A similar situation exists at the other extreme. For example, using the same numerical values of the mean and standard deviation from Equ. (5–4), suppose we wanted to find the probability of selecting a bar at random whose yield strength is greater than $y* = 5{,}000{,}000$ psi. Physically, we know that for this stock, there should be zero chance of finding such an unrealistically strong bar. However, the mathematical definition tells us that $Pr(y > y*) > 0$ for all finite values of $y*$. While the mathematical model conflicts with reality for values of $y*$ much greater than the mean, we will see again that the magnitude of the error is very small.

The Standard Normal Distribution

For a given set of numerical values for the parameters $\bar{y}$, s_y, and $y*$, the integral in Equ. (5–2) can be evaluated to get a numerical value for $Pr(y < y*)$. This can be accomplished readily with a variety of commercially available software packages. However, for many situations it is desirable to simplify the evaluation of this integral. This can be accomplished by introducing a new independent variable z defined as

$$z = \left(\frac{y - \bar{y}}{s_y} \right) \tag{5–5}$$

This variable is frequently called the standard normal variable, or the z variable. The associated probability distribution (called the standard normal distribution, or the z distribution) is obtained by substituting Equ. (5–5) into Equ. (5–1). Note from Equ. (5–5) that

$$dz = \frac{dy}{s_y}$$

After simplification, Equ. (5–1) can be expressed in terms of the new variable z as

$$Pr(z < z*) = \frac{1}{\sqrt{2\pi}} \int_{-\infty}^{z*} e^{-(z^2/2)} dz \tag{5–6}$$

where

$$z* = \left(\frac{y* - \bar{y}}{s_y} \right) \tag{5–7}$$

By comparing this equation with Equ. (5–1) we see that the z distribution is a normal distribution whose mean value is 0 and whose standard deviation s_z is unity. Thus, by virtue of the transformation described in Equ. (5–5), any normal distribution with parameters, $\bar{y}$, s_y, and $y*$ can be converted into a z distribution with $\bar{z} = 0$, $s_z = 1$, and $z*$ given by Equ. (5–7). One benefit of having $s_z = 1$ is that the numerical value of z then can be interpreted as, "the number of standard deviations from the mean." The graph of the z, or standard normal, distribution is shown in Figure 5–2.

The numerical values of $Pr(z < z*)$ for a given $z*$ can be easily calculated from built-in functions incorporated in many popular commercial software packages. The usefulness of the standard normal distribution stems from the fact that it only requires

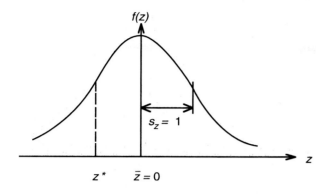

Figure 5–2.   The Standard Normal Distribution

specification of a single parameter, z^*, to evaluate the integral in Equ. (5–6). For convenience, an abbreviated listing for the values of $Pr(z < z^*)$ is provided in Table 5–1.

Since the z distribution is symmetrical, we only need to tabulate numerical values to one side of $z^* = 0.00$. For example, suppose we wish to evaluate $Pr(z < 1.00)$. Since Table 5–1 provides information only for values of $z < 0$, we first apply Equ. (5–3) to get

$$Pr(z < 1.00) = 1 - Pr(z > 1.00) \qquad (5\text{--}8)$$

Then we take advantage of symmetry (see Fig. 5–3) to write

$$Pr(z > 1.00) = Pr(z < -1.00)$$

This allows us to write Equ. (5–8) as

$$Pr(z < 1.00) = 1 - Pr(z < -1.00)$$

Now we can use Table 5–1 to get

$$Pr(z < 1.00) = 1 - 0.159 = 0.841$$

We can also use the table in reverse to find the value of z^* associated with a specified value for $Pr(z < z^*)$. If we are given, for example, that $Pr(z < z^*) = 0.001$, Table 5–1 tells us that $z^* = -3.00$.

We will apply this technique to our problem involving the yield strength of our structural material. Consider first the example discussed earlier where we want $Pr(y < y^*)$ when $y^* = 0$ for the normal distribution defined by Equ. (5–4). Substituting Equ. (5–4) to Equ. (5–5) yields $z^* = -8.33$. From Table 5–1 we see that $Pr(z < -8.33) < 0.001$. This confirms our earlier statements that the probabilities of a

TABLE 5–1.   ABBREVIATED TABLE OF CUMULATIVE DISTRIBUTION FUNCTION FOR THE STANDARD NORMAL DISTRIBUTION

z^*	−3.00	−2.75	−2.50	−2.25	−2.00	−1.75	−1.50	−1.25	−1.00	−0.75	−0.50	−0.25	0.00
$Pr(z < z^*)$	0.001	0.003	0.006	0.012	0.023	0.040	0.067	0.106	0.159	0.227	0.308	0.401	0.500

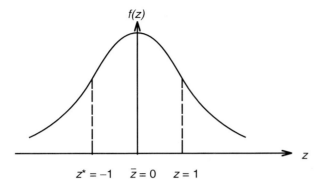

Figure 5–3. Symmetry of Standard Normal Distribution About $z = 0$

normally distributed random variable taking on values that are extremely small or extremely large relative to the mean are negligible.

Now for a more realistic example. We wish to find the probability of a randomly selected bar from the stock defined by Equ. (5–4) having a yield strength less than 40,000 psi. Using Equ. (5–5) with $y^* = 40,000$ leads to $z^* = -1.67$. Table 5–1 then provides $Pr(z < -1.67) = 0.0475$. Thus, there is only a 4.75% chance that the selected bar will have a yield strength less than 40,000 psi.

Table 5–1 also can be used in conjunction with Equ. (5–5) to find $Pr(y > y^*)$. Suppose we want to know what the chances are of selecting a bar from the same stock whose yield strength is greater than $y^* = 55,000$ psi. From Equ. (5–5) we obtain $z^* = 0.83$. With this value, Table 5–1 provides

$$Pr(y < 55,000) = 0.7967$$

Now using Equ. (5–3) we find

$$Pr(y > 55,000) = 0.2034$$

So there is a 20% chance of the selected bar having a yield strength greater than 55,000 psi.

Importance of both mean and standard deviation.

Because the properties of a probability distribution depend on both its location and spread, a complete picture of the properties of a random variable requires that both the mean and standard deviation are considered. The significance of this can be appreciated by considering the probability distributions for the yield strength of two different materials shown in Figure 5–4. If we are interested in selecting the material which has the smaller $Pr(y < y^*)$, it is not clear that we should select the one with the higher $\bar{y}$. To illustrate, let's use Equ. (5–4) to define the distribution for material 1 and we'll specify $y^* = 40,000$ as we did before. We calculated $Pr(y < 40,000) = 0.0475$, that is, there is only a 4.75% chance that a randomly selected bar from this population will have a yield strength less than 40,000 psi. Now suppose we drew a second rod at random from a different population, one characterized by $\bar{y}_2 = 55,000$ psi and $s_{y_2} = 12,000$ psi. Since the second rod is drawn from a population with a higher mean yield strength

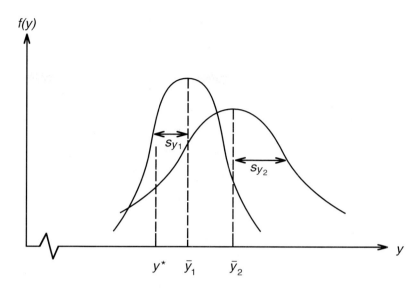

Figure 5–4.   Probability Distributions for the Yield Strength of Two Different Materials

than the population from which the first rod was drawn, you might be tempted to say that $Pr(y_2 < 40{,}000) < Pr(y_1 < 40{,}000)$. However, since $s_{y_2} > s_{y_1}$ the opposite may in fact be true. Using the probability tables, it turns out that the yield strength of the second rod has a 10.6% chance of falling below 40,000 psi. Thus, for these two specific distributions, the yield strength of the rod associated with the higher mean value has a higher probability of falling below the specified value of 40,000 psi. As we will see, this has major implications for design strategies.

5.3. PROBABILITY OF FAILURE

Engineering systems can fail in many ways. Columns can buckle, circuit wires can short, and gear teeth can be stripped. For any given system, the mode of failure has to be specified before the probability of that type of failure occurring can be assessed. In this section we will use the failure of rods by yielding in tension to illustrate the important concepts and techniques associated with probability of failure.

Consider a slender rod of cross-sectional area A subject to a tensile load P (see Fig. 5–5). Basic strength of materials theory tells us that the tensile stress in the rod is

$$\sigma = \frac{P}{A} \qquad (5\text{–}9)$$

The rod will fail by yielding if σ exceeds the yield strength; failure occurs if

$$y < \sigma \qquad (5\text{–}10)$$

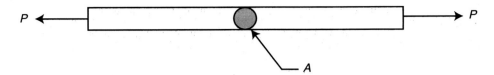

Figure 5–5. Slender Circular Rod of Cross-Sectional Area A Under Axial Tension P

For purposes of a numerical example, let's assume the values of P and A are

$$P = 20,000 \text{ lb.} \tag{5–11}$$

$$A = 0.5 \text{ in.}^2$$

then

$$\sigma = \frac{20,000 \text{ lb.}}{0.5 \text{ in.}^2} = 40,000 \text{ psi}$$

Let's further assume that the rod is made from the same material discussed in the previous section; the yield strength of a material is assumed to be a normally distributed random variable with $\bar{y} = 50,000$ psi and $s_y = 6,000$ psi. We will approach this problem from two perspectives. First, we'll use the factor of safety approach. Then we'll use the probabilistic approach.

The safety factor.   The factor of safety (FS) approach handles the uncertainty in the yield strength by specifying how far the stress shall be below the nominal (mean) yield strength;

$$FS = \frac{\bar{y}}{\sigma} \tag{5–12}$$

In this case, with both $\bar{y}$ and σ known, we calculate the factor of safety as

$$FS = \frac{50,000}{40,000} = 1.25$$

Whether this is a large enough safety factor depends on the particular application and the consequences of a failure. A higher safety factor is warranted in situations where public safety is jeopardized by a failure than when a failure does not lead to catastrophic consequences. In many industries and settings, the required safety factor is established by a design code or safety regulation. It is not unusual for safety factors of 1.5, 2.0, or higher to be specified.

If, instead of being given a numerical value for A, a minimum required safety factor of, say $FS = 1.25$, was part of the problem formulation, we could solve for A so as to satisfy the safety factor requirement. We accomplish this by combining Equs. (5–9) and (5–12) to get

$$A = \frac{P}{\sigma} = \frac{P}{\bar{y}/FS} = \frac{FS \cdot P}{\bar{y}}$$

Substituting in the numerical values yields

$$A = \frac{1.25 \cdot 20{,}000 \text{ lb.}}{50{,}000 \text{ psi}} = 0.50 \text{ in.}^2$$

Hence, for a given load P and a given nominal yield strength $\bar{y}$, the safety factor dictates the size of the rod, A. However, this approach is not an absolute guarantee against failure. We saw in the last section that for the material defined by Equ. (5–4) there is a 5% chance that the yield strength of a randomly selected rod would fall below 40,000 psi, thereby leading to failure as defined by Equ. (5–10).

The factor of safety approach can never be a direct measure of safety because it relies solely on $\bar{y}$ and completely ignores s_y. Nevertheless, the approach continues to be used because of its inherent simplicity, especially in situations where experience has demonstrated it to be an effective guard against failure. But we turn now to a more sophisticated approach in which the properties of the known random variable y are explicitly linked to the probability of failure.

Designing to a Specified Probability of Failure

The probability that the yield strength of the selected rod is less than 40,000 psi was found in the last section to be $Pr(y < 40{,}000 \text{ psi}) = 0.0475$. Therefore, the probability of the rod failing by yielding in tension is 4.75%. We now treat the cross-sectional area A as the design parameter whose value is to be determined so that there is no more than a 0.005 probability of a randomly selected rod yielding under the 20,000 lb. tensile load. We will continue to assume that A is deterministic[3] and the yield strength is a normally distributed random variable defined by Equ. (5–4).

The failure criterion is still given by Equ.(5–10) but now we are told that we want

$$Pr(y < \sigma) = 0.005. \tag{5–13}$$

Our design task is to select a value for the design parameter A that is consistent with this equation. To clarify the role of A, we use Equ. (5–9) to write Equ. (5–13) as

$$Pr\left(y < \frac{P}{A}\right) = 0.005 \tag{5–14}$$

where, as we discussed earlier, $\sigma = P/A$ plays the role of y^*, that is,

---

[3]By deterministic variable, we mean that the vendor can supply a rod with a cross-sectional area that has precisely the specified value of A. This is in contrast to the yield strength for which, as a random variable, all we can specify is the mean and standard deviation of the rod population from which the vendor draws.

$$y^* = \frac{P}{A} \qquad (5\text{-}15)$$

Next, we use the probability tables for the standard normal distribution in the reverse manner as they were used in the previous example. First, we can rewrite Equ. (5–13) in terms of the z variable as

$$Pr(z < z^*) = 0.005$$

Then interpolating from Table 5–1 (for more accuracy use a spreadsheet or other software) we get

$$z^* = -2.575 \qquad (5\text{-}16)$$

Substituting Equs. (5–11), (5–15), (5–16), and (5–5) into Equ. (5–7) leads to

$$-2.575 = \left(\frac{\dfrac{20,000}{A} - 50,000}{6,000} \right) \qquad (5\text{-}17)$$

or

$$A = 0.579 \text{ in.}^2 \qquad (5\text{-}18)$$

Thus, if a large number of rods were selected randomly from a stock in which the cross-sectional area of each rod is $A = 0.579$ in.^{2}, we would expect that 0.5% of the rods will yield when subjected to a 20,000 lb. tensile load.

But suppose now that the vendor can only furnish rods with values of A in increments of 0.1 in.^{2}. The next available standard size larger than the required size is $A = 0.6$ in.^{2}, so that is the preferred design. The probability of failure of this design is then found by calculating the stress from Equ. (5–9) as

$$\sigma = 20,000/0.6 = 33,300 \text{ psi}$$

If the yield strength is below this value, failure occurs [see Equ. (5–10)]. The associated value of the standard normal variable is determined from Equ. (5–7) as

$$z^* = (33,300 - 50,000)/6,000 = -2.78$$

so the probability of failure of the preferred design is 0.27%, which is less than the maximum permissible failure rate of 0.5%.

Random Variables as Design Parameters

The design task we just accomplished involved selecting the appropriate value of the design parameter A. It turned out that there was a single, correct answer ($A = 0.579$ in.^{2})—one of the rare times within this book that we will be able to arrive at such a direct, unequivocal, unique answer to a design task. The luxury of being able to reach such an unambiguous conclusion is short-lived, because we now want to reformulate the problem of the rod in tension in a slightly different way.

Under the original formulation, the probabilistic considerations came from us specifying the random character of the yield strength; the design parameter A which we were asked to solve for was assumed to be deterministic in nature. Now, consider the situation where all the specified characteristics are deterministic and the sought after design parameter is assumed to be a random variable. For this version of the problem, we'll assume the same magnitude of the applied load ($P = 20,000$ lb.) as given before in Equ. (5–11). Now we will assume that the yield strength is also deterministic, with a value $y = 50,000$ psi (the value previously associated with $\bar{y}$). Our job now is to find A, such that the probability of failure is 0.005, under the assumption that A is a normally distributed random variable. We formulate the problem [using the counterpart to Equs. (5–9) and (5–10)] as determining the random variable A such that

$$Pr\left(A < \frac{P}{y}\right) = 0.005 \qquad (5\text{–}19)$$

To emphasize that A is now the random variable, we redraw Figure 5–1 as Figure 5–6. From the probability table (Table 5–1), we get, as before

$$z^* = -2.575$$

Therefore,

$$-2.575 = \left(\frac{\dfrac{20,000}{50,000} - \bar{A}}{s_A}\right) \qquad (5\text{–}20)$$

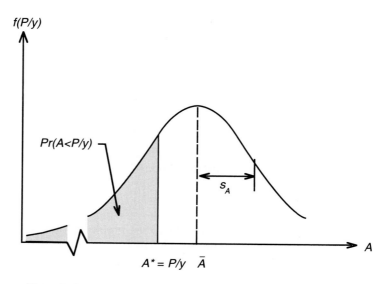

Figure 5–6.   Probability Distribution for Random Design Parameter A

This equation has two unknowns, $\overline{A}$ and s_A. These are the two parameters needed to completely define the normally distributed random variable A. In the absence of additional information or assumptions, there are an infinite number of solutions. We can assign a value to either $\overline{A}$ or s_A and then use Equ. (5–20) to solve for the other. For example we can say that we want this rod to be selected from a population whose mean cross-sectional area is $\overline{A} = 0.5$ in.^{2}. Then the required standard deviation is, from Equ. (5–20),

$$s_A = \left(\frac{0.4 - 0.5}{-2.575} \right) = 0.0388$$

Geometric tolerances and standard deviations. When the desired design parameter is a random variable that describes a geometric feature of the design, we can interpret the standard deviation as an indication of the tolerances associated with fabricating the component. Assuming that the design parameter is normally distributed, the tolerances then must be symmetric about the mean; of the form $\overline{A} \pm$ tol. Generally, the tolerances (tol.) can be thought of as representing a value that is three standard deviations away from the mean,[4] so

$$s_A = \frac{\text{tol.}}{3} \tag{5–21}$$

The two most frequent approaches to establishing tolerances are: 1) specifying the tolerances as a percent of the mean value; and 2) specifying the tolerances in absolute terms.

Taking the first of these approaches, let's assume that the tolerances are tol. = $0.10\overline{A}$ so that

$$s_A = 0.0333\overline{A}$$

Then Equ. (5–20) takes the form

$$-2.575 = \left(\frac{0.400 - \overline{A}}{0.0333\overline{A}} \right)$$

This yields

$$\overline{A} = 0.3143$$

As an example of the second approach, let's assume that the tolerances are given as tol. = 0.100. Substituting this directly into Equ. (5–20) provides

$$\overline{A} = 0.1425$$

So we see that when the design parameter is a random variable, the required mean value of the parameter is very sensitive to the specified fabrication tolerances. This illustrates the importance of considering fabrication (part of what we called "imple-

---

[4]Middendorf, p. 424.

mentation" in the nine-step design process model introduced in Chapter 2) as an integral part of the design process. Smaller tolerances are generally associated with a more expensive fabrication process, but allow us to select smaller mean values, thereby saving on material costs. These kinds of economic trade-offs are discussed in more detail in Chapter 8.

5.4. FUNCTIONS OF A RANDOM VARIABLE

In Section 5.3 our design variable was A, the cross-sectional area of the rod. That was somewhat artificial since it is customary for vendors to specify the size of circular rods by their diameter d, rather than by their cross-sectional area. This departure from reality is of no great consequence when we treat A as deterministic; all we have to do is substitute

$$A = \frac{\pi d^2}{4} \qquad (5\text{--}22)$$

into Equ. (5–17) and solve for d. However, things get a little more complicated when we treat A as a random variable. Specifically, if A is a random variable, then d must also be a random variable. Since the design task in Section 5.3 was to select values for $\overline{A}$ and s_A, here we are looking for $\overline{d}$ and s_d. That means we need to express $\overline{A}$ in terms of $\overline{d}$ and s_A in terms of s_d. To do this we apply the general equations presented in Appendix 5A.

First, lets deal with the means. Applying Equ. (5–26) to Equ. (5–22) provides

$$\overline{A} = \overline{\left(\frac{\pi d^2}{4}\right)} = \frac{\pi}{4}\,\overline{(d^2)} \cong \frac{\pi}{4}\,\overline{d}^2 \qquad (5\text{--}23)$$

Then using Equ. (5–27) with Equ. (5–22) gives

$$s_A \cong \left|\frac{\partial A}{\partial d}\right|_{d=\overline{d}} s_d$$

or

$$s_A \cong \frac{\pi}{4}\,|2d|_{d=\overline{d}}\ s_d = \frac{\pi}{2}\,\overline{d}s_d \qquad (5\text{--}24)$$

Returning to Equ. (5–20) and replacing $\overline{A}$ and s_A by Equs. (5–23) and (5–24), we arrive at

$$-2.575 = \left(\frac{0.400 - \dfrac{\pi}{4}\,\overline{d}^2}{\dfrac{\pi}{2}\,\overline{d}s_d}\right) \qquad (5\text{--}25)$$

Once we express s_d in terms of the tolerances, we can solve Equ. (5–25) for d.

Appendix 5A: Means and Standard Deviations of Functions of Random Variables

Let x be a given random variable whose mean and standard deviation is known; and let y be another variable that is defined in terms of x as

$$y = f(x)$$

The relationship between the mean and standard deviation of y in terms of the mean and standard deviation of x are given as[5]

$$\bar{y} \cong f(\bar{x}) \tag{5--26}$$

and

$$s_y \cong \left| \frac{dy}{dx} \right|_{x=\bar{x}} s_x \tag{5--27}$$

for

$$s_x \ll \bar{x}$$

    More generally, if y is a function of n independent random variables $x_1, x_2, \ldots, x_n$, then

$$\bar{y} \cong f(\bar{x}_1, \bar{x}_2, \ldots, \bar{x}_n) \tag{5--28}$$

and

$$s_y^2 \cong \left(\frac{\partial y}{\partial x_1} \right)_{x_1=\bar{x}_1}^2 s_{x_1}^2 + \left(\frac{\partial y}{\partial x_2} \right)_{x_2=\bar{x}_2}^2 s_{x_2}^2 + \cdots + \left(\frac{\partial y}{\partial x_n} \right)_{x_n=\bar{x}_n}^2 s_{x_n}^2 \tag{5--29}$$

for all

$$s_{x_i} \ll \bar{x}_i$$

For example, let

$$y = x_1 + x_2$$

where x_1 and x_2 are independent random variables. Then from Equ. (5–28)

$$\bar{y} = \bar{x}_1 + \bar{x}_2$$

and from Equ. (5–29)

$$s_y^2 \cong s_{x_1}^2 + s_{x_2}^2$$

or

$$s_y \cong \sqrt{s_{x_1}^2 + s_{x_2}^2}$$

For convenience, Table 5–2 lists several of the more common relationships.

---

[5] Siddall, p. 240.

TABLE 5–2. MEANS AND STANDARD DEVIATIONS
OF FUNCTIONS OF RANDOM VARIABLES

y	$\bar{y}$	s_y
$x_1 \pm x_2$	$\bar{x}_1 \pm \bar{x}_2$	$\sqrt{s_{x_1}^{\,2} + s_{x_2}^{\,2}}$
$x_1 x_2$	$\bar{x}_1 \bar{x}_2$	$\sqrt{\bar{x}_2^{\,2} s_{x_1}^{\,2} + \bar{x}_1^{\,2} s_{x_2}^{\,2}}$
$\dfrac{x_1}{x_2}$	$\dfrac{\bar{x}_1}{\bar{x}_2}$	$\dfrac{\sqrt{\bar{x}_2^{\,2} s_{x_1}^{\,2} + \bar{x}_1^{\,2} s_{x_2}^{\,2}}}{\bar{x}_2^{\,2}}$

5.5. MULTIPLE RANDOM VARIABLES

In this section we consider another variation of the rod in tension problem. In Section 5.3 we examined the case where the yield strength was random and the design parameter was deterministic, and the case where the yield strength was deterministic and the design parameter was random. Here we will treat the design variable A as deterministic and the yield strength as random, but now we will also assume that the load (which previously was a deterministic quantity, $P = 20{,}000$ lb.) is a normally distributed random variable defined by

$$P = 20{,}000 \text{ lb.} \quad s_P = 2{,}000 \text{ lb.} \qquad (5\text{--}30)$$

This formulation of the problem will reveal the complexities associated with dealing with multiple random variables.

Determining the Probability of Failure

We'll start with the relatively straight-forward problem of determining the probability of failure of a rod subjected to the randomly distributed load characterized by Equ. (5–30). We'll assume that the rod has a (deterministic) cross-sectional area $A = 0.5$ in.^{2}, and is taken from the same stock described by Equ. (5–4) ($\bar{y} = 50{,}000$ psi and $s_y = 6{,}000$ psi). In this case, we have two random variables (y and P), each defined by their own probability density function. See Figure 5–7 for the two distributions.

Dealing with these two probability density functions is awkward. We can simplify the solution by introducing new random variables whose probability density functions are an appropriate combination of the probability density functions of the original random variables y and P.

The first step is to use the definition of stress in the rod [see Equ. (5–9)] to define the new random variable σ in terms of the random variable P as

$$\sigma = \frac{P}{A} \qquad (5\text{--}31)$$

Since A is deterministic, the mean and standard deviation of the new random variable are found from Equs. (5–26) and (5–27) to be

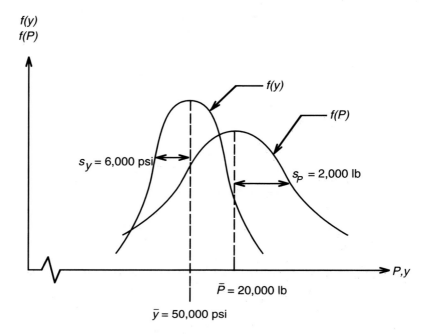

Figure 5–7. Probability Density Functions for P and y

$$\bar{\sigma} = \frac{\bar{P}}{A} \tag{5-32}$$

and

$$s_\sigma = \frac{s_P}{A} \tag{5-33}$$

From the given data, the numerical values of σ and s_σ are

$$\bar{\sigma} = 40,000 \text{ psi} \tag{5-34}$$

and

$$s_\sigma = 4,000 \text{ psi} \tag{5-35}$$

We now introduce an additional random variable Q defined as

$$Q = y - \sigma$$

An appropriate name for Q is "the margin of safety" since incipient failure occurs when $Q = 0$ and values of $Q > 0$ represent the margin of safety. Now y and σ are independent random variables since the yield strength of a rod randomly selected from stock is not influenced by the magnitude of the load P to be applied to the rod, and vice-versa.

Since y and σ are normally distributed, Q is also normally distributed with a mean and standard deviation related to the means and standard deviations of the original random variables according to Table 5–2,

$$\overline{Q} = \overline{y} - \overline{\sigma} \tag{5–36}$$

and

$$s_Q = (s_y^2 + s_\sigma^2)^{1/2} \tag{5–37}$$

Substituting in the numerical values from Equs. (5–5), (5–34), and (5–35), Equs. (5–36) and (5–37) become

$$\overline{Q} = 10{,}000 \text{ psi} \tag{5–38}$$

and

$$s_Q = 7{,}210 \text{ psi} \tag{5–39}$$

The Q distribution is shown in Figure 5–8. The onset of failure is associated with

$$Q^* = 0 \tag{5–40}$$

To determine the probability of failure, that is,

$$Pr(Q < 0)$$

we introduce the standard variable z. We get

$$z^* = \left(\frac{Q^* - \overline{Q}}{s_Q} \right) \tag{5–41}$$

But with Equs. (5–38), (5–39), and (5–40), the above expression is evaluated as

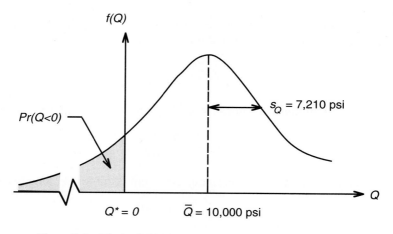

Figure 5–8.   The Probability Distribution for the Margin of Safety

$$z^* = \left(\frac{0 - 10,000}{7,210}\right)$$

or

$$z^* = -1.39$$

Table 5–2 then provides

$$Pr(Q < 0) = 0.0823$$

Designing to a Specified Probability of Failure

As in Section 5.3, we now solve the inverse problem: Given the random variables y and P, what (deterministic) cross-sectional area A is needed in order for the probability of failure to be less than 0.005? Because of the awkwardness of dealing with multiple random variables, we will start the analysis by using the margin of safety Q.

From Table 5–1, the z^* associated with $Pr\,(z < z^*) = 0.005$ is

$$z^* = -2.575$$

Then Equ. (5–41) with $Q^* = 0$ becomes

$$-2.575 = \frac{-\overline{Q}}{s_Q} \tag{5–42}$$

We now need to write this expression in terms of the design variable A. The first step is to invoke Equs. (5–36) and (5–37) so that Equ. (5–42) becomes

$$2.575 = \frac{\overline{y} - \overline{\sigma}}{(s_y^2 + s_\sigma^2)^{1/2}} \tag{5–43}$$

Now $\overline{\sigma}$ and s_σ can be expressed in terms of A using Equs. (5–32) and (5–33) so that the above expression is transformed into

$$2.575 = \frac{\overline{y} - \overline{P}/A}{[s_y^2 + (s_P/A)^2]^{1/2}} \tag{5–44}$$

Inserting the numerical values for $\overline{y}$, $\overline{P}$, s_y, and s_P into Equ. (5–44) and squaring gives

$$6.63 = \frac{(50,000 - 20,000/A)^2}{[(6,000)^2 + (2,000/A)^2]} \tag{5–45}$$

which, after clearing, becomes a quadratic equation for A^2. The two solutions are

$$A_1 = 0.268$$

$$A_2 = 0.617$$

Inserting A_1 into the right hand side of Equ. (5–44) yields –2.575 so this is not an acceptable solution.[6] Hence the desired solution is $A_2 = 0.617$.

Random design variable.   If we now let A be a random variable, then we have to back up to Equ. (5–43) and express $\bar{\sigma}$ and s_σ in terms of $\bar{P}$, s_P, $\bar{A}$, and s_A. Using Table 5–2, we can write

$$\bar{\sigma} = \frac{\bar{P}}{\bar{A}}$$

and

$$s_\sigma^2 = \frac{\bar{A}^2 s_P^2 + \bar{P}^2 s_A^2}{\bar{A}^4}$$

Substituting these into Equ. (5–43) yields

$$2.575 = \frac{\bar{y} - \dfrac{\bar{P}}{\bar{A}}}{\left(s_y^2 + \dfrac{\bar{A}^2 s_P^2 + \bar{P}^2 s_A^2}{\bar{A}^4}\right)^{1/2}} \qquad (5\text{–}46)$$

This simplifies considerably if we assume that $s_A = k\bar{A}$ where k is some numerical constant. We get

$$2.575 = \frac{\bar{y} - \dfrac{\bar{P}}{\bar{A}}}{\left(s_y^2 + \dfrac{\bar{A}^2 s_P^2 + \bar{P}^2 k^2}{\bar{A}^2}\right)^{1/2}} \qquad (5\text{–}47)$$

For any specific value of k, this can be solved for $\bar{A}$. Let's use the numerical values already introduced: $\bar{y} = 50,000$ psi, $s_y = 6,000$ psi, $\bar{P} = 20,000$ lb., $s_p = 2,000$ lb., and select $k = 0.1$. We square Equ. (5–47) then solve to get the two roots $\bar{A} = 0.237, 0.647$. Substituting each root back into Equ. (5–47) confirms that the solution is $\bar{A} = 0.647$.

5.6. RELIABILITY

Our discussion earlier in this chapter of the probability of failure of a rod under tension was in the context of a system whose loading and performance was independent of time. We turn our attention now to situations in which the performance of the sys-

---

[6]If we had solved Equ. (5–43) directly for A, we would have only obtained one solution, A_2. But by squaring Equ. (5–43) to eliminate the radical and solving Equ. (5–44), we obtained the two solutions A_1 and A_2. Note that if we square both sides of Equ. (5–43) after we substitute in the numerical value of $A_1 = -2.575$, the equality holds. In other words, both A_1 and A_2 are solutions to Equ. (5–44) but only A_2 is a solution to Equ. (5–43).

tem is a function of time. This may occur because the load on the rod changes with time, as it would if the rod was a piston rod in an automobile engine or part of a bridge truss subject to the ever-changing loads from vehicular traffic; or a tiny electrical connection subject to tension as the result of thermal expansion. In these situations time becomes the independent variable for our analysis. Also, for many other components and systems, we are interested in the probability of failure within a specified nonchronological period of analysis. The period of analysis might be the number of accumulated miles for automobile tires, or the number of takeoff and landing cycles for airplanes. To simplify the following discussion, we will use the word "time" to indicate the period of analysis. We begin this discussion by defining reliability. Then we will look at other indicators of performance related to failure: per-unit failure rate, and the mean time to failure.

We designate the probability of a system failing by time t as $F(t)$ and express it as the integral of the probability of failure density function $f(t)$ (see Fig. 5–9).

$$F(t) = \int_0^t f(\tau)d\tau \qquad (5\text{–}48)$$

It turns out that the complexity of many analyses can be reduced by using the complement of failure probability, rather than failure probability itself, as the dependent variable. For this reason, we define reliability $R(t)$ as the complement of failure probability; the probability of a system not failing by time t. Since the events of "failing by time t" and "not failing by time t" cover all possibilities, we can write

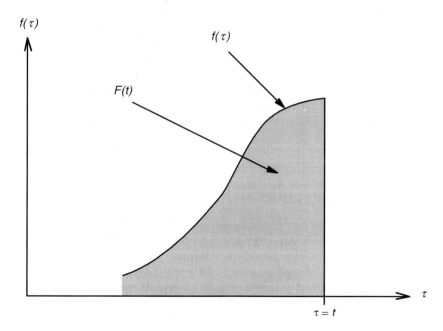

Figure 5–9. Failure Probability

$$R(t) + F(t) = 1 \tag{5-49}$$

We saw in the previous sections how the rather straight-forward engineering analysis and design of a deterministic system quickly became very complicated when we tried to account for probabilistic considerations. The analytical approach becomes even more difficult if we incorporate time as another independent variable. So, in the real world, the reliability of a device or system is much more likely to be determined empirically from operational experience or test data than from the kind of analytic approach discussed earlier in this chapter.

Per-Unit Failure Rate

To get some insight on how a numerical value of $R(t)$ might be determined from test data or observations of service history, consider two different populations P_1 and P_2 of the nominally same device. Let P_1 consist of 10,000 transistors in service or undergoing tests. Suppose that we observe that these transistors are failing at the rate of 50 failures/hour. Now suppose P_2 consists of 1,000 of the same kind of transistors experiencing the exact same service/test conditions as P_1. Because there are only 1/10 as many P_2 transistors as there are P_1 transistors, we would expect the failures in P_2 to occur at the rate of 5 failures/hour.

This description of performance is not very useful, since the quantitative measure of failure activity for the transistors depends on the size of the population. It would be nice if we could quantify the rate at which the transistors are failing in a manner that is independent of the size of the particular population group. This can be done by dividing the rate at which the failures are occurring by the size of the population. If we do that in this case, we can describe what is happening in both P_1 and P_2 as transistors failing at "a per-unit failure rate" of 0.005.

We can envision these two groups of transistors P_1 and P_2 experiencing these service/testing conditions simultaneously. On the other hand, we can also imagine that we start out with the population P_1 and at some time later the failures have reduced the population to P_2. This leads us to define the per-unit failure rate $h(t)$ as the ratio of the rate at which failures are occurring at time t to the size of the population at time t. Mathematically, it is convenient to write this as

$$h(t) = \frac{dN_f(t)/dt}{N_s(t)} \tag{5-50}$$

where $N_f(t)$ is the number of these objects that have failed by time t, and $N_s(t)$ is the number of objects which have survived to time t. As shown in Appendix 5B, we can rewrite the right hand side of this equation in terms of the reliability as

$$h(t) = -\frac{dR/dt}{R} \tag{5-51}$$

As also detailed in Appendix 5B, we can integrate Equ. (5–51) to get $R(t)$ as

$$R = e^{-\int h(t)dt} \tag{5-52}$$

This is the desired relationship. It allows us to express reliability in terms of an empirically determined per-unit failure rate.

For many products, the per-unit failure rate $h(t)$ consists of three stages which together take the form of what is known as the "bathtub" curve shown in Figure 5–10.

Early in a product's lifetime, the per-unit failure rate declines as product defects are rectified by improvements in design, manufacturing, and inspection procedures. During the mature phase of a product's lifetime, the only failures that occur are random and tend to be from unexplained causes. Such failures typically occur at a constant or linearly increasing per-unit failure rate. As the system approaches the end of its useful lifetime, it begins to wear out and the per-unit failure rate starts increasing rapidly. Let's look more closely at the longest part of the useful lifetime, the mature life span.

Constant Per-Unit Failure Rate

Consider the simplest case, where the per-unit failure rate is a constant. Let

$$h(t) = \lambda$$

where λ is a constant determined from test data or other empirical evidence. Then Equ. (5–52) reduces to

$$R(t) = e^{-\lambda t} \tag{5–53}$$

Thus, for a system that has a constant per-unit failure rate, the reliability decreases exponentially with time.

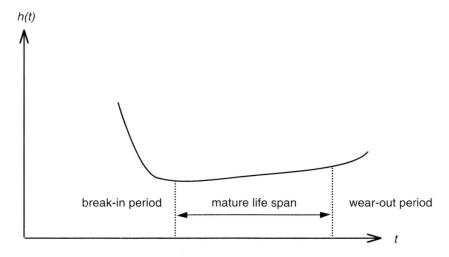

Figure 5–10.   Typical Per-Unit Failure Rates During System Life Cycle

Consider as an example that 2,000 items are being tested for 500 hours. Early observations indicate that failures are occurring at a constant per-unit failure rate of $\lambda = 2 \times 10^{-3}$ hr^{-1}. How many objects will survive the 500 hours?

From Equ. (5–53) we calculate the reliability at 500 hours as

$$R(500) = e^{-(2 \times 10^{-3} \, \text{hr}^{-1} \times 500 \, \text{hr})} = e^{-1} = 0.37$$

Now we can use Equ. (5–10) to get the number of surviving objects as

$$N_s(500) = N_o R(500) = 2,000 \times 0.37 = 740$$

Mean time to failure.   Another important descriptor of performance is the mean time to failure ($MTTF$). This provides the average time an object will be in service before failure occurs. By definition, $MTTF$ is the expected value of $f(t)$, the failure probability density function, or

$$MTTF = \text{Expected Value of } f(t) = \int_{-\infty}^{+\infty} t f(t) dt \qquad (5\text{–}54)$$

As shown in Appendix 5B, we can use Equ. (5–60) plus the fact that t cannot be negative, to write this in terms of the reliability as

$$MTTF = -\int_{o}^{\infty} t \frac{dR}{dt} dt \qquad (5\text{–}55)$$

For the special case of a constant per-unit failure rate λ, we can substitute Equ. (5–53) into the above equation to express $MTTF$ as

$$MTTF = \frac{1}{\lambda}$$

Thus, for systems with constant per-unit failure rates, the mean time to failure is the inverse of the per-unit failure rate. For the above example problem in which the objects had a per-unit failure rate of $\lambda = 2 \times 10^{-3}$ hr^{-1}, the mean time to failure is

$$MTTF = \frac{1}{2 \times 10^{-3} \, \text{hr}^{-1}} = 500 \text{ hr}$$

Appendix 5B: Derivation of Key Equations

In this Appendix we provide the detailed steps in deriving Equs. (5–51), (5–52), and (5–55).

Derivation of equation (5–51).   To explore the relationship between the per-unit failure rate $h(t)$ and the reliability $R(t)$, we start by letting the total number of nominally identical objects at the beginning of the utilization/testing period, that is, at $t = 0$, be N_o.

Then

$$N_s(t) + N_f(t) = N_o \qquad (5\text{–}56)$$

In the limiting case of a very large population, we can express the probability of failure and the reliability at any time t in terms of the number of objects as

$$F(t) = \lim_{N_o \to \infty} \frac{N_f(t)}{N_o}$$

and

$$R(t) = \lim_{N_o \to \infty} \frac{N_s(t)}{N_o}$$

In practice, when we are dealing with large but finite populations, we don't insist on the limiting condition and just define

$$F(t) = \frac{N_f(t)}{N_o} \quad \text{and} \quad R(t) = \frac{N_s(t)}{N_o} \tag{5–57}$$

Now lets rewrite Equ. (5–50) to eliminate N_f. We do this by using Equ. (5–56) to rewrite Equ. (5–50) as

$$h(t) = \frac{d[N_o - N_s(t)]/dt}{N_s(t)}$$

Dividing numerator and denominator by N_o leads to

$$h(t) = \frac{d\left[1 - \left(\dfrac{N_s}{N_o}\right)\right]/dt}{\left(\dfrac{N_s}{N_o}\right)}$$

which, after invoking Equ. (5–55) reduces to

$$h(t) = \frac{d[1 - R]/dt}{R}$$

which simplifies to Equ. (5–51).

Derivation of equation (5–52).   We first rewrite Equ. (5–51) as

$$\frac{dR}{R} = -h(t)dt \tag{5–58}$$

This integrates to

$$\ln R = -\int h(t)dt$$

or

$$R = e^{-\int h(t)dt} \tag{5–59}$$

This is labeled as Equ. (5–52) in the body of this section.

Derivation of Equation (5–55). Differentiating both sides of Equs. (5–48) and (5–49), we get

$$\frac{dF(t)}{dt} = f(t) \tag{5–60}$$

$$\frac{dR(t)}{dt} + \frac{dF(t)}{dt} = 0 \tag{5–61}$$

Substituting Equ. (5–61) into Equ. (5–60) gives us the relationship between $f(t)$ and $R(t)$ as

$$f(t) = -\frac{dR(t)}{dt} \tag{5–62}$$

Inserting this into Equ. (5–54) yields Equ. (5–55).

5.7. SYSTEM RELIABILITY

While our discussion so far has focused on the reliability of a group of nominally identical objects, much of the application of reliability concepts occurs in the analysis and design of systems that consist of many components, each one of which has its own reliability. So we now turn our attention to examining the relationship between overall system reliability and the reliability of individual components. Not surprisingly, that relationship depends on how the components are connected together to form the system. We focus on the two extreme cases of series systems and parallel systems. The analogy with electrical circuits will be obvious.

Series System

If failure of a single component causes the entire system to fail, we call such a system a series system. For a series system with n components, the system reliability is related to the component reliabilities R_i according to

$$R_{\text{system}} = R_1 \times R_2 \times \ldots R_i \times \ldots R_n \tag{5–63}$$

Note that since each of the component reliabilities is less than one, the reliability of a series system is always less than the reliability of the least reliable component. If we consider the special case in which every component has a constant per-unit failure rate, then from Equ. (5–53) we have

$$R_i(t) = e^{-\lambda_i t}$$

and we can write Equ. (5–63) as

$$R_{\text{system}} = e^{-\lambda_1 t} \times e^{-\lambda_2 t} \times \ldots e^{-\lambda_i t} \times \ldots e^{-\lambda_n t}$$

which simplifies to

$$R_{system} = e^{-\Lambda t} \qquad (5\text{-}64)$$

where Λ is the per-unit failure rate of the system, defined as

$$\Lambda = \lambda_1 + \lambda_2 + \dots \lambda_i + \dots \lambda_n \qquad (5\text{-}65)$$

Thus, a series system in which every component has a constant per-unit failure rate has a system per-unit failure rate that is also constant, and is equal to the sum of the per-unit failure rates of all the components in the system.

Parallel System

A parallel system is one that will not fail unless every component in the system fails. For a parallel system with n components, the system reliability is related to the component reliabilities R_i according to

$$R_{system} = 1 - (1 - R_1)(1 - R_2) \dots (1 - R_i) \dots (1 - R_n) \qquad (5\text{-}66)$$

Consider the special case of a two component parallel system in which both components have a constant per-unit failure rate, that is,

$$R_{system} = 1 - (1 - e^{-\lambda_1 t})(1 - e^{-\lambda_2 t})$$

This expands to

$$R_{system} = e^{-\lambda_1 t} + e^{-\lambda_2 t} - e^{-(\lambda_1 + \lambda_2)t}$$

This demonstrates that parallel systems have variable per-unit failure rates, even if all their components have constant failure rates.

It can be shown that the reliability of a parallel system is always greater than the reliability of the most reliable component. The reason why we don't design many systems as parallel systems in order to ensure their high reliability is that it costs too much. Placing two components in parallel is akin to establishing a redundancy. Clients are only willing to pay the extra costs of these redundancies in situations where the cost of failure is very high. Certainly, in situations where failure presents a threat to human life, multiple redundancies are clearly justified, and in many cases required by safety regulations.

5.8. CLOSURE

In principle, the techniques for probabilistic design discussed in this chapter allow us to design a system to achieve any desired probability of failure. The design considerations can account for random variations in loading, material properties, dimensions, or other relevant parameters. Such a design approach is much more sophisticated than the traditional approach of using safety factors to account for these uncertainties.

However, this powerful tool is much more complicated to use than safety factors. Even the simple problem of a slender cylindrical rod under tension becomes considerably more complicated when several random variables are present. For a more complicated system and loading, the problem can become quite formidable, especially if multiple failure modes are possible (that is, buckling and yielding). Also, we made several simplifying assumptions regarding the form of the probability distributions (normal), the independence of the random variables, and the relative size of the standard deviation to the mean ($s_x \ll \bar{x}$). In the latter part of this chapter we saw how reliability theory allows us to use test data or actual usage history to estimate the probabilities of failure of more complicated systems.

Like several other topics covered in this book, our treatment of the material is introductory. Entire books are written just on reliability theory. All we've been able to do here is familiarize you with the basic concepts and make you aware of the usefulness of the techniques. At least you now have enough background to tackle a more sophisticated coverage of the subject.

5.9. REFERENCES

BENJAMIN, J. R. and C. A. CORNELL. 1970. *Probability, Statistics, and Decision for Civil Engineers.* New York: McGraw-Hill Book Co.

MIDDENDORF, WILLIAM H. 1990. *Design of Devices and Systems, 2nd Edition.* New York: Marcel Dekker, Inc.

MODARRES, M. 1993. *What Every Engineer Should Know About Reliability and Risk Analysis.* New York: Marcel Dekker, Inc.

RAO, S. SINGIRESU. 1992. *Reliability-Based Design.* New York: McGraw-Hill, Inc.

SIDDALL, J. N. 1983. *Probabilistic Engineering Design: Principles and Applications.* New York: Marcel Dekker, Inc.

5.10. EXERCISES

1. Aluminum tubes of diameter $d = 1.5''$ and length $L = 20''$ will be used as simply supported columns subject to an axial load. The load on each column has a mean value of 1,000 lbs. and a standard deviation of 50 lb. What should the thickness of the tube be so that the probability of the column buckling is less than 0.0001? The tolerance on the wall thickness is $\pm 0.001''$; the tube diameter and length can be treated as deterministic variables.

2. A two-bar symmetric pin-connected truss shown below is subject to a nominally vertical load of 1,000 lbs. acting at point A. Misalignment of the load occurs randomly about the vertical axis of symmetry according to a normal distribution with a mean of $\bar{\theta} = 0°$ and a standard deviation of $s_\theta = 10°$. Select the cross-sectional area of the steel legs ($y = 30,000$

psi) so that the probability of failure of the truss is 0.001. All parameters except for the orientation of the load are deterministic.

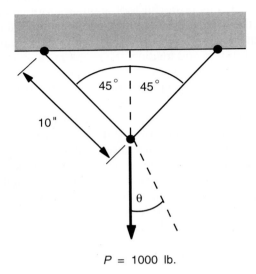

$$P = 1000 \text{ lb.}$$

3. A helical tension spring has a natural frequency f given by

$$f = \frac{1}{4}\sqrt{\frac{kg}{W}}$$

where k is the spring constant given by $f = d^4G/8D^3N$ and the weight W of the spring is $W = \pi^2d^2DN\rho/4$, N is the number of turns, D is the diameter of the spring, d is the wire diameter, G is the shear modulus, and ρ is the density. To avoid resonance, the natural frequency of the spring should be 20 times higher than the frequency ω of the applied force. If the only random variable is the diameter d of the spring wire, obtain an expression for the tolerance on this diameter so that the probability of resonance occurring is <0.0001.

4. The international standard for transportation of perishable foodstuffs requires that the heat transfer rate ($\dot{H} = k/t$) through the insulated walls of a refrigerated container does not exceed 0.4 Watts/m²–°C. If the thermal conductivity k of the insulating material is known to be normally distributed with a mean value of 0.35 Watts/m–°C and a standard deviation of 0.02 Watts/m–°C, what insulation thickness t is required to ensure that there is a 99.5% probability that the standard is met?

5. A steel ($E = 30 \times 10^6$ psi) cantilever beam of length 10 ft is required to support a concentrated load at its free end whose mean value is 500 lbs. and whose standard deviation is 100 lbs. If the cross section of the beam is rectangular with a 2 in. width, how deep should the beam be so that there is a 99.9% probability that its deflection at the free end is less than 3 in.?

6. A straight rod of solid circular cross-section is made of steel ($E = 30 \times 10^6$ psi) and subject to a tensile load P. Both the load and the length of the rod are normally distributed random variables ($\overline{P}, s_P; L, s_L$). Develop an analytical expression to select the cross-section area A of the rod such that the probability of the rod stretching more than 0.5 in. under the load is 1/100. Assume that $s_A = .05\,\overline{A}$.

7. The two-bar truss with pinned joints shown below is subject to a normally distributed load P whose mean value is 1,000 lb. and whose standard deviation is 200 lb. Bars 1 and 2 are circular stock whose normally distributed yield strength in psi are characterized as

	Bar 1	Bar 2
$\bar{y}$	45,000	42,000
s_y	5,000	6,000

The only available stock sizes for bars 1 and 2 are in increments of 0.1 in.^{2} of cross-sectional area. Select the bar diameters so that both 1 and 2 have a probability of yielding of ≤ 0.005. Based on your design choices, what is the probability of the truss failing due to yielding in at least one of its legs?

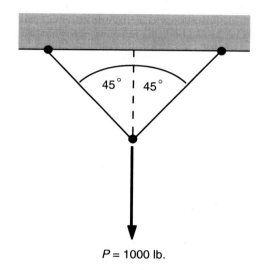

$P = 1000$ lb.

8. An electrical network has three resistors in series as shown. The resistance of each resistor is a normally distributed random variable characterized by the numbers shown. What is the mean and standard deviation of the resistance of the network?

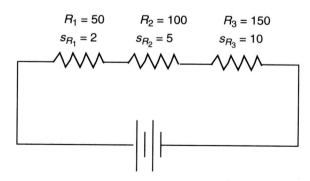

$R_1 = 50$ $R_2 = 100$ $R_3 = 150$
$s_{R_1} = 2$ $s_{R_2} = 5$ $s_{R_3} = 10$

♦ *Source:* Rao, p. 143. Reprinted with permission from The McGraw-Hill Companies.

9. A computer modem has a per-unit failure rate of 0.005 failures/hour. What is the probability that one of these modems will fail during 1,000 hours of operation?

10. The results of testing recently redesigned bus components shows that the axle has a mean "time" between failures of 600,000 miles, the modified electric motor controller has a mean "time" between failures of 740,000 miles, and the anti-lock braking feature has a mean "time" between failures of 650,000 miles. If failure of any one of these components results in failure of the bus, what is the probability that a bus will fail before it accumulates 500,000 miles? Assume that the failure of these three components are the only remaining causes of bus failure, and that the failure rates are constant.

11. A mechanical system consists of three components whose probabilities of failure are 0.0040, 0.0020, and 0.0030 respectively. If failure of all three components is required before the system fails, what is the probability of system failure?

12. A system is composed of parts A, B, and C. A has a reliability of 0.95; B has a reliability of 0.99; and C has a reliability of 0.90. These parts can be assembled in either of the two configurations shown. What arrangement of the three parts yields the highest system reliability?

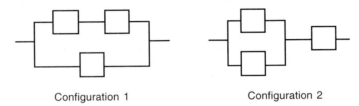

Configuration 1   Configuration 2

13. A gear box is being designed to have a reliability of at least 99.98%. Three vendors offer gear sets which meet the functional requirements of the gear box, and have reliabilities and costs shown in the table. What combination of gear sets will achieve the desired reliability at the least cost? You can use as many gear sets from each vendor or combination of vendors as needed.

Vendor	Reliability	Cost ($)
a	0.95	50
b	0.93	40
c	0.90	35

14. A simplified model of a thermal power plant is shown below along with the per-unit failure rate of each component. What is the reliability of the power plant at 1,000 hours?

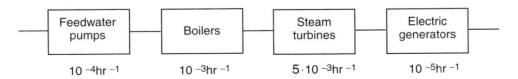

♦ *Source:* Rao, p. 185. Reprinted with permission from The McGraw-Hill Companies.

15. A component has a reliability of 0.90. How many identical components do we have to add in parallel to the first to increase the overall reliability to 0.999?

CHAPTER 6

Concept Generation

6.1. OVERVIEW

Concept generation (Step 5 in the design process) is where possible design solutions are first envisioned. This is where engineers use creativity and imagination to develop approaches to achieve the design objectives while satisfying the constraints.

The conceptualization step of design requires a creative, unconfined, uninhibited mindset dedicated to conceiving new designs. This is the deepest part of the design swamp, but it is also the richest because it contains numerous unusual and exotic plants and animals. If we think of each one of these elusive species as a design idea waiting to be realized, this can be the most exciting and fulfilling aspect of the engineering design experience. In order to maximize the value of this experience, we must be willing to prepare ourselves and put forth the effort, just as we need to have the right equipment and preparation to fully explore and appreciate the intricate ecology of the wetlands.

We begin this chapter by examining the opportunities for creative concepts to be incorporated into different kinds of engineering design assignments. Then we examine the meaning of creativity and the process of creative thinking. Next we identify the barriers to creative thinking and provide a set of techniques designed to help overcome those barriers. This is followed by a discussion of several tools for eliciting creative design ideas from individuals and groups. Finally, we introduce several approaches to fostering new concepts by modifying, combining, or otherwise organizing design alternatives.

6.2. LEVELS OF CREATIVE DESIGN

Not all engineering design assignments require the same degree or type of creative thinking. In this section, we discuss one approach to classifying design problems according to the opportunities they present for innovative and creative solutions.[1] While most real design activities involve combinations of the categories described here, it is still useful to consider them one at a time.

Selection Design

For this category of problems, design involves selecting standard components from a vendor and assembling them in a straight-forward manner to achieve the design objectives. The components frequently are selected from vendor's catalogs that describe all the relevant characteristics of the components. These standard components may include gears, valves, actuators, fasteners, sensors, motors, pumps, extruded shapes, structural beams of various cross-sectional sizes and shapes, electric circuitry, springs, switches, and bearings. You would be surprised at the rich and varied array of components and systems that are available "off the shelf." In general, it's a good idea to assume that a standard component exists to meet your needs; you should postpone a commitment to design your own until a search of vendors' literature fails to reveal what you are looking for. Because "selection design" is a mix and match type activity, there is less opportunity for innovative approaches than with other types of design problems. Most creative opportunities in this category are associated with selecting unusual combinations of components.

Configuration Design

This class of design problems was discussed in Section 1.6. It generally involves standard components, similar to "selection design," but the design challenge is to locate and arrange the components to improve performance or reduce size. An example of this is the design challenge of locating the jet engines on a commercial aircraft. The most popular configurations consist of the engines suspended beneath the wings, or cantilevered off the side of the rear fuselage, or mounted in the tail assembly. Even though many of the components are standard "off the shelf" items, there generally are areas in the configuration design category for modifying the size or shape of the components in order to take advantage of particularly attractive configuration opportunities.

Parametric Design

Parametric design refers to problems in which the primary design challenge is to vary the performance or design parameters to achieve the design objectives. These parameters may be imbedded in equations which express the relationships among the parameters, the objectives, and the constraints that must be satisfied. Boeing uses parametric design to create an entire family of aircraft based on one basic design. Consider

---

[1]This classification is based on one presented by Ullman, pp. 21–25. Although Ullman's classification is not solely for the purpose of assessing levels or amounts of creativity, we find that it is quite useful for this purpose.

the many models of the 737, which is the best selling aircraft in history. The variety of the models was obtained by varying the fuselage length, wingspan, and engine characteristics to create various combinations of range, speed, and carrying capacity. Compared to selection design and configuration design, parametric design provides greater freedom to the engineer to examine different sizes, shapes, and other key parameters.

Original Design

This category refers to the design of objects and systems that are a substantial and fundamental departure from existing products or processes. The push-button telephone is an example of a sudden, major, innovative departure from the rotary dial telephone. Other examples of breakthrough original designs include the personal computer, the microwave oven, the automatic transmission for automobiles, prestressed concrete structures, the elevator, the transistor, and the laser. Because the designs in this category involve radical changes from the existing norm, the creative element is a major, if not the dominant, factor in the design process.

6.3. CHARACTERISTICS OF CREATIVE THINKING

Many people have studied the phenomenon of creative thinking. Some psychologists devote a substantial part of their careers trying to understand what distinguishes the creative person from the individual who rarely, if ever, comes up with a new idea. There are no definitive answers, but there are at least some indications of factors that appear to affect our creativity.

In the following subsections we examine the concept of creativity from a perspective similar to that used in Chapter 1 with our discussion of engineering design. In particular, we first define creativity, then present models of the creative process that describe how it is carried out.

The Nature of Creativity

Creativity is an elusive concept, and there have been many definitions proposed by professionals in the field. Davis organizes definitions of creative thinking into four groups. The first group deals with the concept of combining ideas: for example, "Creative ideas are new combinations of previously unrelated ideas."[2] The second group of definitions emphasizes the originality of the ideas, as typified by this definition attributed to Herbert Fox: "The creative process is any thinking process which solves a problem in an original and useful way."[3] Next are those definitions which emphasize the unexplainable nature of the creative process, such as Carl Jung's assertion that,

[2]Davis, p. 6.
[3]Davis, p. 8.

". . . the creative act . . . will forever elude . . . human understanding."[4] The last group of definitions discussed by Davis are those that propose the idea that the creative act is inherently a product of the unconscious mind. For our purposes, we can be content with the insight provided by this range of definitions. We have no need to select from among them, or to consolidate them into a single composite definition.

Models of the Creative Process

Here we present three widely accepted models of the creative process developed within the professional psychology community.[5] Like the engineering design model discussed in Chapter 1, these are not rigid linear models; the steps may come out of order, be skipped entirely, or be revisited.

We will see that there is substantial overlap between these psychological models of the creative process and our model of the engineering design process. Under our model of engineering design, creativity is part of concept generation, and concept generation in turn is just one of nine steps in the design process. However, as we will see, the creative process models include elements that correspond to steps in our design model other than the concept generation step. We will handle this by presenting the creativity models in their entirety and indicate where they overlap those other elements of our engineering design model. Our objective is to strip away the redundancies and arrive at a model of the creative process that fits comfortably within Step 5 of our design process model.

Wallas model.   This is a five-step model originally proposed in 1926.

- Preparation: This stage includes clarifying and defining the problem. It corresponds to Steps 1 and 2 in our model of the design process.
- Incubation: This is the formative stage which takes place while the mind is relaxed and the individual is engaged in an unrelated activity which frees up the conscious and perhaps unconscious mind to be in a receptive mode.
- Illumination: This is the conscious recognition of the new idea, the so-called "Eureka" phenomenon.
- Verification: This is the reality check to determine whether the idea has merit. It corresponds to what we called the Evaluation Step in our engineering design model.

If we strip away those stages that are accounted for elsewhere in our engineering design model, we are left with a two-stage (incubation and illumination) model of the creative process.

---

[4]Ibid.
[5]Davis, pp. 39–44.

Fabun model. First proposed in 1968, this model uses the four stages from Wallas and adds three more.

- Desire: This deals with a thinker's equilibrium being disturbed by a problem. The individual then becomes motivated to restore the equilibrium by solving the problem. This sounds like Step 1 in our model of engineering design.
- Preparation: This is the same as in the Wallas model. But with the addition of the Desire stage, the Preparation stage now fits better with our Step 2. So the combination of the Desire and Preparation stages corresponds to what we call Problem Formulation in Chapter 2.
- Manipulation: This is where the mind actively manipulates materials or ideas in an exploratory fashion. This may include what we refer to as Step 4: Gathering Information. But it goes beyond that and represents the active effort that usually precedes the Incubation stage.
- Incubation: This is the same as in the Wallas model.
- Intimation: This is the feeling that you get when you realize that you are on the right track and are making progress.
- Illumination: This is the same as in the Wallas Model
- Verification: This is also the same as the Wallas Model.

Removing the overlap with our engineering design process model leaves us with a four-stage model: Manipulation, Incubation, Intimation, and Illumination.

Creative education foundation model. Proposed in 1976 by the Creative Education Foundation (CEF), this model consists of five stages. Each step involves both a divergent phase in which options are generated, and a convergent phase in which those options are winnowed down and combined or modified. Only the best of those ideas are taken forward to the next phase.

- Fact Finding: This consists of collecting facts which might have some bearing on the problem. This appears to be Step 4 of our design process model—Gathering Information.
- Problem Finding: This begins with developing alternative statements of the problem. This is the Problem Formulation phase of our engineering design model—Steps 1 and 2.
- Idea Finding: This is where possible solutions are first presented.
- Solution Finding: This is the evaluation phase. It closely corresponds to a combination of Step 6: Evaluation of Alternatives and Step 7: Decision Making.
- Acceptance Finding: This phase is the counterpart to our Step 8: Implementation.

After stripping away the overlaps, what remains is a single-stage model of the creative process-Idea Finding.

Consolidated Model

We will combine the non-overlapping elements of the three models described above into a consolidated model of the creative process. From the Wallas model, we take the Incubation and Illumination stages. The Fabun model contributes, in addition, the Manipulation and Illumination stages. However, we combine the Manipulation stage with the Idea Finding stage of the CEF model and rename it the Exploration phase. The result is the following four-stage model:

- Exploration: The active and concerted search for new ideas, often terminated in frustration because of the inability to identify sufficiently attractive ideas
- Incubation: The extended period of relaxation during which there is no conscious effort to address the problem
- Intimation: The resumption of purposeful pursuit combined with the feeling that you are beginning to make progress
- Illumination: The breakthrough that produces an attractive new idea

Creative Personalities

Creative geniuses are few and far between. In the field of technology, we instantly think of people like Thomas Edison and Benjamin Franklin. Clearly, they were born with a talent for invention, and we can't expect ordinary people like you and me to learn how to be as creative as they. But that doesn't mean that we can't improve our creative skills beyond their current level.

In Section 6.5 we will discuss particular techniques and exercises for stimulating your creative juices. Before doing that, we want to describe attributes of a general frame of mind that appear to be correlated with increased creative powers.[6] A creative mindset is optimistic, confident, enthusiastic, and endorses no preconceived solutions. Creativity also requires an adventurous spirit and a willingness to take risks. This includes a sense of spontaneity, even impulsiveness. A good sense of humor and even a childlike playfulness are valuable allies in your search for creative and innovative ideas.

Creative time is not the time to be conservative, analytical, and fearful of failure, since these attitudes can only inhibit you. Remember, in this phase of design our focus is on generating new ideas. Assessing the value of those ideas is a subsequent activity. Let's not jump the gun and confine ourselves to ideas that we think are workable. We have plenty of time to discard or modify ideas. So let's do one thing at a time. Let's emphasize the generation of ideas at this point. Our awareness of the attitudes that can help sharpen our creative tendencies can be reinforced by understanding a little bit of the physiology of the human brain.

---

[6]Davis, pp. 24–28.

Modes of Thought

The human brain is a marvelous and mysterious organ. It is the home of our "mind" and the source of all our thinking abilities. Extensive research on the physiological structure of the brain has revealed that different parts of the brain are associated with different modes of thought. One classical approach to characterizing the relationship between brain structure and modes of thought is in terms of the right and left hemispheres of the brain. The thought modes that are associated with each hemisphere are summarized in Table 6–1. Most engineering students probably recognize that left-brain functions dominate their engineering course work while the right-brain functions are more characteristic of humanities courses. More specifically, traditional engineering education, with its emphasis on mathematics, physical sciences, and engineering sciences, is structured to strengthen left-brain capabilities; almost to the point of suppressing any tendencies toward right-brain functions. The problem is that the creative aspect of engineering design requires a right-brain mode of thought. This situation has been described as one that requires a controlled schizophrenia, the ability to switch back and forth between these two modes of thinking.[7] It may be that engineering students, because of their background, have a harder time activating their creative tendencies than others who have not had the benefit of an education dominated by rigorous left-brain activities.

TABLE 6–1.   MODES OF THOUGHT

Left Brain	Right Brain
Deductive	Inductive
Derivative	Integrating
Specific	General
Logical	Imaginative
Hard	Fuzzy
Linear	Nonlinear

6.4. BARRIERS TO CREATIVE THINKING

Now that we have some familiarity with basic concepts of creative thinking and with the attitudes that can improve our creativity, why aren't we more creative? The answer to this question has two parts. First, most of us are continuously immersed in and surrounded by very strong barriers to creative thinking. Some of these are self-imposed, others arise from external factors. Second, overcoming these barriers takes a concerted effort. We devote this section to exploring several of these barriers.[8] As we

[7]Chaplin, C.R., 1989.
[8]Our classification of barriers is drawn from Adams, pp. 13–81.

will see, some of them are pretty powerful. In Section 6.5 we will describe a variety of exercises to overcome these barriers and enhance our creative powers.

Types of Barriers

Some of the existing barriers to creative thinking include perceptual barriers, cultural barriers, environmental barriers, emotional barriers, and intellectual barriers. Once we understand the nature of these barriers, we can focus on reducing the extent to which they interfere with our mental preparation and inhibit our creative thinking.

Perceptual barriers.   Perceptual barriers prevent a clear understanding of the design problem and/or the opportunities for solutions. Since this barrier is perhaps the one most frequently encountered by design engineers, we devote the next subsection to a more detailed examination of some of the specific ways in which perceptual barriers inhibit our creative abilities. Keep in mind that this kind of barrier can be easier to overcome by devoting proper attention to the Problem Formulation phase of design as discussed in Chapter 2.

Cultural barriers.   Cultural barriers are restrictions imposed by society. They discourage nontraditional approaches that violate societal norms. For example, a female redesigning a men's locker room for greater efficiency might feel that it would be very helpful, perhaps essential, for her to see the present design. This could cause both the designer and the patrons discomfort because of societal conventions. Having the courage to think and act unconventionally will help overcome these barriers.

Environmental barriers.   Anything in your immediate surroundings which inhibits creativity is an environmental barrier. This includes poor lighting, interruptions from frequent telephone calls, background noise, distracting aromas (both pleasant and unpleasant), uncomfortable temperature, or an uncomfortable chair. Taking the time to make your surroundings more conducive to creative thinking will pay handsome dividends. Some people experience their most creative periods late at night just before going to bed, or first thing in the morning. Others get their creative breakthroughs while hiking or sailing. This suggests that finding a time and setting to clear the cobwebs of daily problems from your mind provides a fertile space for new ideas to take root.

Environmental barriers are not just physical. The attitudes of your peers or your supervisors may be inimical to a creative thinking environment. Coworkers or bosses who are overly critical, who have no sense of humor, and who insist on rigid rules of dress and behavior can discourage your creative tendencies. An extreme example of such a situation is depicted in Figure 6–1.

Emotional barriers.   Emotional barriers are feelings which discourage you from considering a solution, or distract you from focusing on the problem. Discomfort with the risk associated with a proposed solution may cause you to discard that solution prematurely. For example, if you lost a relative due to a firearm accident, you

DILBERT® by Scott Adams

Figure 6–1.   An Office Environment That Discourages Creative Thinking (*Dilbert © United Features Syndicate. Reprinted by Permission.*)

may be uncomfortable working on projects dealing with explosives. Recognizing the existence of emotional barriers is an important first step in overcoming them.

  Intellectual barriers.   Intellectual barriers occur when you don't have enough knowledge of a topic to incorporate it into a design. Successful engineers overcome these barriers by remaining up-to-date in their field. You can't know everything, but you can expand your personal knowledge base by interacting with colleagues who have diverse backgrounds.

Types of Perceptual Barriers

Of the five categories of barriers mentioned above, perceptual barriers are the most common and the most difficult to recognize and overcome. For this reason we concentrate here on examining these barriers in more detail. Perceptual barriers manifest themselves when we perceive design problems too narrowly and thus become captive to an unnecessarily constrained perspective regarding possible solutions. Here we will examine five different ways in which this phenomenon can occur.

  Patterns.   Many design assignments involve making incremental changes in prior designs. Much technological progress has been made though these evolutionary developments. The internal combustion automobile engine is an excellent example of the impressive results achieved using the incremental approach over approximately one-hundred years. However, there also are situations that could be improved by making a radical departure from an existing pattern instead of continuing with incremental improvements. Consider the following simple geometrical example of this situation.

  We are given two identical pieces of cardboard, 1 and 2, shown in Figure 6–2(a) and asked to arrange them to form a simple geometrical shape. The obvious result is

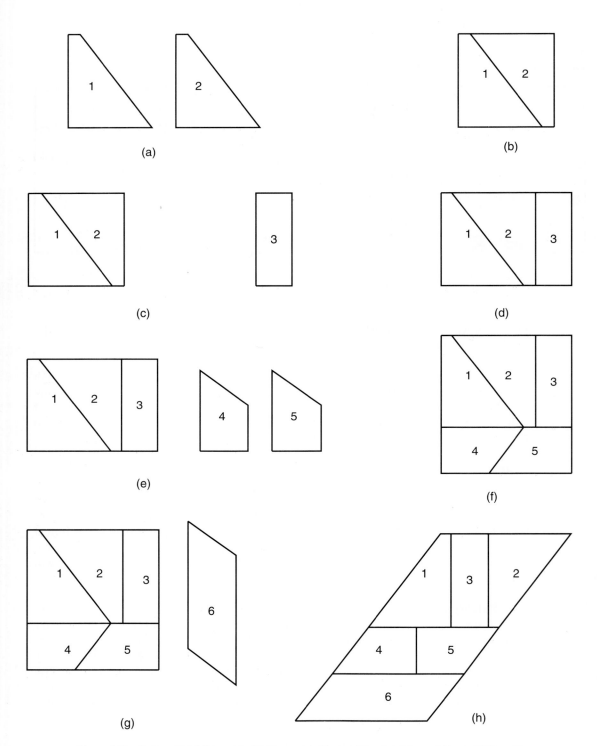

Figure 6–2. Pattern Establishment with Geometric Shapes *(Reprinted from deBono, 1970, p. 31. Copyright © 1970 by Edward de Bono. Reprinted with permission from Harper Collins Publishers, Inc.)*

the rectangle shown in Figure 6–2(b). Next we are given piece 3 as shown in Figure 6–2(c) and asked to use it together with pieces 1 and 2 to create another simple geometrical shape. The easy response is shown in Figure 6–2(d). The process continues with pieces 4 and 5 in Figure 6–2(e) and (f). In each stage of the activity we were able to incorporate the new piece by making incremental changes.

However when we are presented with piece 6 in Figure 6–2(g) the incremental approach can no longer provide a solution. The existing shape has to be decomposed and reassembled in a completely different way in order to accommodate piece 6 [see Figure 6–2(h).

Many people who easily achieved the desired result up through step (f) find it extremely difficult to solve the challenge posed in step (g). We succumbed to two pattern syndromes in the activities leading up to (g). First, we assumed that each new shape was to be formed by adding the new piece to the existing shape without disturbing the existing shape. Second, we assumed that the desired shape had to be a rectangle. The more stages that we went through applying the solution strategy based on these assumptions, the more committed we became to the pattern, and the harder it was to make non-incremental changes in the existing shape to obtain the new shape.

While we used a simple geometrical puzzle to illustrate this type of barrier, the phenomenon of existing patterns of thought serving to inhibit our creative tendencies is well known, and can even permeate entire organizations. The corporate "way of doing things" may serve a company well for a period of time, but can paralyze the firm when new circumstances require a rethinking of basic design philosophy. The Keuffel & Esser company dominated the market for slide-rules but was unable to adapt to the era of pocket calculators. Several mainframe and mini-computer manufacturers went out of business because they failed to appreciate the significance of the desktop personal computer.

Boundaries.   Another barrier to letting loose our creative juices is our tendency to interpret situations too literally and impose artificial boundaries on the range of solution possibilities. Perhaps the most well-known geometric puzzle that illustrates the implied boundary syndrome is the nine-dot problem shown in Figure 6–3. The

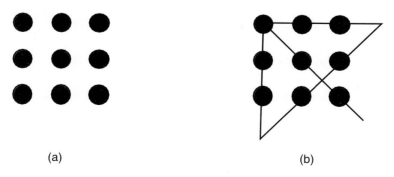

(a)   (b)

Figure 6–3.   Example of Perceived Boundary Constraint

challenge presented in Figure 6–3(a) is to draw four connected straight line segments that pass through all nine dots without lifting your pencil from the paper. Many people find this to be a difficult challenge, and after several tries will even claim that it is impossible (try it on an unsuspecting friend). The solution, shown in Figure 6–3(b), is easy once we free ourselves of the self-imposed boundary that we cannot go outside the implied square formed by the dots.

A more challenging puzzle in the same category is shown in Figure 6–4. The task is to place your pencil point at point 1, and without lifting the pencil from the paper, draw a continuous curve that crosses each of the 16 straight line segments (e.g.,   is a segment) just once, ending at point 2. See the Appendix to this section for a solution (but try to solve the problem before you peek).

As illustrated in both of these examples, many people take for granted that certain restrictions or boundaries exist. These perceived constraints can serve to severely limit the possibilities for creative solutions.

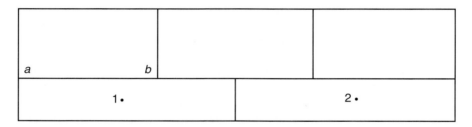

Figure 6–4. A Second Example of Perceived Boundary Constraint *(Reprinted from Baker, p. 12. Copyright © 1962 by Samm S. Baker. Reprinted by permission of Curtis Brown, Ltd.)*

Illusions. The fact that we don't always perceive objects correctly is the basis for many optical illusions. Consider the geometric shapes shown in Figure 6–5. In (a),

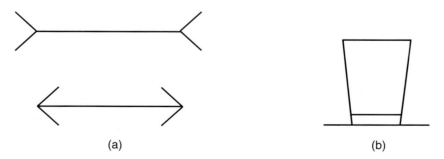

(a)   (b)

Figure 6–5. Optical Illusions Involving Perceived Line Lengths *(Reprinted from Baker, p. 79. Copyright © 1962 by Samm S. Baker. Reprinted by permission of Curtis Brown, Ltd.)*

Figure 6–6. Optical Illusion Involving Alternative Interpretations *(Reprinted from Lumsdaine, p. 37, with permission of the McGraw-Hill Companies)*

there are two parallel lines terminated by arrow heads. We perceive the upper line as being longer than the lower line even though they are the same length. In (b) we perceive the height of the top hat as being longer than the width of the brim. Again, they are both the same length. The lesson to be learned from this phenomenon is that when we act too quickly on our first impression of an object or situation, we may misinterpret the situation or at least ignore alternative explanations.

Another class of optical illusions are those in which our perception of an image can involve alternative interpretations. A classical illusion in this category is shown in Figure 6–6. Are we looking at two people, up close and face-to-face, or is it a fancy goblet?

Lenses.   Another kind of perception barrier arises when we look at situations with the wrong lens or focus too much on details. When this occurs we often miss the larger picture (we can't see the forest for the trees). Again, we will use a simple example, widely cited in the creative thinking literature, to illustrate this phenomenon. Figure 6–7 depicts an object that most people do not recognize, even after concerted effort, and even after being told that the object is being viewed with a close-up lens.

The same scene, viewed with a wide-angle lens, is shown in Figure 6–8. This version provides more information on the surroundings to help establish a context for the original object. Prematurely focusing on project details at the expense of exploring the broader context is a common syndrome in the early stages of the design process, and can severely limit the opportunities for creative solutions.

Meanings.   A perceptual barrier arises because we tend to adopt conventional roles for objects. We severely limit our options if we fail to consider the possibilities

Figure 6–7.   View of Object With a Close-Up Lens *(Reprinted from Parnes, p. 50).*

that become available if we adopt new interpretations for the role of familiar objects. The objects we refer to here can also include words, since relying only on the conventional meaning of words can also unnecessarily restrict our opportunities for creative solutions. A class of word problems that provides many good examples of this includes the following:[9] "What large and well-known city in the United States is half golden and half silver?" The answer is Denver. If you had trouble obtaining the solution, it may be that you interpreted the phrase "half golden and half silver" as describing a physical characteristic rather than a literal formula for constructing the name. The key to success with this puzzle may be your willingness to explore alternative meanings of "is".

Barrier Removal

Now that we understand the nature of these barriers to creative thinking and the negative effects they can have, we can design strategies for overcoming them so that creative solutions to design problems can be sought. There are a number of specific tech-

---

[9]Baker, p. 34. *Copyright © 1962 by Samm S. Baker.* Reprinted by permission of Curtis Brown, Ltd.

Figure 6–8.   Wider View of Object Using a Different Lens

niques that can help you in your effort to develop new ideas and nurture them into workable solutions. Several such techniques are described in Section 6.6. Before we turn to those, consider the following humorous, but quite profound, example of overcoming three of the perceptual barriers described above. To fully appreciate the editorial cartoon displayed in Figure 6–9, we must provide some background information.

During the late 1980s, Seattle constructed the first segment in a rail-based transit system, a tunnel that went from one end of downtown to the other. Recognizing that it would be quite a few years before the planned rail system could be completed, the tunnel was designed to be used by buses in the interim. However, since the tunnel was not equipped with a ventilation system to handle the exhaust fumes from the diesel engines, only buses powered by electric motors would be able to use the tunnel. Although Seattle had a large fleet of electric buses, they operated on close-in routes that had been strung with overhead electric wires many decades ago. The newer, longer routes between the suburbs and downtown were too expensive to electrify. In order to make the downtown tunnel accessible to suburban bus commuters, the transit agency ordered a fleet of more than 200 new dual-mode buses. These buses would operate under diesel power until the last stop before the tunnel, at which point they would connect their trolley poles to the overhead wires and switch over to the onboard electric motors. The concept was quite creative—these would be the first dual-mode buses in the U.S..

Things went well with the design of the dual-mode buses until the first prototype bus weighed in at 5,000 pounds heavier than expected. The extra weight was espe-

Figure 6–9. Overcoming Barriers to Creative Thinking on Multi-mode Bus Design *(Reprinted from the Seattle Post-Intelligencer of Feb. 13, 1989, with permission from the Seattle Post-Intelligencer.)*

cially significant because it meant that the buses violated a city ordinance regarding maximum permissible vehicle weight on city streets.[10] What a predicament—the transit tunnel was almost ready to open, but the special buses bought solely for the purpose of using the tunnel would be breaking the law each time they did. The dilemma was the lead story on local radio and television news and on the front page of the local newspapers. The editorial cartoonist for the daily morning newspaper captured the dilemma and proposed ways to solve it that beautifully illustrate several of the key points we've addressed in this chapter.

Figure 6–9 reveals the cartoonist's three creative design alternatives for dealing with the overweight dual-mode buses. His proposed Option A is an example of overcoming the "meanings" type of barrier presented by the conventional meaning of the word "overweight." The municipal ordinance imposes a weight limit on each axle, not

---

[10]Street pavements are designed to accommodate a certain range of vehicle loads. Loads that exceed design capacity can cause pavements to crack, buckle, and settle.

on the entire vehicle. Thus, the proposed solution is to add an additional axle, thereby spreading the total vehicle weight over more axles. Option B overcomes a "boundary" type barrier associated with our assumption that the buses were limited to being dual-mode. Once the cartoonist broke free of that constraint, his idea for a triple-mode bus was forthcoming. Finally, most people were confined in their thinking because the "lens" type barrier focused all our attention on the bus. The cartoonist took a step back to see the larger picture and realized that it was the excess weight of the bus and passenger system that was the problem. This allowed him to propose a solution, Option C, that was passenger-oriented. Clearly, the cartoonist used humor to make his point. That doesn't mean we can't benefit in a serious way from the in sight that his sense of humor allowed him to have.

Appendix 6A: Solution to Perceived Boundary Problem

Figure 6–10 shows a solution to the problem posed earlier in this section (see Fig. 6–4). The key step in the solution is to draw the curve so that part of it coincides with line segment *cd* of the original diagram. The problem cannot be solved without such a coincident segment. The constraint was that the curve could not cross a line segment more than once; it said nothing about not having part of the curve coincide with one of the line segments. You say that's not fair . . . it's a trick question. Actually, it makes the point we want to make quite well. Since the goal of the problem is to intersect each of the line segments, most people subconsciously constrain their thinking to consider intersection as the only form of contact between the original diagram and the curve to be constructed. Tangency, and colinearity, are implicitly excluded from our conscious thought processes. This artificial boundary is self-imposed. We have nobody to blame but ourselves.

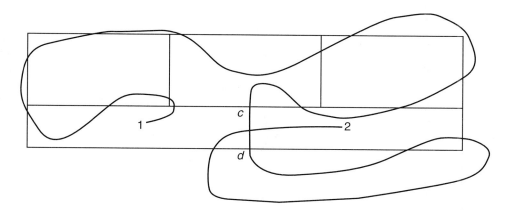

Figure 6–10.   Solution to Perceived Boundary Problem

6.5. TECHNIQUES FOR GENERATING ALTERNATIVES

Understanding the nature of creative thinking and the barriers that serve to inhibit it are a good start towards enhancing our creative capabilities. In this section we take the next step by learning several techniques that are designed to help stimulate our ability to generate design alternatives. Before discussing specific techniques, we want to emphasize two extremely important principles that should govern all your concept generation activities.

Lateral Thinking

A comment attributed to Thomas Edison was his response to a question about how prolific he was as a source of new inventions. He said, "It's easy to obtain one-hundred patents if you also have 5,000 unsuccessful inventions." A fundamental principle of the concept development phase of design is to concentrate on generating a large number of alternatives. The corollaries to this principle are: (a) resist the urge to pursue any one of the concepts in detail; and (b) avoid critiquing any of the concepts.

All this amounts to is separating Step 5 in the design process (Conceptualizing Alternative Approaches) from Step 6 (Evaluating the Alternatives). This sounds easy, but in practice it takes a great deal of will power and discipline to resist, what for most engineering students, is an innate urge to analyze. The importance of postponing the analysis activities is underlined by the phrase "lateral thinking" to emphasize the broad search for multiple design concepts; as opposed to "vertical thinking" which conveys the idea of an in-depth examination of individual alternatives.

Perseverance

A myth about creativity is displayed in the cartoon of the light bulb going off above somebody's head to signal the instant creation of a new idea. Much closer to the truth is that many new ideas are the result of a gradual evolution of thinking about existing concepts in new ways. While light bulbs occasionally do flash, they usually are the end result of a long process of searching. Thomas Edison is alleged to have said, "Invention is 95% perspiration and 5% inspiration."

Many people find that an extended, conscious, dedicated search for new ideas is often unsuccessful, but that it prepares the necessary background for a new idea that is born during a period of relaxation. The key factor appears to be that these break-throughs are more likely to occur if they follow an extended, intensive, conscious effort. This is consistent with our model of the creative process discussed in Section 6.3.

Mental Push-ups

Athletes engage in regular physical exercise between athletic events as a way to stay in shape. Though football players never do push-ups in a game situation, they engage in a regular and rigorous routine of push-ups and similar exercises in order to keep in top physical condition. It should not be surprising then that a regular routine of mental exercises can keep your creative thinking powers in top form. Here we offer several versions of "mental push-ups" to keep you in peak creative condition.

Making lists.   A good warm-up exercise is to make lists of twenty different uses for common objects such as: a wooden pencil, a spoon, a computer diskette, a burned out light bulb, a staple remover, an empty salt shaker, or a garden rake. After the first obvious use, and maybe one or two others, most people find it difficult to accumulate a list of twenty. Nobody said it would be easy. The important thing is to make the effort. If ideas are still not forthcoming, take a break to allow the incubation period to set in. If you still find it too difficult, you can, in the spirit of push-ups, start with a shorter list of uses and gradually work your way up to twenty. The point of this kind of exercise is to give you practice in thinking about using conventional objects in unconventional ways, which is an important attribute of creative thinking.

Word games.   Puns, anagrams, and word-building games such as Scrabble™ are another class of good mental exercises that can help keep your creative juices flowing. For practice try to identify a one-syllable, six-letter word that can have two of its letters switch positions to make another one-syllable word that rhymes with the original word.[11]

One of my favorite kind of word games are cryptic crosswords. These are crossword puzzles in which each clue is itself a puzzle, containing a synonym, hidden word, pun, anagram or other mischievous element in addition to the definition. Half the challenge is figuring out which part of the clue is the definition. Here are some sample clues from a cryptic crossword puzzle; the numbers in parentheses are the number of letters in the answers.[12] The answers are given in the Appendix to this section.

- Making coffee in copper kettle (4)
- Buddies strike back (4)
- Shish kebab item is more crooked? (6)
- Award boy crackers in play area (8)

The search for alternative meanings in each of these clues is a direct assault on the "meanings" barrier to creative thinking described in Section 6.4. Some people might describe this particular activity as the equivalent of one-handed push-ups on your finger-tips!

Solving puzzles.   Classical board games like chess and checkers are excellent exercises to foster lateral thinking, since they put a premium on examining a wide range of alternatives before making each move. Less time-consuming activities that

---

[11]Reprinted from *Games,* June 1996, p. 43, with permission from Games Publications, Inc.

[12]Reprinted from *Games,* June 1996, p. 34, with permission from Games Publications, Inc.

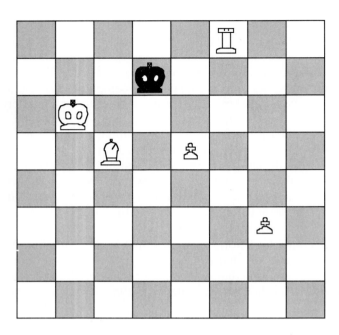

Figure 6–11. Chess Puzzle: White Check-
Mates in One Move *(Based on a puzzle in
Brecher, p. 105.)*

don't require physical playing pieces and a board include chess and checker puzzles.
These present the reader with a diagram of a given configuration, usually involving
only a few pieces, and the challenge is to achieve an objective, usually within a specified number of moves. Figure 6–11 is a typical puzzle of this genre.[13] Warning: this
particular chess puzzle has an added twist that challenges your agility in overturning
the "boundaries" barrier to creative thinking. If you can't quite make it, try peeking at
the clue in the footnote. If that still doesn't do it, use the two clues imbedded in the
sentence before the previous sentence. If you're still stumped, try solving the puzzle in
Figure 6–12 first.

A related class of puzzles that are easy to set up are those in which a set of
toothpicks, matches, coins, or other common objects are arranged in a given geometric pattern. The challenge is to transform the given pattern to a prescribed new pattern by moving only a certain number of the objects. A classic puzzle from this category is shown in Figure 6–12. It consists of eight toothpicks and a coin arranged to
look like a fish. The challenge is to move the coin and only three toothpicks to make
the fish swim in the opposite direction. Success with this kind of puzzle is usually associated with overcoming self-imposed restrictions on what moves are considered.

Magic tricks and other games.   Many magic tricks base their success on illusions or other mechanisms to convince you to misinterpret what you are seeing. Observing magicians work, and trying to figure out "how they do it" can be a very rigorous form of mental push-ups designed to overcome the "meanings" barrier to creative

---

[13]Hint: the pawn(s) are crucial to the solution.

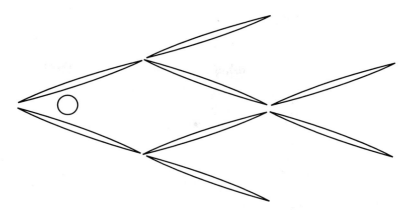

Figure 6–12.   Toothpick and Coin Geometric Puzzle *(Reprinted from Gardner, p. 64, with permission from Dover Publications, Inc.)*

thinking. Card games, such as bridge, pinochle, and some versions of solitaire have a high strategy/luck ratio and are also good material for mental push-ups because they demand a high degree of lateral thinking.

Procedural Techniques

Now that we have practiced our "mental push-up" exercises diligently, it is time to implement our game strategies. These are approaches to use in actual design situations. The first set that we will discuss provide a structure to enhance the creative productivity of design groups.

Brainstorming.   Brainstorming is a structured group-oriented technique for conceiving design alternatives. It consists of a group of individuals letting their imaginations run wild, but in accordance with certain procedural rules. The hope is that the group members will inspire and support each other; as a result the group will be able to conceptualize design alternatives that are more elegant than any that the individuals could have achieved independently.

The objective of a brainstorming session is to generate as many design concepts as possible. Individuals should be encouraged to mention all their ideas: the more outrageous, the better. Understandably, many people are reluctant to suggest design concepts that they may not have thoroughly thought through. In particular, many people may not be willing to risk eliciting negative reactions from other members of the group. Certainly, the possibility that another group member will deride your suggestion as totally impractical is enough to intimidate many individuals. In order to guard against such a stifling environment, a crucial brainstorming rule is that no criticism of individuals or ideas is allowed. This means, among other things, no comments such as, "that won't work," or "it's interesting, but impractical," are allowed. Further, no facial expressions or body lanquage that convey negative reactions are permissible. The success of a brainstorming sesssion depends on every participant feeling that they are in a supportive and sympathetic atmosphere in which all ideas are welcome.

Another important brainstorming rule is that ideas may not be pursued in detail. Once the basic gist of the idea is clear, the group should resume the search for other ideas. Group members must resist the urge to inquire about details of specific ideas. There will be plenty of opportunities to do that at a later time. The emphasis must be kept on generating as many ideas as possible, without worrying about their details, or even their validity. This is the "lateral thinking" notion mentioned in Section 6.4.

One of the other problems that frequently occurs in brainstorming sessions is that one or two individuals dominate the session, either by force of personality, or because they are much better than the other group members in generating ideas. Another problem is that if there is no designated discussion leader or session organizer, the activity can become chaotic at one extreme or die from lack of ideas at the other. The following organizational suggestions are offered as a way to overcome these problems.

First, you can establish a process to encourage group members to take turns offering ideas. This can be done by moving around the group, with each person getting a turn. During his/her turn, an individual can either (a) offer one idea, or (b) pass. An individual who passes during any cycle still can offer an idea in a subsequent cycle (their creative juices may have been stimulated by an idea offered by one of the other group members). This process continues until every person in the group passes when their turn comes.

Second, a brainstorming session should be recorded on paper, videotape, or some other medium. One person should be assigned the task of session recorder. This person, who may or may not be part of the brainstorming group, is responsible for making sure that a record is kept of all the ideas. If the recorder is taking notes and is unsure how to capture the essence of an idea in a few words, he/she should ask the individual who offered the idea for clarification. This clarification should occur before moving on to the next idea.

Consider having one person serve as session coordinator; their responsibility is to make sure the procedures are followed and the brainstorming rules are observed. Because of the discipline required to squelch any hints of negativism or vertical thinking, having a strong session facilitator will significantly improve the prospects for a successful and productive outcome.

Brainwriting. This is a variation of the brainstorming approach. Instead of ideas being presented orally, each participant writes their ideas on slips of paper, and the slips of papers are circulated among the group.[14] The brainwriting technique preserves the anonymity of the individuals, and prevents the more vocal personalities from dominating the session. A computer-age twist replaces the slips of papers with personal computers connected via a local area network, while still preserving the anonymity feature. Because the personal interaction is less direct and spontaneous in

---

[14]Kiely, pp. 32–40.

comparison to the traditional brainstorming approach, there may be less synergism of ideas.

Storyboarding.   Storyboarding is another variation of brainstorming. The main feature of this approach is the format and media for recording and organizing the ideas that are generated.[15] In this method, participants write each of their ideas on 3×5 index cards. All cards are then collected and tacked or pasted on a wall. After all the cards have been posted, the group engages in a discussion of the proposed concepts that can lead to many of the cards being grouped into categories. While storyboarding suffers from the same spontaneity weakness as brainwriting during the early part of a session, the ability to move cards around on the wall fosters an exploratory atmosphere for synthesizing and modifying the initial set of ideas.

Sources of Ideas

The techniques just presented describe processes, formats, and procedures to help elicit creative ideas. However, there are many times when, in spite of using the best techniques, the ideas are just not forthcoming. Here are two practical approaches to breaking the impasse and stimulating your creative juices. Both approaches rely on the notion that creative thinking involves making connections between situations that at first glance are not connected.

Random stimulation.   This approach relies on selecting a word at random (say, from a dictionary, or the daily newspaper, or a magazine) and then using word association or a sequence of word associations, to suggest an idea for a design alternative. Here is the result of my use of this approach to generate design options for the automobile bumper. I decided to randomly select a page in my *Webster's New World Dictionary,* and use the word in the upper right corner of the right-hand page. I opened to p. 581 and found "general court-martial". My initial reaction was, "Oh no, now I'm really in trouble." I had a strong temptation to pretend that this first try was just for practice. I also felt a very strong urge to scan the two pages in front of me for a better word. It took a lot of will-power to resist both of these opportunities and stick with this apparently hopeless situation. I made a sincere, but half-hearted effort to find a connection between "general court-martial" and "automobile bumper." The first thing that came to mind was "military"—it was also the last thing that came to mind. I just could not move any further. Mercifully, my phone rang and I realized I had serious work to do, so I gladly set aside this silly little word game that had turned so sour so quickly.

Perhaps six hours later, as I lay down in bed for the evening, the "general court-martial" monster reappeared in my consciousness. This time, I almost effortlessly glided from "general court-martial" to "military" to "tank" to "tank-treads" to the concept of an automobile bumper in the form of a tank-tread rotated 90 degrees so

---

[15]Barr, pp. 42–46.

the flat face of the tread (the part that the tank sits on) served as the contact surface between the bumper and the object with which the car was colliding [see Fig. 6–13(a)]. The idea was that if part of the bumper got damaged in a low-speed collision, we could just advance the tread to bring in a fresh section of undamaged tread. Then it didn't take too long to come up with a variation of the tank-tread idea. Instead of the bumper consisting of one continuous tread that advanced horizontally from one side of the car to the other, the variation consisted of several sections of treads, each one advancing vertically [see Fig. 6–13(b)].

Are these good design concepts for an automobile bumper? Well, maybe . . . and maybe not. That will be determined in the subsequent evaluation phase. Our objective at this point was simply to generate a number of design options. And, the random stimulation method worked!

Several lessons were learned from that experience. First, don't give up until you've allowed for a reasonable incubation period. Second, exert the discipline required to stick with each application of the random stimulation method. Giving up too easily may deprive you of the pleasure of generating some very unusual options. Remember, the stimulus does not have to be a word. Photographs, objects, or paintings, can also be used.

Analogies. Another approach to fostering the generation of concepts is the use of analogies. This technique encourages the participants to see the similarities between the given design situation and another situation. The idea is that it may be eas-

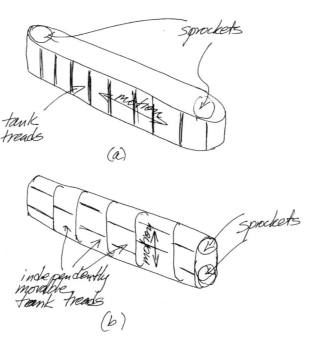

Figure 6–13. Sketches of Bumper Design Concepts

ier to generate ideas for that other situation; some of which may be transferrable to, or adapted for, the present situation. A four-step approach to using analogies to generate design concepts is listed here:[16]

1. State the need.
2. Generate the analogies by completing the phrase, "this situation is like . . ."
3. Solve the analogy.
4. Transfer the analogy to the original problem.

You can use brainstorming or one of the other procedural approaches in Steps 2 and 3 to generate as many analogies as possible and as many solutions as possible to each analogy. For example, consider the automobile bumper design problem introduced in Chapter 1. Some analogies to the situation of having too much damage from low-speed automobile collisions are: too many shin bruises from banging into office furniture; too many bumps on the head when retrieving dropped objects from under tables; too many pencil points breaking; and too many broken eggs in egg cartons at the grocery store.

Let's pursue this approach using the shin bruise analogy. One way to reduce or eliminate the shin bruises is to add padding to furniture legs. We can transfer this solution back to the original bumper problem in the form of adding padded strips to walls in parking lots. This reveals a solution to the bumper problem that doesn't require us to redesign the bumper.

Some people might need assistance in generating the analogies in Step 2. One approach is to invoke personal analogies[17] by asking the person to place themselves in the situation of the object; for example, "How would you deal with the situation if you were the automobile bumper?" Another good source for analogies is the natural world. Try to find situations involving plants and animals that have adapted to similar design challenges presented to them.

Organizing and Combining Ideas

Once ideas start flowing, we want to take advantage of the situation and combine elements from separate ideas to form new ones. In this way, even those suggestions which are impractical may be able to contribute to a new concept if part of the impractical idea is combined with another item.

Osborn's checklist.   You can direct your search for these kind of opportunities by using the following checklist of questions:[18]

---

[16]Folger and LeBlanc, pp. 78–79.

[17]Davis, p. 68.

[18]Folger and LeBlanc, p. 69.

Adapt?	How can this (product, idea, plan, etc.) be used as is? What are other uses it could be adapted to?
Modify?	Change the meaning, material, color, shape, odor, etc.?
Magnify?	Add new ingredient? Make longer, stronger, thicker, higher, etc.?
Minify?	Split up? Take something out? Make lighter, lower, shorter, etc.?
Substitute?	Who else, where else, or what else? Other ingredient, material, or approach?
Rearrange?	Interchange parts? Other patterns, layouts? Transpose cause and effect? Change positives to negatives? Reverse roles? Turn it backwards or upside down? Sort?
Combine?	Combine parts, units, ideas? Blend? Compromise? Combine from different categories?

For designs that have multiple functions or attributes, the previous methods can be used to generate concepts for each function and/or attribute. The methods discussed next are designed to ensure that all combinations of attribute concepts are considered. These approaches help to overcome the natural tendency to focus immediately on only the most logical, most familiar, or most obvious combination.

Morphological box. Another technique for developing new ideas is the morphological box. Each function or attribute is assigned a dimension in an array. The various design options for each attribute or for accomplishing each function are listed along the axis in the respective dimension. Consider the problem discussed in Section 2.5 of designing an electric power plant. One dimension of the design problem is selecting the energy resource which will be converted to electricity. A second dimension is the type of energy conversion technology used. Figure 6–14 shows a two-dimensional morphological box whose two dimensions are "Resource" and "Conversion Technology." There are three categories of conversion technologies and eight options along the resource dimension. In principle, each of the twenty-four possible combinations of a resource with a conversion technology represents a design option. Some of these combinations, such as using the energy in falling water (hydro) as input to a thermal energy conversion system, are not likely to be viable design options. We have placed marks in those cells which we consider to be viable. That reduces the number of options to ten.

Morphological boxes can be constructed at several levels of detail. They can also be used in layers, to pursue selected options in more detail. For example, suppose we wanted to examine the combination of fossil fuels and thermal conversion in more detail. That can be done by zooming in on that one cell in Figure 6–14 (shown with the thick border) and constructing another morphological box, such as the one shown in Figure 6–15.

We can also add a third function, such as emissions control, and construct a three-dimensional morphological box like the one shown in Figure 6–16. In this diagram, we display four emissions control options so we have a total of $3 \times 3 \times 4 = 36$ combinations. Of course, as before, some of these may not be viable design options.

The morphological box technique provides a structure that encourages design engineers to consider all possible combinations of functions and attributes, not just

		Conversion Technology		
		Thermal	Mechanical	Photoelectric
Resource	Fossil fuel	x		
	Nuclear	x		
	Wood	x		
	Wind		x	
	Solar	x		x
	Hydro		x	
	Tidal		x	
	Waves		x	
	Geothermal	x		

Figure 6–14. Two-Dimensional Morphological Box for Designing a Power Plant

the most obvious ones. As such, it is a useful graphical tool for engaging in lateral thinking. Its primary shortcoming is that it is limited to simultaneous consideration of no more than three characteristics.

Morphological charts.   Morphological charts are a modification of the morphological box concept that removes the restriction on the number of attributes or functions that can be depicted. This is accomplished by listing each attribute or function in a single column. The options associated with that characteristic are listed horizontally in the corresponding row. See Figure 6–17 for a morphological chart that corresponds to the three-dimensional morphological box shown in Figure 6–16. Note that

		Thermal Conversion Technology		
		Steam Turbine	Gas Turbine	Diesel Engine
Resource	Coal			
	Natural gas			
	Oil			

Figure 6–15. Two-Dimensional Morphological Box for a Fossil-Fuel Fired Thermal Power Plant

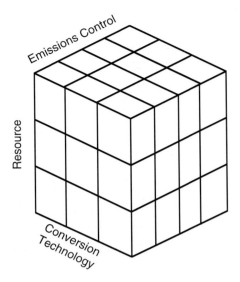

Figure 6–16. Three-Dimensional Morphological Box for a Fossil-Fuel Fired Thermal Power Plant

any number of additional attributes can be accommodated by adding another row for each attribute. Additional options can be easily handled by adding more columns to the chart. Each design option consists of one entry from each row. We have depicted one option (a steam turbine fired by coal and using flue gas desulfurization) by shading the appropriate cells in each row.

Attributes	Options			
Resource	coal	natural gas	oil	
Conversion Technology	steam turbine	gas turbine	diesel engine	
Emissions Control	electrostatic precipitator	flue gas desulfurization	catalyst	pre-combustion fuel cleaning

Figure 6–17. Three-Attribute Morphological Chart for a Fossil-Fuel Fired Thermal Power Plant

Appendix 6B: Solutions to Cryptic Crossword Clues

- Making coffee in copper kettle (4): The definition is "making coffee"; The answer is *perk* a hidden word in cop*per k*ettle. The word "in" is a clue that the answer is a hidden word.
- Buddies strike back (4): The definition is "Buddies"; the answer is "pals", a synonym for buddies; it is obtained by spelling "slap", a synonym for "strike", backwards, i.e., strike back.

- Shish kebab item is more crooked? (6): There are two definitions; "Shish kekab item" and "more crooked". The answer is "skewer", a word that satisfies both definitions ("skewer" means more skewed!).
- Award boy crackers in play area (8): The definition is "play area" but it has to interpreted as "an area in which plays are held". The answer is "Broadway", an anagram of "Award boy". The word "crackers" is a hint that the answer is in the form of an anagram, i.e., "Award boy" is "crackers", or all mixed up.

6.6. CLOSURE

Our examination of concept generation has included a glimpse of the thinking on creative thinking by mainstream psychologists. This exploration of the creative process has provided insights and given us tools that can help us heighten our individual creative powers. Clearly, what we have learned will not convert us into a modern day Thomas Edison or Benjamin Franklin; they were truly extraordinary creative geniuses. But that isn't a fair test. We don't have any reason to expect such dramatic results anymore than we could expect that learning how to do aerobics will convert us into olympic athletes. But what we have learned here can improve our individual capacity for innovative thinking. We also have acquired a capability to improve the organization and activities of design teams, and the environment in which they operate, toward the end of increasing their productivity in developing innovative design ideas.

Perhaps the three most important lessons to take away from this discussion are that: (1) the concept development phase of design should focus on generating a large number of options, and postponing judgment on their viability; (2) it is possible to improve individual and group ability to generate creative solutions to design problems; and (3) to be successful in our efforts to be creative requires a great deal of effort, discipline, practice, and a supportive working environment.

6.7. REFERENCES

ADAMS, JAMES L. 1986. *Conceptual Blockbusting: A Guide to Better Ideas, 3rd Edition*, Reading, MA: Addison Wesley Publishing Co.

BAKER, S. S. 1962. *Your Key to Creative Thinking: How to Get More and Better Ideas*. New York: Harper and Row Publishers.

BARR, V. Nov. 1988. The Process of Innovation: Brainstorming and Storyboarding. *Mechanical Engineering*.

BRECHER, E. 1996. *The Ultimate Book of Puzzles, Mathematical Diversions, and Brainteasers*. New York: St. Martin's Griffin.

CHAPLIN, C. R. November 1989. *Creativity in Engineering Design—The Educational Function*. London: The Fellowship of Engineering.

DAVIS, G. A. 1983. *Creativity is Forever*. Dubuque, Iowa: Kendall/Hunt Publishing Co.

DE BONO, E. 1967. *New Think: The Use of Lateral Thinking in the Generation of New Ideas*. New York: Basic Books, Inc.

DE BONO, E. 1970. *Lateral Thinking: Creativity Step by Step*. New York: Harper and Row Publishers.

FOGLER, H. SCOTT and S. E. LEBLANC. 1995. *Strategies for Creative Problem Solving*. Englewood Cliffs, N.J.: PTR Prentice Hall.

FRENCH, M. J. 1985. *Conceptual Design for Engineers, 2nd Edition*. New York: Springer-Verlag.

GARDNER, MARTIN. 1988. *Perplexing Puzzles and Tantalizing Teasers*. New York: Dover Publications, Inc.

KIELY, T. January 1993. The Idea Makers. *Technology Review*. Vol. 96, No. 1.

LUMSDAINE, E. and M. LUMSDAINE. 1990. *Creative Problem Solving: An Introductory Course for Engineering Students*. New York: McGraw-Hill Book Co.

PARNES, SIDNEY J. 1967. *Creative Behavior Workbook*. New York: Charles Scribner's Sons.

ULLMANN, D. G. 1992. *The Mechanical Design Process*. New York: McGraw-Hill, Inc.

VON OECH, R. 1986. *A Kick in the Seat of the Pants*. New York: Harper Collins Publishers.

VON OECH, R. 1990. *A Whack on the Side of the Head*. New York: Warner Books, Inc.

WALKER, D. J., B. K. J. DAGGER, and R. ROY. 1991. *Creative Techniques in Product and Engineering Design: A Practical Workbook*. Cambridge, England: Woodhead Publishing Limited.

6.8. EXERCISES

1. Cross out six letters from the following sequence so that the remaining letters, without altering their sequence, will spell a familiar English word: BSAINXLEATNTEARS.

♦ *Source:* Reprinted from von Oech, 1990 p. 114, with permission from Warner Books.

2. (a) What is the next shape in the sequence of shapes shown below?

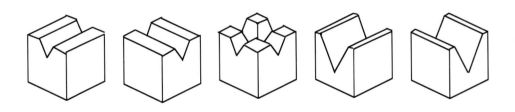

(b) Complete the set shown below

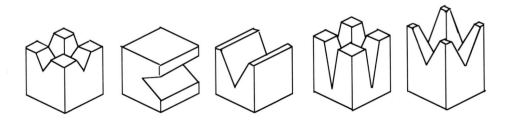

♦ *Source:* Adapted from Walker, et. al., p. 22, with permission from Woodhead Publishing Ltd.

3. As an exercise in creative thinking, find the missing words that make each of the five following equations true. Each equation contains the first letter of the missing words. If unsuccessful at first, don't give up. Many people report that answers come to them after the
exercise had been set aside, particularly at unexpected moments when their minds are relaxed.

(a)   7 = W of the W
(b)   12 = S of the Z
(c)   13 = S on the A F
(d)   76 = T in the B P
(e)   9 = P in the SS

4. Part of the skill of creative thinking is being able to find unusual uses for, or give unusual
interpretations to, common objects. Use your powers of creative thinking to describe five
completely different physical objects or systems that are suggested by the abstract arrangement of the geometrical shapes shown below.

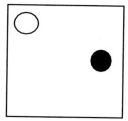

♦*Source:* Reprinted from Walker, et al., p. 23, with permission from Woodhead Publishing
Ltd.

5. One barrier to creative thinking is the assumptions we make regarding the existence of
boundaries that constrain our thinking. Here is an exercise to encourage you to break free
of such boundaries. Draw six connected straight lines that pass through all 16 dots shown
without lifting your pencil from the paper.

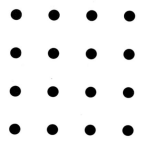

6. This exercise illustrates the value of lateral thinking; the systematic identification of alternative solutions before committing yourself to an apparent "best" solution. You have three
containers: an eight-liter container, a five-liter container, and a three-liter container. The
eight-liter container is filled with liquid; the other two are empty. Is it possible, without the
use of any additional equipment, to separate the eight liters of liquid into two four-liter
parts? If so, design a strategy for achieving this separation with the minimum number of
steps.

7. This is an exercise in overcoming cultural, environmental, intellectual, and emotional blocks to creative thinking. Assume that a steel pipe is embedded in the concrete floor of a bare room as shown below. The inside diameter is 0.06″ larger than the diameter of a ping-pong ball (1.50″), which is resting at the bottom of the pipe. You are one of a group of six people in the room, along with the following objects:

    100′ of clothesline
    A carpenter's hammer
    A chisel
    A box of Wheaties
    A file
    A wire coat hanger
    A monkey wrench
    A light bulb

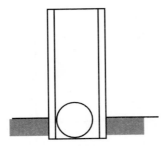

Develop five distinctly different preliminary designs for getting the ball out of the pipe without damaging the ball, tube, or floor.

◆ *Source:* Adams, p. 54. Reprinted with permission from Addison-Wesley Longman Inc.

8. A barrier to creative design is the tendency to be too critical of unusual approaches before fully exploring their advantages. Shown below is a sketch of a proposed design for a coffee cup involving a thin, hollow stem. Describe five advantages of this design.

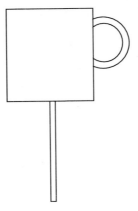

◆ *Source:* Adapted from von Oech, 1986, p. 90. Copyright © 1986 by Roger von Oech. Reprinted with permission from Harper Collins Publishers, Inc.

9. Use the minimum number of words to unambiguously describe the following shape. Your description must be clear enough to allow another person to duplicate the shape (except for size) solely on the basis of your description. Employ the concept of lateral thinking to develop at least three fundamentally different approaches to the problem before finalizing your solution.

◆ *Source:* de Bono, 1967, p. 37. Copyright © 1967 by Edward de Bono. Reprinted with permission from Basic Books, a division of Harper Collins Publishers, Inc.

10. As an exercise in creative thinking, find the missing words that make each of the five following equations true. Each equation contains the first letter of the missing words. If unsuccessful at first, don't give up. Many people report that answers come to them after the exercise had been set aside, particularly at unexpected moments when their minds are relaxed.

(a)   57 = H V
(b) 3600 = S in an H
(c)   3 = M in a T
(d)   29 = D in F in a L Y
(e)   54 = C in a D (with Js)

11. One of the barriers to creative engineering design is the difficulty in breaking free of the patterns which we easily fall into. This is particularly evident when a sequence of engineering designs evolve as small refinements of previous designs. Sometimes it is necessary to make a sudden break from the preceding pattern in order to arrive at a solution. Consider the following activity:

> Place four identical blocks (these may be matchboxes, paperback books, cereal boxes, etc.) on a flat horizontal surface. The problem is to arrange them in certain specified ways. These ways are specified by how the blocks touch each other in the arrangement. For two blocks to be regarded as "touching", any part of any flat surface must be in contact—a corner or an edge does not count.

The specified arrangements are as follows:

a. Arrange the blocks so that each block is touching two others.
b. Arrange the blocks so that one block is touching one other, one block is touching two others, and another block is touching three others.
c. Arrange the blocks so that each block is touching three others.
d. Arrange the blocks so that each block is touching one other.

Submit neat sketches to illustrate your solutions to parts (a) through (d).

◆ *Source:* de Bono, 1970, p. 97. Copyright © 1970 by Edward de Bono. Reprinted with permission from Harper Collins Publishers, Inc.

12. One aspect of creativity is using conventional objects in unconventional ways. As a mental "push-up" exercise to stimulate your creative powers, list fifteen different uses for a spoon.

13. One aspect of creativity is using conventional objects in unconventional ways. As a mental "push-up" exercise to stimulate your creative powers, list fifteen different uses for a paper clip.

14. You work for an office equipment manufacturer and have been asked to design a new line of ball point pens. Identify four attributes of this product and three alternative design approaches for each attribute. Use any three attributes to construct a morphological box for use in identifying alternative design concepts.

15. As chief design engineer for an aircraft engine manufacturer you are asked to explore the possible concepts for an engine suitable for use on large commercial aircraft beginning ten years from now. Conduct a broad examination of options and a more focused study of one approach by constructing a two-stage, morphological chart. With careful labeling, no accompanying text will be necessary.

16. One aid to creative thinking is to use words selected at random to suggest new design ideas. Use this approach to generate new design ideas for a writing instrument. In particular, pick the first and last words defined on pages 50 and 100 of any convenient dictionary (a total of four words). Explain the relation between each word and one of the design concepts.

17. Here are some more cryptic crossword clues, similar to the examples provided in Section 6–5. Find the answers and explain the connection between each clue and the corresponding answer.
    - Toss aside proposals (5)
    - The Lone Ranger's sidekick has a choice word for Canadian city (7)

◆ *Source:* Reprinted from *Games World of Puzzles,* July 1997, p. 57 and *Games,* April 1997, p. 35, with permission from Games Publications, Inc.

18. You want to use the minimum number of words to unambiguously describe the following shape. Your description must be clear enough to allow another person to duplicate the shape (except for size) solely on the basis of your description. Employ the concept of lateral thinking to develop at least three fundamentally different approaches to the problem before finalizing your solution.

◆ *Source:* Reprinted from de Bono, 1970, p. 66. Copyright © 1970 by Edward de Bono. Reprinted with permission from Harper Collins Publishers, Inc.

19. Use no more than two cuts to divide the square and rearrange the pieces to form an L-shaped figure.

Draw sketches to illustrate three fundamentally different strategies for accomplishing this task such that the two legs of the L have the same thickness.

♦ *Source:* Reprinted from de Bono, 1970, p. 79. Copyright © 1970 by Edward de Bono. Reprinted with permission from Harper Collins Publishers, Inc.

20. Below are what remains of three, common, related one-word proper nouns after all their consonants have been removed. What are the original words?

1. AAAA   2. EEEE   3. IIII

♦ *Source:* Reprinted from *Games*, Aug. 1996, p. 60, with permission from Games Publications, Inc.

CHAPTER 7

Project Planning

7.1. OVERVIEW

In this chapter we examine project planning—Step 4 in our nine-step model of the design process. We begin with a discussion of the purpose of the planning activity and the benefits of making it an integral part of the design process. Several of the more widely used planning techniques such as Gantt Charts, Critical Path Method, and Program Analysis and Review Technique, are then described. We conclude by examining several key variations and combinations of these standard methods.

As with many other topics covered in this book, our treatment of project planning is introductory in nature. Entire books are devoted to the subject, and many engineering schools offer courses devoted to this subject. Also, there are many commercially available software packages that allow the techniques described in this chapter to be applied to extremely large projects without having the engineer burdened with tedious calculations and diagram construction.

7.2. PLANNING CONCEPTS AND PRINCIPLES

Planning how to undertake a design activity is an integral part of the engineering design process. In fact, project planning is listed as Step 3 in the nine-step model of the design process described in Chapter 1. At a very minimum, a plan for an engineering design project consists of a list of tasks needed to complete the project, an estimate of the duration of each task and a sequential ordering of those tasks based on their logical relationships to each other. More detailed project plans may also include cost estimates for each task and personnel assignments for each task. Several widely used planning tools and formats are covered in the subsequent sections of this chapter. In this section we describe the main features of engineering project plans and discuss the role of planning in the design process.

Benefits of Planning

Many engineering design projects are extremely complex, involve many people, and extend over a considerable time period. Dividing a large complicated project into a series of smaller, manageable tasks gives structure to the project and helps minimize wasted efforts. A systematic planning effort includes these additional benefits:

1. A project plan provides a framework for communicating with clients and coworkers. While there may be general agreement regarding the major features of a project, it is easy for the many parties in a large project to have different perceptions of some of the specifics that have not been explicitly addressed. This may lead to individual parties making different assumptions about how those specifics are going to unfold; the proverbial "devil is in the details." By developing a plan that documents your expectations that, say, task G will take X weeks, will require M engineering hours, can't begin until task Z is completed, and has to be finished before task Q begins, you have taken a huge step to minimize any misunderstandings arising out of conflicting expectations. It is not unreasonable that the project definition will get modified as a result of issues, relationships, and constraints that first get revealed during the planning effort.

2. A project plan helps allocate resources, including project funds, personnel, facilities, and equipment. It also identifies the lead times for acquiring materials and components from outside sources. As an example of one of these dimensions, it is not unusual for engineers to be engaged simultaneously on more than one project. On one hand, you don't want to over-commit an individual; on the other hand, planning can keep individuals fully engaged in productive activities by assigning them to several projects if necessary (see Figure 7–1). So the planning may have to evolve simultaneously for several projects, with a particular emphasis on these cross-project linkages.

3. A plan also serves as a benchmark to measure progress during the project and helps to identify adjustments needed in the remaining tasks to meet the deadlines and stay within budget. Project plans should never be viewed as rigid. They are, at best, an informed projection of future activities. They should be considered as dynamic entities, meant to be modified on a regular basis as the project unfolds. About the only thing you can say for sure about a project plan is that it is obsolete as soon as it is developed.

Planning Challenges

Prospective clients frequently require that you submit a proposal to them describing your approach to meeting their need before they give you the go-ahead on the project. Many clients solicit such proposals from several competing design firms before making hiring decisions. Having a detailed, credible plan can make the difference between getting hired or not. This is tricky. If your plan leads to a higher cost and/or longer project duration, your firm can lose out to a competing company that promised

DILBERT® by Scott Adams

Figure 7–1. Keeping Engineers Busy With Multiple Projects (*Reprinted with permission from United Media.*)

to complete the work sooner and at less cost. On the other hand, if you are unrealistic in promising a quicker result at less cost and you cannot deliver on those promises, your company will lose money on that project. You can't stay in business very long if your planning efforts regularly put your firm in such a predicament.

One of the most common statements made by new design engineers when faced with this planning task is, "How am I supposed to know how long this will take and how much it will cost? I've never done this before!" That is a natural reaction. Early in your career, your design team might rely on a more senior engineer to do the planning effort. Certainly, practical experience on prior design projects is very valuable for planning future projects. But every design project is, to some extent, a venture into the unknown. The design must be different in some way from anything that already exists or else your client will use the existing version rather than asking you to design something new. So every design has an unknown element to it that introduces uncertainties into planning. The challenge is to have confidence in your professional judgment so that you are not paralyzed by the uncertainties. Remember, you're in the swamp and you must keep moving.

The amount of effort you devote to the planning task should be based on the overall size of the project. For a two-year project with a budget of $10,000,000 it may be appropriate to spend a month and $100,000 on planning. On the other hand, if you must complete the design within two weeks at a cost of $5,000, you probably will be better off keeping the planning effort very informal and with a low cost.

It has been said that there are six real stages of any engineering design project:[1]

- Enthusiasm
- Disillusionment

[1]Dieter, p. 452.

- Panic
- Search for the guilty
- Punishment of the innocent
- Praise and honors for the nonparticipants

Probably every design engineer has been involved with projects that consisted of at least several of these phases. When you find yourself in such a situation (and believe me, you will), such encounters should be chalked up to experience and serve to improve your insight so as to reduce the extent to which future projects have these characteristics.

7.3. GANTT CHARTS

Perhaps the most widely used project planning tool is the Gantt chart, named after Henry Gantt who developed the concept in about 1917. In its simplest form, a Gantt chart is a list of project activities along the vertical axis and a chronological time scale along the horizontal axis for displaying the schedule and durations for each task. In this section we use the automobile bumper project from Chapter 1 to introduce the features of the Gantt chart. We then illustrate its application to a much more complicated real-world project.

Gannt Chart for Automobile Bumper Project

The planning information for the automobile bumper project listed in Table 1–2 is presented in Figure 7–2 in the form of a Gantt chart. The project extends through the month of July with the starting and finishing dates for each task from Table 1–2 forming the basis for the horizontal bars associated with each task in Figure 7–2. Note that no work is scheduled for this project on July 18 and 19. This may be because all project personnel are needed full-time on another project during those two days.

Gantt charts are easy to construct and interpret, and provide a good chronological overview of the project. There are many additional elements that can be added to

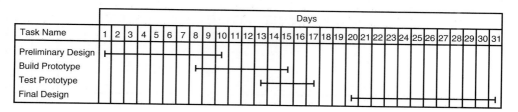

Figure 7–2.   Gantt Chart for Automobile Bumper Design Project

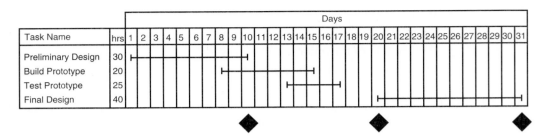

Figure 7–3. Enhanced Gantt Chart for Automobile Bumper Project

the basic Gantt chart shown in Figure 7–2. An enhanced Gantt chart shown in Figure 7–3 includes diamond symbols to identify project milestones, such as dates when progress reports are due. Also included in Figure 7–3 is a column that lists the duration of each task in hours. These values identify the amount of effort or resources required to carry out each task, and supplement the chronological durations represented by the horizontal bars.

Gantt charts are also very well suited for displaying project planning at various levels of detail. For example, each of the tasks shown in Figure 7–2 can be subdivided into several subtasks that are displayed on a separate Gantt chart. For complex systems, there can be many such layers of Gantt charts.

Gantt charts are also very well suited for keeping track of progress during a project. For example, Figure 7–4 shows cross-hatched bars to indicate actual progress through the 18th on the project whose plan was presented in Figure 7–3. As indicated, the preliminary design task was completed on schedule along with the first progress report (indicated by the cross-hatched diamond symbol). But the prototype construction took two days longer than expected and that led to a corresponding delay in starting the testing phase. In turn, that caused the prototype testing activity to spill over onto the 18th. This may require an adjustment in the other project that we were scheduled to be working on during the 18th.

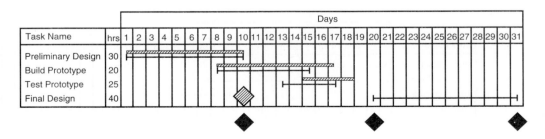

Figure 7–4. Keeping Track of Project Progress on a Gantt Chart

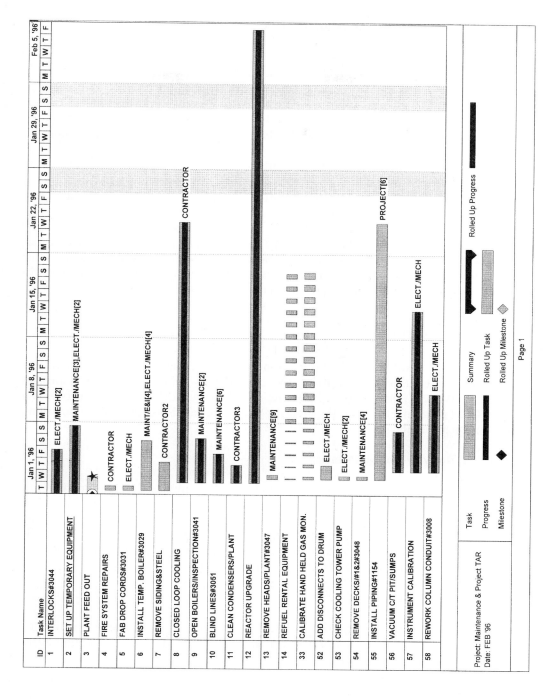

Figure 7-5. Gantt Chart for Chemical Plant Maintenance and Project Shut-down

Gantt Chart for Plant Shut-Down Project

Figure 7–5 shows part of a Gantt chart for the regular maintenance shut-down of a large processing plant owned by a major chemical company. To take advantage of the plant shut-down, several process and equipment modifications are incorporated into the project. Because of concerns over revealing proprietary information, we cannot reveal either the location of the plant or the name of the company. In addition, we have edited some of the information displayed in Figure 7–5 to avoid revealing the plant and company identity.

The complete Gantt chart for this project consists of ten sheets, involving 238 tasks. The sheet displayed in Figure 7–5 is a retrospective view of part of the project, representing the actual schedule and duration of the activities. The term "rolled up" in the legend at the bottom of Figure 7–5 signifies that the chart displays task durations and activities as they were modified over the six-week course of the project. Note that tasks ID #15 thru 32 and #34 thru 51 are missing from the chart. That is because these tasks, originally part of the plan for this project, were rescheduled after the project started and now appear on subsequent sheets of the chart. The labels to the right of each bar indicate the personnel that worked on each task.

7.4. CRITICAL PATH METHOD

The Critical Path Method (CPM) is a graphical network diagram approach to project planning that focuses on identifying the potential major bottlenecks in a project schedule. While Gantt charts emphasize the chronological relationships between tasks, CPM diagrams focus on the logical precedence relationships while submerging chronological details.

Establish Activity Characteristics

Consider the problem of constructing the network diagram and determining the critical path for the hypothetical project that consists of the activities shown in Table 7–1. Each activity is characterized by the estimated time required for its execution, and by listing all other activities that must be completed before that activity can begin. This is not a trivial step, but we will assume that it has already been completed so that we can focus here on applying the CPM method to this project.

Notation and Conventions

Two project elements are required in the CPM method: activities and events. An activity is an ongoing effort on a project activity. Every activity has an initiating event (typically the "start" event) and a closing event (typically the "stop" event). Even though events may represent discrete elements such as a decision point, delivery of a report or product, or a meeting, they are assumed to consume no time. The primary role of events in CPM diagrams is to separate activities.

TABLE 7–1.   PROJECT ACTIVITIES AND THEIR
PRECEDENCE RELATIONS

Activity	Duration	Preceded by
A	3	
B	3	A
C	4	
D	1	C
E	3	B, D
F	2	A, B, D
G	2	C, F
H	4	G
I	1	C
J	3	E, G
K	5	F, H, I

Graphically, activities are represented in a network diagram by directed line segments that point in the direction from the start event to the stop event of the activity. The name of the activity and the time required to complete the activity are associated with the directed line segment. Such a line segment is displayed in Figure 7–6(a). The directed line segment is not a vector, so its length and orientation can be chosen for convenience and to facilitate construction of the network diagram. All events are represented by circles, as indicated in Figure 7–6(b); they typically are not labeled.

There are several rules and conventions for assembling a network diagram from a set of activities and events. All networks must begin with a single "begin" event and end with a single "end" event. Consecutive activities must be separated by events. No pair of events can be directly connected by more than one activity with no intervening events. If an activity R must precede several otherwise unrelated activities S and T, the relationship is depicted as shown in Figure 7–7(a). If several otherwise non-related activities R and S must both precede activity T, the relationship is depicted as shown in Figure 7–7(b).

Dummy activities.   Sometimes precedence relationships require us to introduce an imaginary "dummy" activity into a CPM diagram (depicted by a dashed line) to indicate the appropriate relationships. To see how this can occur, consider the logi-

(a) Activities   (b) Events

Figure 7–6.   Conventions for Depicting Activities and Events in CPM Diagrams

Figure 7–7. Depiction of Activity Precedence Relationships.

(a)   (b)

cal relations between several hypothetical subtasks listed in Table 7–2 for the automobile bumper problem.

Clearly, we cannot build the bracket until we have designed it. Let's assume that the bracket will be bolted to the bumper, and the first step in the assembly process is to drill the mounting holes in the bumper. Obviously the holes cannot be drilled until the bumper is built. Also, the bracket design must be complete so that we know where to drill the holes in the bumper and what size the holes should be. We encounter problems in trying to depict these relationships.

Figure 7–8(a) properly shows that the bumper must be built and the bracket must be designed before we drill the holes in the bumper. However, Figure 7–8(a) is incomplete because it doesn't include the activity of building the bracket. If we try to add that activity to Figure 7–8(a), and also reflect the fact that the bracket has to be designed before it can be built, we end up with Figure 7–8(b). But this incorrectly indicates that the bumper must be built before the bracket is built. We can resolve the problem by introducing a dummy activity as shown in Figure 7–8(c). With the addition of the dummy activity, the proper relationships can be depicted.

Some precedence relations will require that you insert more than one dummy activity. For example, the following set of relationships require three dummy activities, as shown in Figure 7–9:[2]

- *A* precedes *D*
- *A* and *B* precede *E*
- *B* and *C* precede *F*

A dummy activity is also needed to distinguish between two activities that share the same start event and the same end event, such as activities *S* and *R* in Figure 7–10.

TABLE 7–2. PRECEDENCE RELATIONS FOR
SEVERAL SUBTASKS OF BUMPER PROJECT

Activity	Preceded by
Design bracket	
Build bracket	Design bracket
Build bumper	
Drill holes in bumper	Build bumper, design bracket

[2]Meredith and Mantel, p. 273.

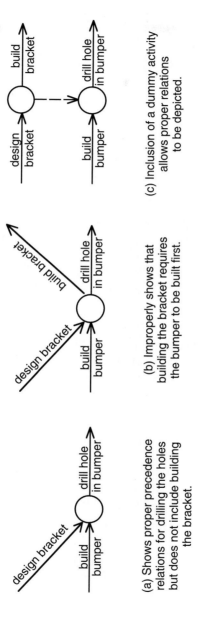

Figure 7-8. Use of Dummy Activity to Depict Precedence Relations

(a) Shows proper precedence relations for drilling the holes but does not include building the bracket.

(b) Improperly shows that building the bracket requires the bumper to be built first.

(c) Inclusion of a dummy activity allows proper relations to be depicted.

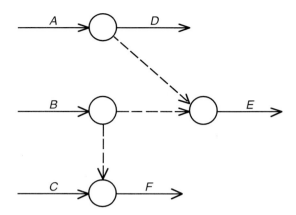

Figure 7–9. Multiple Dummy Activities

Dummy activities have zero duration but their inclusion increases the number of alternative paths between the begin and finish events. However, you should use them sparingly, including them in a network diagram only when necessary to display the precedence relations. You should not use them to close gaps in the network diagram that otherwise could be closed by modifying the location of events and the length and orientation of activity line segments.

Network Construction

We start constructing the network diagram for the project defined in Table 7–1 by placing a circle to represent the begin event. Then line segments are drawn for each activity that does not require any preceding activities. In this case, activities A and C are the only two activities that are in this category. The tails of these line segments must emanate from the begin event. However, their length and orientation are selected tentatively for convenience. We can always rotate these line segments and adjust their lengths later so that the appropriate combinations of activities meet at their common closing events.

We terminate activities A and C in events, then construct line segments emanating from these events representing those activities that require either A or C as preceding activities. According to Table 7–1, B is the only activity for which A is the only

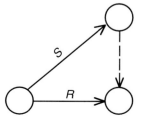

Figure 7–10. Using Dummy Activity to Depict Concurrent Activities

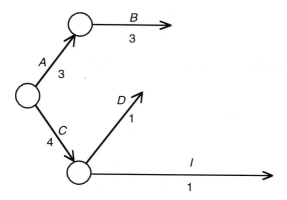

Figure 7–11.   Partial Network Diagram of Project Activities

preceding activity, and *D* and *I* are the only activities for which *C* is the only preceding activity. See Figure 7–11 for the partial network diagram depicting the relationships discussed to this point. Continuing in this manner until line segments are constructed for all activities, we then complete the diagram by connecting to the finish event all activities that have not yet been assigned an intermediate terminating event. The completed network diagram for the project defined in Table 7–1 is depicted in Figure 7–12. Note the dummy activity connecting activity *G* to activity *J*. Also note that we located the events so that none of the activity line segments crossed each other.

Determining the Critical Path

The critical path is defined as that path of activities from begin to finish of the network diagram for which delay in any activity along that path will cause the completion of the entire project to be delayed. For networks that have a small number of alternative paths from begin to finish, the critical path can most easily be identified as the

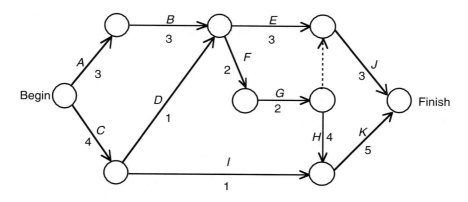

Figure 7–12.   Network Diagram of Project Activities

TABLE 7–3. ALTERNATE PATHS
AND THEIR LENGTHS

Path	Length
A-B-E-J	12
A-B-F-G-dummy-J	13
A-B-F-G-H-K	19
C-D-E-J	11
C-D-F-G-dummy-J	12
C-D-F-G-H-K	18
C-I-K	10

longest path from the begin event to the finish event. For the project depicted in Figure 7–12, the alternative paths and their lengths are listed in Table 7–3. We see that the critical path is *A-B-F-G-H-K* and its length is nineteen weeks.

For networks with many possible paths from begin to finish, an alternative, more efficient, approach to determining the critical path involves a forward sweep through the network diagram to determine the "earliest start" (*ES*) for each activity, and a backward sweep through the diagram to determine the "latest start" (*LS*) for each activity. Besides helping to identify the critical path, the *ES* and *LS* values have other uses that will be discussed later.

Earliest start. The earliest start (*ES*) of an activity is the earliest time the activity can start. It is found by tracing a forward path (from tail to head of each activity) from the begin event of the network to the tail of the selected activity. When several paths are possible, use the longest path (as determined by the sum of their durations). The length of this longest path is the earliest start (*ES*) for the selected activity. For example, Figure 7–12 shows that there are two paths from the begin event to the tail of activity *F*. The path involving activities *A* and *B* has a length of 6; the path involving activities *C* and *D* has a length of 5. Therefore, the *ES* for activity *F* is 6, as indicated in Table 7–4.

For activities that are further along a path, we can take advantage of the *ES* calculations already completed for preceding activities. So, the *ES* for activity *G* in Figure 7–12 is clearly 8, obtained by adding the duration of 2 for activity *F* to the already determined *ES* of 6 for activity *F*. This result is also listed in Table 7–4.

Project duration. The *ES* procedure described above is continued until we have calculated the *ES* for all activities that terminate in the finish event. When the duration of each of those activities are added to their respective *ES* times, the largest of the resulting sums is defined as the project duration. Again referring to Figure 7–12, the two activities that terminate in the finish event are *J* and *K*. Their *ES* times are listed in Table 7–4 as 10 and 14 respectively for *J* and *K*. Adding their durations (also listed in Table 7–4) to their *ES* values, the largest sum (14 + 5 = 19 versus 10 + 3 = 13) is associated with activity *K*. Thus the project duration for this project is 19.

TABLE 7–4. TOTAL FLOAT CALCULATIONS

Activity	Duration	Earliest Start	Latest Start	Total Float
A	3	0	0	0
B	3	3	3	0
C	4	0	1	1
D	1	4	5	1
E	3	6	13	7
F	2	6	6	0
G	2	8	8	0
H	4	10	10	0
I	1	4	13	9
J	3	10	16	6
K	5	14	14	0
Project Finish	19			

Latest start.   The latest start (LS) of an activity is the latest time the activity can start and still have the project be completed within the project duration time. To find the LS for each activity, trace a backwards path (from head to tail of each activity) from the finish event of the network to the tail of the selected activity. Make sure you reach the tail of the selected activity via the head of that activity. When several paths are possible, use the longest path (as determined by the sum of their durations). The project duration minus the length of this longest path is the LS for the selected activity. We illustrate this procedure by calculating LS for activity F. Starting at the finish event in Figure 7–12, one path backward to the tail of F is K-H-G-F; its length is 13. Another is J-dummy-G-F; its length is 7. Subtracting the longer of these paths from the project duration of 19 gives us a LS of 6 for activity F.[3] Since the procedure for finding LS involves moving backward through the network, the LS column in Table 7–4 is filled in generally from the bottom up.

Total float and critical path.   The "total float "(TF) for each activity tells how much that activity can be delayed while still allowing the entire project to be completed on time. It is determined for each activity as the difference between the latest possible starting time for that activity and the earliest possible starting time for that activity:

$$TF = LS - ES \qquad\qquad (7\text{--}1)$$

The critical path is defined as the sequence of activities for which $TF = 0$. The calculations for finding the critical path for the project depicted in Figure 7–12 are shown in Table 7–4. The critical path found using this approach consists of activities A-B-F-G-H-K, which is the same critical path as found in Table 7–3.

---

[3]Note that the path J-E to the tail of F is not eligible for the LS calculation because it doesn't reach the tail of F by going through F itself.

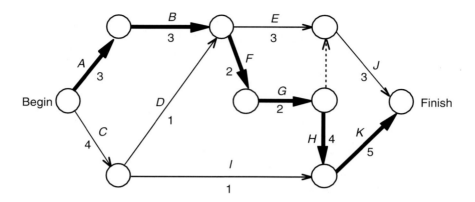

Figure 7–13. Critical Path Depicted on the Network Diagram

Once the critical path has been determined, it is conventional to update the network diagram to highlight the critical path, such as shown in Figure 7–13.

Any delay in an activity on the critical path will force the project finish to exceed 19. On the other hand, a delay in a noncritical activity can be accommodated up to the value of that activity's total float without affecting the project finish.

7.5. PROGRAM EVALUATION AND REVIEW TECHNIQUE

One of the disadvantages of the CPM method is the difficulty of making accurate estimates for the duration of each activity. We can overcome this limitation by using the Program Evaluation and Review Technique (PERT), which explicitly incorporates uncertainties regarding activity durations into the project schedule. PERT uses the same graphic convention as CPM, but replaces the single estimate for activity duration by a set of three estimates for activity duration. These estimates are based on the assumption that a Beta probability distribution function can represent the uncertainties associated with activity duration.[4] The Beta probability distribution is a skewed finite probability density function whose left terminus, mode, and right terminus can be specified by three parameters. As applied to PERT the three parameters are as follows:

- t_s: shortest time within which this activity can be completed assuming everything goes right. This represents the left terminus of the probability density function.
- t_m: most likely time that will be required to complete the activity. This is the mode (or peak) of the probability density function.
- t_l: the longest time it will take this activity to be completed assuming everything goes wrong. This represents the right terminus of the probability density function.

[4]Although rarely addressed in introductory treatments of PERT, additional assumptions are needed to derive the PERT formulas from the Beta distribution. See Meredith and Mantel, p. 275.

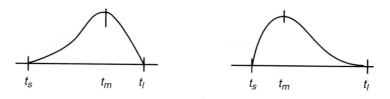

(a) skewed left Beta distribution   (b) skewed right Beta distribution

Figure 7–14.   Typical Beta Distributions for Activity Durations

Typical shapes of the Beta probability distribution are shown in Figure 7–14(a) and (b).

Determining the Critical Path

Let's apply PERT to the example problem solved in the previous section, as modified to incorporate estimates for the three parameters t_s, t_m, t_l, for each activity. In this case, we choose the time duration estimates used in the previous section to be our values for t_m. We supplement those values by estimates for t_s and t_l for each activity. We display the numerical values in Figure 7–15.

The first step in determining the critical path is to calculate the expected duration t_e, or mean, of each activity. This is found from a weighted average of the mode t_m, and the midpoint $(t_s + t_l)/2$, of the Beta distribution, with the mode weighted twice as much as the midpoint.[5] This yields

$$t_e = \frac{[2t_m + (t_s + t_l)/2]}{3} = \frac{t_s + 4t_m + t_l}{6} \qquad (7\text{--}2)$$

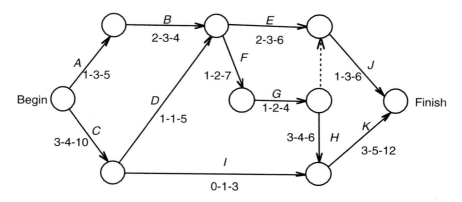

Figure 7–15.   Network Diagram for PERT Example Problem

[5]Shtub, p. 311.

TABLE 7–5.   DETERMINATION OF CRITICAL PATH FOR PERT EXAMPLE CALCULATION

Activity	Expected Time	Variance	Earliest Start	Latest Start	Total Float
A	3.00	0.44	0.00	0.50	0.50
B	3.00	0.11	3.00	3.50	0.50
C	4.83	1.36	0.00	0.00	0.00
D	1.67	0.44	4.83	4.83	0.00
E	3.33	0.44	6.50	14.84	8.34
F	2.67	1.00	6.50	6.50	0.00
G	2.17	0.25	9.17	9.17	0.00
H	4.17	0.25	11.34	11.34	0.00
I	1.17	0.25	4.83	14.34	9.51
J	3.17	0.69	11.34	18.17	6.83
K	5.83	2.25	15.51	15.51	0.00
Project Duration	21.34				

We show the results of the expected time calculations in Table 7–5. The expected times for each activity are then used to identify the earliest start, latest start, total float, and the critical path using the same approach as used in the CPM method. See Table 7–5 for the results. Also included in Table 7–5 is a column labeled Variance; it is calculated using Equ. (7–6) and will be used in a subsequent calculation.

The critical path consists of activities for which the total float $= 0$. Therefore the critical path for this project is *C-D-F-G-H-K*. This is a different critical path from that found for the same project using CPM. This is not surprising since the determination of the PERT critical path depends on the values of t_s and t_l for each task in addition to t_m, while the CPM critical path is determined only by the t_m values.

The expected time to complete the project is the sum of the expected durations for each activity on the critical path. In this case, the expected completion time for the project is

$$T_e = 4.83 + 1.67 + 2.67 + 2.17 + 4.17 + 5.83 = 21.34$$

(We'll use upper case T to identify project parameters and lower case t to represent individual activity parameters.)

Time to Project Completion

One of the main advantages of PERT is that it allows us to estimate the probability that the project will be completed by any specified time T_s. For this example problem, suppose we want to know the probability that the project will be completed in twenty weeks:

$$Pr(T < T_s) = ?$$

when

$$T_s = 20 \tag{7-3}$$

To use the probability table, we have to convert the random variable T to the standard variable z. The value of z associated with T_s is

$$z_s = \frac{(T_s - T_e)}{\sigma_T} \tag{7-4}$$

From Table 7–5, we know that

$$T_e = 21.34 \tag{7-5}$$

We next calculate the standard deviation of the time to project completion (σ_T) by assuming that the variance of the project completion time is the sum of the variances of the completion times for each activity on the critical path.[6] In this case

$$\sigma_T = (\sigma_C^2 + \sigma_D^2 + \sigma_F^2 + \sigma_G^2 + \sigma_H^2 + \sigma_K^2)^{1/2} \tag{7-6}$$

We assume that the standard deviation of the Beta distribution for each activity is one-sixth of the range of possible durations for each activity. We thus calculate the variance of each activity's duration time as

$$\sigma^2 = \left(\frac{t_l - t_s}{6}\right)^2 \tag{7-7}$$

The numerical values of the variances for each activity as calculated by Equ. (7–7) are listed in the Variances column of Table 7–5. Substituting the appropriate numbers from Table 7–5 into Equ. (7–6) yields

$$\sigma_r = (1.36 + 0.44 + 1.00 + 0.25 + 0.25 + 2.25)^{1/2}$$

or

$$\sigma_r = 2.36 \tag{7-8}$$

Inserting Equs. (7–3), (7–5), and (7–8) into Equ. (7–4) yields

$$z_s = -0.57$$

Using this value of z_s, Table 5–1 yields

$$Pr(z < -0.75) = 0.285$$

Hence there is a 28.5% chance that this project will be completed within 20 weeks.

---

[6]To be exact, the variance of the sum is equal to the sum of the variances only if the distributions are normal. However, we take advantage of the property that the variance of the sum of a large number of distributions approaches the sum of the variances, regardless of the shape of the distributions.

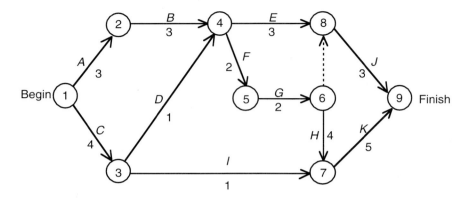

Figure 7–16.   CPM Network Diagram with Nodes Numbered

7.6. VARIATIONS AND COMBINATIONS OF APPROACHES

In this section we examine several modifications to the project planning techniques described earlier in this chapter.

Calendarized Network Diagram

The network diagram is a powerful technique for identifying the critical path. However, unlike the Gantt chart, it does not display the chronological relationships between the activities. This can be rectified by "calendarizing" the CPM network diagram.[7] We'll illustrate this technique by applying it to the CPM network diagram shown in Figure 7–12 (the method can just as easily be applied to PERT network diagrams). For convenience, we reproduce Figure 7–12 as Figure 7–16 and add identifying numbers to each node.

We have to modify several of the CPM conventions adopted earlier in order to depict the chronological relationships. First, the length of each activity line segment now has to be proportional to the duration of that activity. Second, all activities other than dummy activities must be depicted by horizontal line segments. In order to maintain the appropriate activity node connections, we will use horizontal and vertical dashed lines to extend activity line segments and to connect them to their terminating nodes.

In addition, nodes from which multiple activities emanate will be displayed separately for each emanating activity; the numbers assigned to each node will identify which of them have been split. The resulting calendarized version of Figure 7–16 is shown in Figure 7–17. Note that nodes 1, 3, and 4 have been split. All horizontal dashed lines represent the total float time for the associated activity. The vertical dashed lines, except for the one connecting nodes 6 and 8 serve only to connect the

[7]Meredith and Mantel, p. 285.

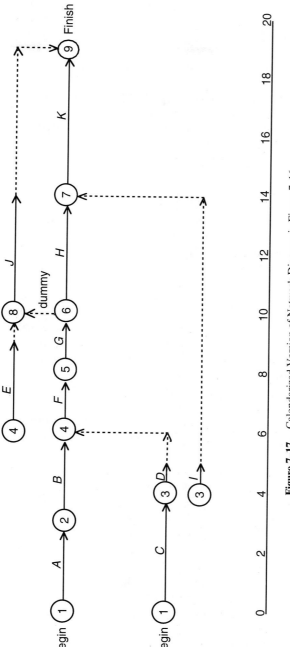

Figure 7-17. Calendarized Version of Network Diagram in Figure 7-16

corresponding activity with its terminating node. The vertical dashed line connecting nodes 6 and 8 is the dummy activity. All dummy activities in calendarized CPM diagrams are vertical, consistent with the definition of dummy activities as activities that have zero duration.

Figure 7–17 contains the chronological display feature inherent in Gantt charts while retaining the advantages of the CPM method, in particular the feature of clearly displaying the critical path. In fact, we can identify the critical path immediately from the calendarized diagram as the only path from begin to finish that does not contain a horizontal dashed line. Further, since the line segment lengths now designate the activity durations, the numerical values of task duration can be omitted from the segments themselves in favor of the master duration scale shown at the bottom of Figure 7–17.

Activities on Nodes

The convention introduced in Section 7.4 for network diagrams is that activities are represented by directed line segments and events are represented by nodes. We call this the activities-on-arrows (AOA) approach. An alternate approach is to represent activities by nodes and represent events by directed line segments. This formulation is called the activities-on-nodes (AON) method. Its application to the project represented in Figure 7–16 in its AOA form is shown in Figure 7–18. The numbers on the directed line segments in Figure 7–18 identify the nodes connecting the activities, whereas in Figure 7–16 the numbers on the directed line segments identified the duration of the tasks. Also, the begin and finish events in Figure 7–16 are now the begin and finish activities. This helps to define the two ends of the network in preparation for determining the critical path.

Note that there are no dummy activities in the AON version of the network. This is the case in general, and is one of the advantages of the AON approach. On the other hand, Figure 7–16 consists of nine nodes and 12 directed line segments while Figure 7–18 contains 13 nodes and 17 directed line segments.

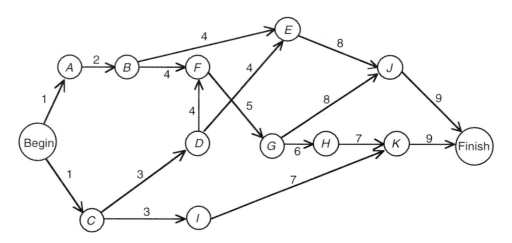

Figure 7–18.   Activity-on-Node (AON) Form

If we were to construct Figure 7–18 directly from Table 7–1, there would be no need to number the events. If, in addition, we add the activity durations to each activity, we can transform Figure 7–18 into Figure 7–19.

The critical path can be determined from Figure 7–19 in the same manner as it was found using Figure 7–12. Though most textbooks on project planning feature the AOA approach, both the AOA and AON approach are used in commercial project planning software packages. It appears that personal preference is the major factor that determines which of the two formulations is used on a given project.

Almost Critical Activities

The CPM and PERT methods appropriately focus attention on the critical path and the activities on the critical path. However, we must not lose sight of the fact that the values for the activity durations are only estimates, and they may change. With regard to Figure 7–12, if activity C were to have a duration of 6 instead of 4, then both C and D would become part of the critical path, replacing activities A and B. In a more complicated project, a near-critical path might very well have no activities in common with the critical path. It is prudent to be alert as a project progresses to the possibility that deviations from estimated activity durations may cause shifts in the critical path.

After using a network diagram approach to determine the critical path for a project, you may decide to reexamine some of your key assumptions regarding precedence relations and durations, and thereby restructure the project.

Estimates for activity durations in particular are keyed to assumptions regarding the amount of resources needed and available to carry out the activities. In many cases, you can reduce the duration if you devote more resources to the activity. You might even be willing to pay a premium (for example, overtime wages) to significantly reduce the duration of certain activities. Thus, these project planning tools can be an integral

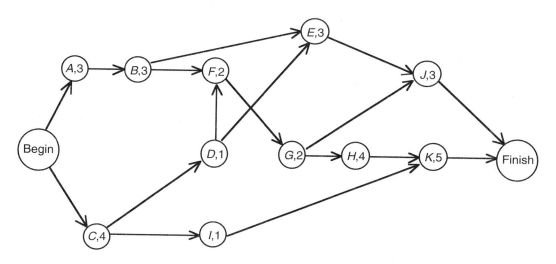

Figure 7–19.   Refined Activity-on-Node (AON) Form

part of a broader project budgeting and cost control effort. In particular, you can use the earliest start and latest start data to bracket anticipated budget requirements.[8]

7.7. CLOSURE

Our examination of project planning in this chapter began with a discussion of the purpose and benefits of planning. We then introduced several of the more widely used planning techniques. Gantt charts, perhaps the most popular project planning tool, are particularly well-suited for displaying chronological relationships between tasks and for combining scheduling with resource allocation. The critical path method (CPM) and PERT help to identify potential bottlenecks in a project, with PERT also allowing us to explicitly incorporate uncertainty over activity duration into the planning process. The "calendarized" version of CPM combines some of the best features of CPM with those of the Gantt chart.

Commercially available software packages allow the techniques described in this chapter to be applied to extremely complicated projects. However, we must not become so enamored by the power of project planning softward that we lose sight of the fact that the toughest part of project planning is to envision the activities that will be required. estimate their duration, and identify the precedence relationships among them. No software can substitute for astute engineering judgment in that phase of the planning effort. Finally, no project plan should ever be confused with a rigid prescription for future activities. A plan is always, at best, a tentative estimate of what we think might happen. All project plans should be reviewed regularly during the course of the project and modified as necessary and as experience dictates.

7.8. REFERENCES

DIETER, G. E. 1991. *Engineering Design: A Materials and Processing Approach, 2nd Edition.* New York: McGraw-Hill, Inc.

GOULD, F. J., GARY D. EPPEN, and CHARLES SCHMIDT. 1988. *Quantitative Concepts for Management, 3rd Edition.* Engelwood Cliffs, NJ: Prentice Hall.

JEWELL, THOMAS K. 1986. *A Systems Approach to Civil Engineering Planning and Design.* New York: Harper & Row, Inc.

MEREDITH, JACK R., and SAMUEL J. MANTEL, JR. 1989. *Project Management: A Managerial Approach, 2nd Edition.* New York: John Wiley and Sons.

SHTUB, AVRAHAM, JONATHAN F. BARD, and SHLOMO GLOBERSON. 1994. *Project Management: Engineering , Technology, and Implementation.* Engelwood Cliffs, NJ: Prentice Hall.

VAUGHN, R. C. 1985. *Introduction to Industrial Engineering, 3rd Edition.* Ames, Iowa: The Iowa State University Press.

WALTON, JOSEPH W. 1991. *Engineering Design: From Art to Practice.* St. Paul, MN: West Publishing Company.

[8]Gould, Eppen, and Schmidt, pp. 448–454.

7.9. EXERCISES

1. An automatic welding machine used in a manufacturing plant is to be rebuilt during a two-week plant shutdown and then reinstalled in a new location in the same plant. The machine must be dismantled to determine which replacement parts to order. The following activities are required (not necessarily listed in sequential order) and estimates of their duration are listed. The activities must occur from 7 a.m. to 3:30 p.m. on weekdays. Construct a Gantt chart for this project.

Activity	Description	Duration (hrs)
A	Move machine to the repair shop	4
B	Order and delivery of replacement parts (delivery by 10 a.m. the following day if ordered by 2:30 p.m.)	4
C	Remove existing foundation on which the machine is mounted	6
D	Repair components requiring no replacement parts	6
E	Repair components that require replacement parts	14
F	Reassemble machine	3
G	Test machine	2
H	Build new foundation	6
I	Move rebuilt machine to new location	2
J	Dismantle machine	2

◆ *Source:* Reprinted from Vaughn, p. 315, with permission from Iowa State Univ. Press.

2. Set up the project activity network diagram for the following relationships.

Activity	Duration	Precedence Relationships
A	5	A is the first activity
B	4	B, C, and D depend on A
C	5	
D	2	
E	7	F and E depend on B
F	6	
G	8	G and H depend on B and C
H	5	
I	2	I and J depend on D and H
J	6	
K	8	K depends on F, G, and I

◆ *Source:* Jewell, p. 357.

3. Consider the five activities A, B, C, D, and E for which the following precedence relationships hold:

- *B* and *C* precede *D*
- *A* and *B* precede *E*

Draw the network diagram for these activities using one or more dummy activities as needed to show the proper relationships.

4. Identify the critical path for the project defined in Exercise 2.

5. Construct the part of a project planning network diagram that represents the following relationships between five tasks of a design project:

Task	Preceding Tasks
A	—
B	—
C	—
D	*A, B, C*
E	*A, B*

6. Consider the project whose network flow diagram is shown below. What is the total float for activity *E*?

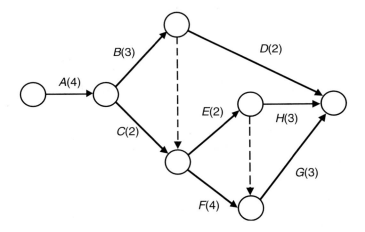

7. As field engineer for a mechanical equipment firm, you are supervising the replacement of a steam boiler and piping distribution system. The activities, durations, and relationships for this project are indicated below. Construct the activity network and identify the critical path.

Activity	Description	Time (weeks)	Preceding Activities
A	Remove old boiler and controls	4	—
B	Install new external steam lines	5	—

Activity	Description	Time (weeks)	Preceding Activities
C	Install new internal steam lines	4	A
D	Construct new footings	5	A
E	Install new thermostats	6	A
F	Install new boiler	4	D,C
G	Install new boiler controls	7	F,B
H	Test boiler and lines for leaks	4	G,E
I	Check thermostat	3	H
J	Install system insulation	2	I
K	Test final system	1	J

♦ *Source:* Walton, p. 70. © 1991. Reprinted with permission from PWS Publishing Co., Boston, a division of International Thomson Publishing, Inc.

8. Determine the critical path for the project described in Exercise 1.

9. Consider the PERT network shown below where the notation *o-m-p* for each task refers to the optimistic, most likely, and pessimistic estimates for the time to complete that task. What is the probability of completing this project within 46 weeks?

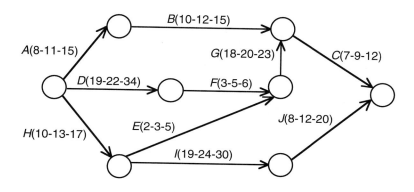

10. The table below lists the expected time (t_e) to complete tasks *A* through *J* of an engineering design project. Also listed are the variances (*var*) of each of those expected times, and the latest start (*LS*) and the earliest start (*ES*) times of each task. What is the probability that this project can be completed within a year (52 weeks)?

Task	t_e	var	ES	LS
A	30.17	26.69	0.00	8.33
B	12.17	0.69	30.17	38.50
C	9.17	0.69	50.67	50.67
D	25.67	13.44	0.00	0.00
E	14.50	1.36	13.17	16.00

(continued)

Task	t_e	var	ES	LS
F	4.83	0.25	25.67	25.67
G	20.17	0.69	30.50	30.50
H	13.17	1.36	0.00	2.83
I	24.17	3.36	13.17	23.00
J	12.67	4.00	37.33	47.17
End			59.83	59.83

11. With respect to Exercise 2, what is the probability that the project will be finished within 30 weeks? Consider the estimates for activity duration in Exercise 2 to be estimates of most likely duration. Assume that the optimistic estimate for the duration of each task is half the estimate of the most likely duration, and the pessimistic estimate for the duration of each task is twice the estimate of the most likely duration.

12. Consider the PERT network shown below where the notation *o-m-p* for each task refers to the optimistic, most likely, and pessimistic estimates for the time to complete that task. What is the probability that this project will be completed in 46 weeks?

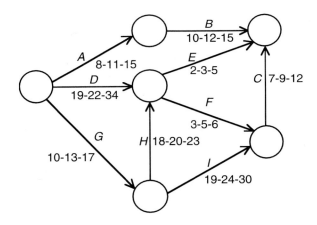

13. Set up an AON network diagram for the project described in Exercise 2.

CHAPTER 8

Engineering Economics

8.1. OVERVIEW

Step 6 in the design process is Evaluating the Alternatives. This is where you use much of your engineering science background to estimate the performance of each design option. Depending on the nature of the design project, you may apply specialized techniques for analyzing electric circuits, bridge structures, chemical reactions, or vibrating springs. In this chapter we concentrate on a set of analytical tools that are used across all engineering disciplines for estimating the economic performance of proposed designs. Unlike the inviolate laws of physics and chemistry that govern the physical behavior of engineering systems, there are a variety of approaches to analyzing a system's economic performance.

The importance of engineering economics to engineering design is underscored by the half-facetious definition of an engineer as someone who can do for $1 what any fool can do for $2.[1] Many talented crafts people with no formal engineering background are capable of building complex technical systems. They rely on experience, intuition, creative talent, and even trial-and-error. Engineers, however, can utilize their professional skills to systematically approach design challenges and arrive at solutions in a much more efficient manner.

Some engineering students question the need to include economics within the scope of an engineering design education. Notwithstanding the inclusion of economics within the ABET definition of engineering design (see Sec. 1.4), the following comments are sometimes offered by engineering students: "I'll get this stuff when I take an economics course—why do we also need to cover it in engineering?"; "Once I'm in industry, I'll just ask the economists or accountants to do the analysis for me." With regard to the first statement, most college-level economics courses deal with larger

[1]Dieter, p. 370.

macroeconomic issues (such as inflation, balance of payments, and money supply) and very rarely focus on the details of comparative project evaluation. With regard to the second comment, many firms do not employ accountants or economists to do this kind of analysis; it is the engineer's responsibility. Even if there were specialists to whom you could turn, you need to know enough about the subject to interact with them. Finally, if you are still not convinced, engineering economics is one of the topical areas included in the Fundamentals of Engineering Examination so you'll need to know it in order to obtain your professional engineer's license.

8.2. ECONOMIC DECISION RULES

In this section we will examine a number of different perspectives that can be used to compare the costs of a set of design alternatives. In order to keep our focus on understanding the methodology, we will examine a relatively simple hypothetical design situation. Later in this chapter, after having introduced the appropriate techniques, we will see how they are applied to a real, much more complex, design study. The context for our hypothetical comparison will be your responsibility as a design engineer to select the electric motors to drive the conveyor belts in a new factory. You have narrowed the choice down to two alternatives: Ajax motors or Blaylock motors.

Ajax motors will cost $30,000 to purchase. The Blaylock motors can be purchased for $20,000. During the expected five-year life of the motors, the Ajax motors will cost $1,000 annually for maintenance; maintenance costs for the Blaylock motors are expected to be $2,000 per year. In addition, the Blaylock motors will have to be rebuilt at the end of the third year at a cost of $3,000. At the end of the fifth year, the Blaylock motors are worthless but the Ajax motors have a salvage value of $4,000.

The Ajax motors are also more energy efficient and, based on the current electricity rates of your local utility, will cost $3,000 a year to run in comparison to the $3,500 annual cost for running the Blaylock motors. Electric rates are expected to increase five percent each year over the next five years. Because the Ajax motors are more reliable, the conveyor belts won't have as much downtime, so annual production output will be $500 higher than with the Blaylock motors.

This problem contains the essential features of the classical engineering economics problem: how to choose between one option with high initial cost and low operating costs (such as the Ajax motors); and another option with low initial cost but high operating costs (such as the Blaylock motors). Before any comparison can be made in a specific situation such as the one described, the engineer has to decide how that comparison is going to be made. In other words, a decision rule, or basis for making the choice, must be adopted. In many companies or government agencies, a particular decision rule is adopted as a matter of company or agency policy; the engineer's role is to apply the rule in the appropriate manner. However, in the consulting engineering field, different clients might prefer different rules; some clients may not even be sure what rule they prefer, or may not even be aware that it makes a difference. In those situations, the engineer's role includes working with the customer to select the best decision rule for the given situation.

In the remainder of this section, we introduce four different decision rules in their most basic form. In subsequent sections, we will consider refinements to these rules and additional considerations which may be needed to conduct a more sophisticated comparison.

Lowest Initial Cost

One approach is to make choices solely on the basis of the initial cost. Such a "lowest initial cost" decision rule might be preferred in situations where the funds available for investment are restricted. In our example problem, application of the "lowest initial cost" rule means that the Blaylock motors (with an initial cost of $20,000) are preferred over the Ajax motors (with an initial cost of $30,000).

Lowest Life-Cycle Cost

Another approach is to treat all costs equally, regardless of when they occur over the expected life of the product or project. That is, the decision rule is to tally up all costs that occur "downstream," or after purchase, and add them to the initial cost. Then select the option with the lowest total cost. To apply the "lowest life-cycle cost" rule to the example problem, we arrange the given data as shown in Table 8–1.

In Table 8–1 we've listed all expenditures as positive numbers and all income as negative numbers. Because the annual maintenance expenditures and annual productivity benefits are the same each year, the five-year totals for those categories are entered as lump sums. However, since the electricity costs change each year, each year's value is entered as a separate item. Since the total life-cycle cost of the Ajax motors is less than that of the Blaylock motors, the Ajax motors are the preferred alternative according to the "lowest life-cycle cost" decision rule.

TABLE 8–1. COMPONENTS OF LIFE-CYCLE COST

	Ajax ($1000)	Blaylock ($1000)
Initial cost	30.00	20.00
Rebuilding at end of 3rd year	—	3.00
Salvage value	−4.00	—
Maintenance (5 years)	5.00	10.00
Productivity benefit (5 years)	−2.50	—
Electricity		
Year 1	3.00	3.50
Year 2	3.15	3.68
Year 3	3.31	3.86
Year 4	3.47	4.05
Year 5	3.65	4.25
Total costs	45.08	52.34

Average Annual Rate of Return

A rate of return decision rule separates initial costs from downstream expenses. This rule examines the extra investment needed to purchase the more expensive option. In this case, the Ajax motors are the more expensive option and the incremental investment needed to purchase them rather than the Blaylock motors is calculated from the first row in Table 8–1 as

$$\text{Incremental Investment} = \text{Initial Cost of More Expensive Option}$$

$$- \text{Initial Cost of Less Expensive Option} \qquad (8\text{–}1)$$

$$= (30.00 - 20.00)(\$10^3) = 10.00\ (\$10^3)$$

The downstream benefits that flow from making this additional investment are then calculated. For this problem, these benefits are found by comparing the Ajax and Blaylock motors for each of the remaining rows in Table 8–1. The results are as follows:

$$\text{Downstream Benefits} = \text{Avoidance of Rebuilding Cost} + \text{Income from Salvage}$$

$$+ \text{Savings on Maintenance Costs over Five Years}$$

$$+ \text{Increased Revenue from Productivity Enhancement}$$

$$+ \text{Annual Savings on Electricity Costs} \qquad (8\text{–}2)$$

$$= [3.00 + 4.00 + (10.00 - 5.00) + 2.50 + (3.50 - 3.00)$$

$$+ (3.68 - 3.15) + (3.86 - 3.31) + (4.05 - 3.47)$$

$$+ (4.25 - 3.65)]\ (\$10^3)$$

$$= 17.26\ (\$10^3)$$

Since these benefits occur over a five year period, the average annual benefit is

$$\text{Average Annual Benefit} = \text{Downstream Benefits/Project Duration}$$

$$= 17.26\ (\$10^3)/5$$

$$= 3.45\ (\$10^3)$$

Then the average annual rate of return is calculated as

$$\text{Average Annual Rate of Return} =$$

$$[\text{Average Annual Benefit/Incremental Investment}]\ 100$$

$$= \left[\frac{\$3,450}{\$10,000}\right] 100$$

$$= 34.5\%$$

Whether or not this 34.5% average annual return on the incremental $10,000 invest-ment is attractive enough to justify making the investment depends on what other in-vestment alternatives are available. For example, investing an extra $10,000 in an ad-vertising program or expanded production capacity may provide higher rates of return. Thus, use of an annual average return rule implies that another option is avail-able to serve as a benchmark for determining an acceptable rate of return.

Payback Period

The average annual rate of return rule averages out the downstream benefits, regard-less of when they occur. The payback period rule focuses only on those downstream benefits which occur chronologically up to the point of "paying back" the incremental initial investment required for the more expensive option over the one with the lower first cost. We invoke this rule by calculating the total benefits at the end of each year until those benefits exceed the value of the incremental investment. For the example problem, we have

$$\text{Benefits in 1st Year} = [(\text{maint})_1 + (\text{prod})_1 + (\text{elec})_1]$$

$$= (1.00 + 0.50 + 0.50) (\$10^3) = 2.00 (\$10^3)$$

This $2,000 benefit received during the first year only partially recoupes the $10,000 incremental investment needed to realize these benefits. So we now consider the ben-efits derived in the second year in addition to those already received in the first year.

$$\text{Total Benefits in 2 Years} = [(\text{yr 1}) + (\text{maint})_2 + (\text{prod})_2 + (\text{elec})_2]$$

$$= (2.00 + 1.00 + 0.50 + 0.53) (\$10^3) = 4.03 (\$10^3)$$

We continue with similar calculations for the third and fourth years:

$$\text{Total Benefits in 3 Years} = [(2 \text{ yrs}) + (\text{maint})_3 + (\text{prod})_3 + (\text{elec})_3 + (\text{rebld})]$$

$$= (4.03 + 1.00 + 0.50 + 0.55 + 3.0) (\$10^3)$$

$$= 9.05 (\$10^3)$$

$$\text{Total Benefits in 4 Years} = [(3 \text{ yrs}) + (\text{maint})_4 + (\text{prod})_4 + (\text{elec})_4]$$

$$= (9.05 + 1.00 + 0.50 + 0.58) (\$10^3) = 11.13 (\$10^3)$$

Thus, the payback period is ~ 3.5 years. Like the rate of return decision rule, use of the payback period decision rule requires that another option be available to serve as a benchmark for determining an acceptable payback period.

8.3. PRESENT VALUE

Among the decision rules examined in the previous section were two extreme cases for handling future expenditures and receipts. The "lowest initial cost" decision rule totally ignored all downstream transactions. On the other hand, the "lowest life-cycle cost" rule treated all downstream transactions on an equal basis as initial costs. In this section we develop a more general approach that incorporates both the lowest initial cost rule and the lowest life-cycle cost rule as special cases.

Time Value of Money

We begin by posing the following question: If I owed you $100, would you rather I give you the money now or a year from now? Even if we removed any element of uncertainty about the delayed payment option (for example, I promise to open an escrow bank account in your name), most people have a preference for receiving the money now. There may be many reasons underlying this preference, but they are equivalent to the opportunity most people have to put that money to work for them (such as depositing it in a bank or buying mutual funds) so that a year from now it would be worth more than $100.

We can quantify the value any person puts on receiving the money now rather than a year from now by asking the question in a slightly different way. Instead of me giving you the $100 now, how much would I have to give you a year from now in order for the delayed payment option to be equally appealing to you (assuming again that we've eliminated all uncertainty from the delayed payment option)? Some individuals might be willing to take a delayed payment of $105. Others might insist on receiving $200 a year from now in exchange for surrendering the immediate receipt of $100. Each individual, in arriving at a quantitative response to the question, is consciously or unconsciously estimating how much return they could get a year from now if they took the $100 now and invested it in the best available investment opportunity.

The annual rate of return, or annual interest rate i, that will convert a present amount P to a future amount F a year from now can be found from

$$F = P(1 + i) \tag{8–3}$$

or

$$i = \frac{F}{P} - 1$$

In the example where an individual required $105 ($F = \$105$) a year from now in order to have the delayed payment option be equivalent to receiving $100 today ($P = \$100$), that person's decision is equivalent to saying that they can invest the $100 at an interest rate of

$$i = \frac{\$105}{\$100} - 1$$

or

$$i = 0.05$$

Present worth factor.   This line of reasoning can be generalized to allow us to convert any future transaction (expenditure or receipt) to an equivalent present amount. If the downstream transaction F occurs n years in the future, Equ. (8–3) takes the form

$$F = P(1 + i)^n \tag{8–4}$$

The most common application of this concept in engineering economics is to convert a future transaction F into an equivalent present value P. So we rewrite Equ.(8–4) in the form

$$P = F(1 + i)^{-n} \tag{8–5}$$

When written in this form, we call i the discount rate. Hence, by definition, the value of the discount rate is equal to the interest rate associated with the best available investment opportunity.

   Equation (8–5) suggests that we define the factor T that transforms the amount F which occurs n years from now into an equivalent present value P using the discount rate i as

$$T_{FP,i,n} = (1 + i)^{-n} \tag{8–6}$$

so that Equ. (8–5) becomes

$$P = F T_{FP,i,n} \tag{8–7}$$

We refer to $T_{FP,i,n}$ as the present worth factor. The notation indicates that this factor transforms F into P using numerical values for the i and n parameters.

   We now return to the task of choosing between the Ajax and Blaylock motors. Assume that your company's accountant has told you to use 20% for the discount rate when calculating present value ($i = 0.20$). Consider the $3,000 item listed in Table 8–1 as the cost to rebuild the Blaylock motors three years from now ($n = 3$). Substituting these values for i and n into Equ. (8–6) gives us

$$T_{FP,0.20,3} = (1 + 0.20)^{-3} = 0.577$$

Then Equ. (8–7) with $F = \$3,000$ provides

$$P = (\$3,000)(0.577) = \$1,730$$

Thus, the $3,000 cost that we expect to incur three years from now to rebuild the Blaylock motors is equivalent to a $1,730 expenditure today. Specifically, if we invested the $1,730 today at an interest rate of $i = 0.20$, that investment will provide us with $3,000 in three years, exactly what we need to rebuild the motors.

   This result, together with a similar calculation for the salvage income received five years from now ($n = 5$) are depicted in the indicated rows of Table 8–2. Note that the columns labeled "Future/Annual Value" in Table 8–2 are working columns to display the value of transactions before they are converted to their equivalent present values.

   Uniform series present worth factor.   We now turn our attention to the maintenance and productivity transactions listed in Table 8–1. Whereas the rebuild and salvage items we just discussed were one-time transactions, the maintenance and

TABLE 8-2. PRESENT VALUE CALCULATION

	Ajax ($1,000)		Blaylock ($1,000)	
	Future/ Annual Value	Present Value	Future/ Annual Value	Present Value
Initial cost		30.00		20.00
Rebuilding at end of 3rd year [from Equ. (8–6), $T_{FP,0.20,3} = 0.577$]	—	—	3.00	1.73
Salvage value [from Equ. (8–6), $T_{FP,0.20,5} = 0.403$]	−4.00	−1.61	—	—
Maintenance [from Equ. (8–8), $T_{UP,0.20,5} = 2.99$]	1.00	2.99	2.00	5.98
Productivity benefit [from Equ. (8–8), $T_{UP,0.20,5} = 2.99$]	−0.50	−1.50	—	—
Electricity [from Equ. (8–9), $T_{GP,0.20,0.05,5} = 3.25$]	3.00	9.75	3.50	11.37
Total present value of costs		39.63		39.08

productivity items reoccur each year. We could apply the present worth factor ($T_{FP,i,n}$) to convert each year's maintenance and productivity transactions to their present values, and then add these converted values to obtain the present value of five years worth of maintenance and productivity transactions. A more efficient approach would be to derive a new factor that can transform the entire series of uniform annual payments U over n years to their present value in one step. This factor is called the uniform series present worth factor and is given by

$$T_{UP,i,n} = \frac{[(1 + i)^n - 1]}{i(1 + i)^n} \tag{8–8}$$

This factor is derived in the Appendix to this section. Both the maintenance and productivity items occur annually over the five year period ($n = 5$), and both are subject to the same discount rate as all other future transactions in this project ($i = 0.20$). The numerical value of this factor is then

$$T_{UP,0.20,5} = \frac{[(1 + 0.20)^5 - 1]}{0.20(1 + 0.20)^5} = 2.99$$

The annual maintenance cost for the Ajax motor is $U = \$1,000$. The present value of five years worth of these costs is

$$P = UT_{UP,0.20,5} = (\$1,000)2.99 = \$2,990$$

Table 8–2 summarizes the application of TUP,i,n to both the maintenance and productivity transactions.

Geometric series present worth factor. The remaining transaction to be converted to its equivalent present value is the electricity payments. These occur annually, but since the annual electricity cost escalates at 5% per year rather than remaining constant, as do the maintenance and productivity transactions, we cannot use $T_{UP,i,n}$. We could rely $T_{FP,i,n}$ to separately discount each year's electricity costs, but is simpler to derive (see the Appendix to this section) a new factor that yields the present value of a set of annual payments that are escalating geometrically. This factor, called the "geometric series present worth factor" and designated as $T_{GP,i,e,n}$ calculates the present value in terms of the first year payment G and the escalation rate e, as well as the discount rate i and period n. The relationship is

$$P = GT_{GP,i,e,n}$$

where

$$T_{GP,i,e,n} = \frac{[(1 + i)^n - (1 + e)^n]}{(i - e)(1 + i)^n} \tag{8-9}$$

The result of using $T_{GP,i,e,n}$ to calculate the present value of the five years worth of electricity costs is shown in Table 8–2.

Application of Least Present Value Decision Rule

All the downstream transactions have now been converted to their equivalent present values. We include the initial investment and add all the items in the Present Value columns of Table 8–2. This gives us the total present value of all costs associated with the Ajax and Blaylock options. Using the decision rule to minimize the present value of costs, the Blaylock motors would be the better choice.

A key factor in this type of analysis is the value of the discount rate. At the beginning of this section we justified the value in terms of the interest that could be earned by investing the available funds in an alternative investment. In a business setting, such as might exist in the context of the Ajax versus Blaylock motor decision, alternative investments might include: expanding the production capacity, increasing the advertising budget, or giving raises to the engineers. In principle, we ought to be able to estimate the expected rate of return on each of those options. The highest rate of return among those alternatives serves as the basis for the discount rate to be used in the motor selection problem.

Under typical business conditions, the rates of return on alternative investments are difficult to estimate with much precision. Some companies have established a corporate-wide discount rate for use in these kinds of projects. In situations where a specific discount rate is not available, there are several options.

One approach is to conduct the comparison over a range of reasonable discount rates to see if the result is sensitive to changes in i. In the Ajax versus Blaylock case we used $i = 0.20$, so we might redo the analysis for $0.17 < i < 0.23$ to see if Blaylock remains the preferred choice for all values of i in that range. Another approach is to note that when $i = 0$, the Ajax motors were the better choice (Table 8–1) but the Blaylock motors are preferred when $i = 0.20$. That means there is some value $0 < i_{equal} < 0.20$ that serves as the border line. If $i < i_{equal}$ then the Ajax motors are the best choice; otherwise Blaylock are the preferred motors. With this approach, we don't need to worry too much about the exact value of i, just its value relative to i_{equal}.

The choice of i, like so many of the subjective judgments that enter into the engineering design process, is subject to politicization. "When the U.S. Federal Highway Administration promulgated a regulation in the early 1970s that the discount rate for all federally funded highways would be zero, this was widely interpreted as a victory for the cement industry over asphalt interests. Roads made of concrete cost significantly more than those made of asphalt, but require less maintenance and less frequent replacement."[2]

Appendix 8: Deriving the Present Worth Factors

Earlier in this section we introduced several factors for transforming a series of transactions into their equivalent present values in a single step. Here we derive the expressions for those transformation factors, starting with the uniform series present worth factor.

Uniform series present worth factor. Consider the need to set money aside to replace an expensive piece of equipment when that equipment reaches the end of its useful life, n years from now. The plan is to invest an amount U at the end of each year at an interest rate i. At the end of the first year, the amount available in the account is simply the amount just invested:

$$F_1 = U$$

At the end of the second year, the amount in the account has earned interest for a year. In addition, we make another of our annual deposits of amount U. So the total amount of money in the account at the end of the second year is

$$F_2 = U(1 + i) + U$$

At the end of the third year, the amount in the account at the end of the second year has earned interest for a year. In addition we make another of our annual deposits of amount U. So the total amount of money in the account at the end of the third year is

$$F_3 = [U(1 + i) + U](1 + i) + U$$

or

$$F_3 = U(1 + i)^2 + U(1 + i) + U$$

At the end of the nth year, the account has accumulated

$$F = U(1 + i)^{n-1} + U(1 + i)^{n-2} + \ldots + U(1 + i)^2 + U(1 + i) + U \quad (8\text{--}10)$$

[2]Shtub, p. 55.

In order to rewrite the right hand side of this equation in a more compact form, we first multiply each term in Equ. (8–10) by $(1+i)$ to get

$$F(1 + i) = U(1 + i)^n + U(1 + i)^{n-1} + \ldots + U(1 + i)^3 + U(1 + i)^2 + U(1 + i) \quad (8\text{–}11)$$

Subtracting Equ. (8–10) from (8–11) leaves

$$F(1 + i) - F = U(1 + i)^n - U$$

which can be solved for F as

$$F = U \frac{[(1 + i)^n - 1]}{i} \quad (8\text{–}12)$$

This leads us to the definition of the future worth uniform series factor (the inverse of which is the sinking fund factor) as

$$T_{UF,i,n} = \frac{[(1 + i)^n - 1]}{i}$$

This is interesting, but not quite what we set out to find. However, we note that Equ. (8–12) gives us the value of our equipment replacement fund n years from now. What we are after is the present value of that future amount. To get it, we simply apply Equ. (8–5) but where F is given in terms of U by Equ. (8–12). This leads us to

$$P = F(1 + i)^{-n} = U \frac{[(1 + i)^n - 1]}{i} (1 + i)^{-n} = U \frac{[(1 + i)^n - 1]}{i(1 + i)^n}$$

We recognize the term on the right hand side that transforms U into P as the uniform series present worth factor introduced earlier as Equ. (8–8).

Geometric series present worth factor.   Let the annual operating cost for the first year of an activity be G, and let e be the annual rate at which this cost increases. At the end of the first, second, and nth years, the respective operating costs are

$$G_1 = G$$

$$G_2 = G(1 + e)$$

$$G_n = G(1 + e)^{n-1}$$

The present value of each of these annual expenditures can be discounted at the rate i from the year in which they occur to get their equivalent present values as

$$P_1 = \frac{G}{(1 + i)}$$

$$P_2 = \frac{G(1 + e)}{(1 + i)^2}$$

$$P_n = \frac{G(1 + e)^{n-1}}{(1 + i)^n}$$

The total present value of all the annual expenditures over the period of n years is

$$P = P_1 + P_2 + \ldots + P_n = G\left[\frac{1}{(1+i)} + \frac{(1+e)}{(1+i)^2} + \ldots + \frac{(1+e)^{n-1}}{(1+i)^n}\right] \tag{8–13}$$

In preparation for simplifying this, we multiply both sides by $(1+e)/(1+i)$ to get

$$P\left(\frac{1+e}{1+i}\right) = G\left[\frac{(1+e)}{(1+i)^2} + \frac{(1+e)^2}{(1+i)^3} + \ldots + \frac{(1+e)^n}{(1+i)^{n+1}}\right] \tag{8–14}$$

Now we can subtract Equ. (8–13) from Equ. (8–14) to obtain

$$P\left(\frac{1+e}{1+i}\right) - P = G\left[\frac{(1+e)^n}{(1+i)^{n+1}} - \frac{1}{1+i}\right]$$

This can be factored as

$$P\left[\left(\frac{1+e}{1+i}\right) - 1\right] = \frac{G}{1+i}\left[\frac{(1+e)^n}{(1+i)^n} - 1\right]$$

which simplifies to

$$P = G\left[\frac{(1+i)^n - (1+e)^n}{(i-e)(1+i)^n}\right] \tag{8–15}$$

The expression in brackets is the geometric series present worth factor introduced in Equ. (8–9) as $T_{GP,i,e,n}$. By inspecting the denominator of Equ. (8–9), it is clear the expression does not apply when $i = e$. To handle that special case, we go back to Equ. (8–13) and set $e = i$ to get n identical terms inside the brackets in the form

$$P = G\left[\frac{1}{(1+i)} + \frac{1}{(1+i)} + \ldots + \frac{1}{(1+i)}\right]$$

or

$$P = G\frac{n}{(1+i)}$$

8.4. ANNUALIZED COSTS

In Section 8.3, we converted the value of all downstream transactions into an equivalent present value. In a slight variation of that approach, we can convert all transactions into equivalent uniform annual costs. Consider the Ajax versus Blaylock motor for which the present value calculations are summarized in Table 8–2. In that table, we used a working column as a temporary holding place for all future and annual transactions before converting them to equivalent present values using the appropriate present value factors. A similar table for the same problem analyzed from an equivalent annual cost basis is provided as Table 8–3. Here our working column is a temporary holding place for all future and present transactions before converting them to equivalent annual values using the appropriate annual value factors.

TABLE 8–3.   ANNUAL VALUE CALCULATION

	Ajax ($1,000)		Blaylock ($1,000)	
	Future/ Present Value	Annual Value	Future/ Present Value	Annual Value
Initial cost [from Equ. (8–16), $T_{PU,0.2,5} = 0.334$]	30.00	10.02	20.00	6.68
Rebuilding at end of 3rd year [from Equ. (8–20), $T_{FU,0.20,3} = 0.194$]	—	—	3.00	0.58
Salvage value [from Equ. (8–20), $T_{FU,0.20,5} = 0.134$]	−4.00	−0.54	—	—
Maintenance		1.00		2.00
Productivity benefit		−0.50	—	—
Electricity [from Equ. (8–21), $T_{GU,0.2,5} = 1.086$]	3.00	3.26	3.50	3.80
Total equivalent annual costs		13.24		13.06

We begin our explanation of the entries in Table 8–3 with the annual maintenance expenses and the annual productivity benefits. Since they are already in the desired form, we enter them directly into the annual value columns. Now let's turn to the initial cost terms. Those, of course, are in the form of present values; and we want to convert them to equivalent annual values.

Capital Recovery Factor

Recall from Equ. (8–8) that we introduced the uniform series present value factor $T_{UP,i,n}$ to transform a uniform annual payment of amount U to an equivalent present value P. Clearly, converting a present value P to an equivalent uniform annual amount U is just the inverse transformation and thus requires application of the inverse factor $T_{PU,i,n}$ defined as

$$T_{PU,i,n} = \frac{1}{T_{UP,i,n}} = \frac{i(1+i)^n}{[(1+i)^n - 1]} \qquad (8\text{–}16)$$

This is known as the capital recovery factor. It transforms an up-front transaction to an equivalent set of uniform annual transactions. This is how home mortgage payments are calculated. For this problem, we can either pay $30,000 up front for the Ajax motors, or borrow the money at 20% interest ($i = 20$) and make five ($n = 5$) annual payments of

$$U = (P)(T_{PU,0.20,5}) = (\$30,000)\frac{0.20(1 + 0.20)^5}{[(1 + 0.20)^5 - 1]} = (\$30,000)(0.334) = \$10,020$$

This result, along with the corresponding result for the Blaylock motors are indicated in Table 8–3.

Sinking Fund Factor

The rebuilding and salvage value categories involve single transactions, as does the initial cost, but they occur downstream rather than at the beginning of the project. We use a combination of our previously derived factors to obtain the factor for converting these transactions into equivalent annual payments. Let's focus first on the salvage value transaction. We can think of converting this to a set of equivalent uniform annual costs in two stages. First we convert the future value F which occurs in the fifth year to an equivalent present value by using the present worth factor defined by Equ. (8–6) to get

$$P = FT_{FP,i,n} = F(1 + i)^{-n} \tag{8–17}$$

where, in this case, $F = -\$4,000$, $i = 0.20$, and $n = 5$. The next step is to use the capital recovery factor defined by Equ. (8–16) to transform the present value P to the uniform annual series U according to

$$U = PT_{PU,i,n} = P\frac{i(1 + i)^n}{[(1 + i)^n - 1]} = F(1 + i)^{-n}\frac{i(1 + i)^n}{[(1 + i)^n - 1]} = F\frac{i}{[(1 + i)^n - 1]}$$

where we have used Equ. (8–17) to substitute for P. This result leads us to define the sinking fund factor as

$$T_{FU,i,n} = \frac{i}{[(1 + i)^n - 1]} \tag{8–18}$$

From an investment perspective, this factor tells us how much money, U, we need to invest each year over n years at an interest rate i in order to realize an amount F in the nth year. We are using the sinking fund factor in this problem from the slightly different perspective of transforming a known income F in the nth year into an equivalent uniform annual income stream, spread out over n years. When applied to the salvage value for the Ajax motors, it means that receiving the $4,000 in the fifth year is equivalent to receiving

$$U = F\frac{i}{[(1 + i)^n - 1]} = 4.00\frac{0.20}{[(1 + 0.20)^5 - 1]} = 0.538$$

or $538 dollars each year over the five year period. This value is displayed in Table 8–3.

   We also want to apply the sinking fund factor to convert the rebuilding expense in the third year to an equivalent uniform series of annual payments over the project lifetime. Here we encounter a slight complication. The salvage value transaction occurred at the end of the project so that the value of $n = 5$ represented both the year in which the transaction occurred and the period of analysis. For the rebuilding expenditure, the cost is incurred in the third year ($n = 3$) of a five year project ($n = 5$). To understand how to handle this, we retrace the two-phase process we used above to arrive at Equ. (8–18).

First we convert the future value F which occurs in the third year to an equivalent present value by using the present worth factor defined by Equ. (8–6) to get

$$P = FT_{FP,i,m} = F(1 + i)^{-m} \tag{8–19}$$

where we use m to identify when the transaction takes place. In this case, $F = \$3,000$, $i = 0.20$, and $m = 3$. The next step is to use the capital recovery factor defined by Equ. (8–16) to convert the present value P to the uniform annual series U over a period of n years according to

$$U = PT_{PU,i,n} = P\frac{i(1 + i)^n}{[(1 + i)^n - 1]} = F(1 + i)^{-m}\frac{i(1 + i)^n}{[(1 + i)^n - 1]} = F\frac{i(1 + i)^{n-m}}{[(1 + i)^n - 1]}$$

where we have used Equ. (8–19) to substitute for P. This result leads us to define a generalized sinking fund factor as

$$T_{FU,i,n,m} = \frac{i(1 + i)^{n-m}}{[(1 + i)^n - 1]} \tag{8–20}$$

Of course, if $n = m$, Equ. (8–20) reduces to Equ. (8–18). The numerical values associated with the rebuilding cost are shown in Table 8–3.

Geometric Series Sinking Fund Factor

The remaining term in the analysis is the electricity cost. This is a sequence of geometrically increasing annual transactions that we want to convert to an equivalent stream of uniform annual payments. To do this, we follow the same two-step conversion process we used above to combine Equ. (8–9) and the inverse of Equ. (8–8). This yields

$$T_{GU,i,e,n} = \frac{i[(1 + i)^n - (1 + e)^n]}{[(1 + i)^n - 1](i - e)} \tag{8–21}$$

The numerical results for $i = 0.2$, $n = 5$, and $e = 0.05$ are shown in Table 8–3.

Application of Least Annual Cost Decision Rule

Summing the entries in the Annual Values columns of Table 8–3 reveals that the Blaylock motors have lower annual costs than the Ajax motors. Therefore, the Blaylock motors are the preferred choice. This is the same conclusion we reached in Section 8.3 using the present value decision rule.

As long as all the parameters (i, n, e, F, etc.) are the same, the least annual cost and least present value decision rules will always lead to the same conclusions. One consideration that may determine which approach to use is the number of conversions that are required. If we count each cost category for each design option, there are nine cost categories in the Ajax/Blaylock problem. Using the present value approach, we performed seven conversions (the only two categories that did not have to be con-

verted were the initial costs of the two motors). Since we performed only six conversions using the annual cost approach, the annual cost method required less algebra for
this problem. In general, examining the nature of each cost category will indicate
which approach will be simpler.

Perhaps more important than the algebraic complexity is the preference of your
client. Some individuals and companies have a preference for expressing the cost
comparison in either the present value or annual cost framework.

8.5. ECONOMIC EVALUATION OF CHITTENDEN LOCKS PROJECT

In this section we describe how the engineering economics concepts discussed in this
chapter are applied to a real engineering design situation. The project we will use for
this purpose is the Chittenden Navigation Locks operated by the U.S. Army Corps of
Engineers. These locks are located in Seattle, WA and are part of the Lake Washington Ship Canal that connects saltwater Puget Sound with freshwater Lake Union and
Lake Washington (see Figure 8–1). In 1997, ASCE designated the locks as a National
Civil Engineering Historic Landmark.

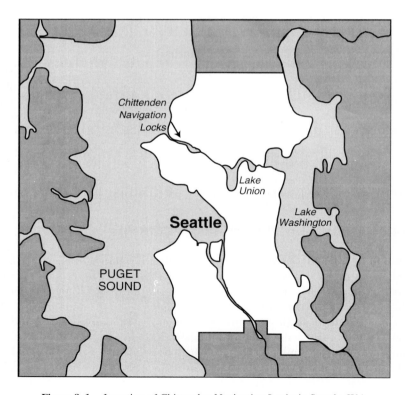

Figure 8–1.   Location of Chittenden Navigation Locks in Seattle, WA

Figure 8–2 is a view looking east over the facility, which consists of a large lock that is 825 ft. long and 80 ft. wide, and an adjacent small lock that is 150 ft. long and 30 ft. wide. Note that several boats are in the small lock while the large lock is empty. Each lock contains several gates that close to allow flooding to lift inbound boat traffic to the higher level freshwater and to allow draining to lower outbound traffic to the saltwater level. The large lock has three gates; any two of which can be operated in tandem to accommodate the amount of boat traffic. As shown in Figure 8–2, the large lock has the western most gate open and the other two gates closed.

Gate Operating Machinery

Each gate consists of two hinged doors, or leaves. When the Chittenden Locks were opened to boat traffic in 1916, cables were used to open and close each leaf. In 1933, the cables on each leaf were replaced by a 20 hp motor that drives a complex gear arrangement and a hinged arm that is attached to the top of the leaf. Part of the engineering drawing for this mechanism is reproduced in Figure 8–3. In Figure 8–3, the

Figure 8–2.   Aerial View of Chittenden Locks *(U.S. Army Corps of Engineers)*

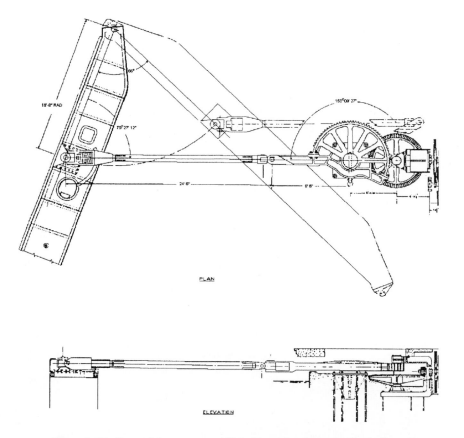

Figure 8–3.   General Arrangement of Machinery for Operating Lock Gates

hinged arm is shown in its fully extended position keeping the gate leaf closed. Note the shock absorber mounted on the end of the arm attached to the leaf. Figure 8–3 also shows in "phantom" lines the retracted position of the arm when the gate leaf is open. Many of the components of the gate opening system have deteriorated over the years and maintenance and repair costs have increased substantially.

Rehabilitation Options

In the early 1990s, the Corps conducted a feasibility study of a major rehabilitation of the large lock gates. The study examined five alternatives:[3]

---

[3]U.S. Army Corps of Engineers, p. 6.

1. Continue operating the locks in their current condition
2. Replace gate opening system by an all-hydraulic system
3. Repair current system
4. Repair current system and replace existing motors with variable speed drive motors
5. Repair current system and add a fluid coupling

Options 1, 3, and 4 are self-descriptive. Option 2, the all-hydraulic system, involves replacing the entire gear system and arm mechanism by a hydraulic piston with associated motors, pumps, and other hydraulic components. Option 5 replaces two sets of gears by a fluid coupling.

A life-cycle cost analysis was conducted for each of the alternatives. The period of analysis was 60 years (the estimated remaining life of the system), and a discount rate of 8% was used. The summary table for the all-hydraulic option (the option that was ultimately selected) is reproduced here as Table 8–4. While the format of the table is similar to that of our tables earlier in this chapter, let's explore the significance of various items in this table.

The items in the "Projected Cost" column are the undiscounted cost estimates of the major capital expenditures. The year in which each expenditure occurs is indicated in the left-most column. The present value of those costs, calculated using Equ. (8–6), are shown in the right-most column.

As shown in Figure 8–2, the large lock has three gates, each with two leaves. In order to keep the large lock functioning during the proposed project, the Corps plans to rehabilitate one gate each year over a three-year period. The cost for each gate and the associated machinery is estimated at $552,000. The cost of the first gate includes an additional $135,000 for engineering design work and preparation of plans and specifications. The total cost for the three gates has a present value of $1,671,000, as summarized toward the bottom of Table 8–4.

Periodic maintenance expenditures include rewinding the motors in years 10 and 20, replacing the motors in year 30, and rewinding the new motors in years 40 and 50. Limit switches are replaced every 10 years, major control components every 20 years, and pumps and tanks also every 20 years. The transformers and magnetic starters are replaced in year 30. These costs have a present value of $42,000.

The total present value of the new gate machinery plus the periodic maintenance items is $1,671,000 + $42,000 = $1,713,000. Using Equ. (8–16), this is converted into an equivalent annual cost of $138,000 over the life of the project. Adding in the routine annual maintenance costs of $8,000 provides a final annual cost estimate of $146,000. See Section 8.10 for a more detailed breakdown of the cost estimates for the all-hydraulic option.

Tables similar to Table 8–4 are included in the Corps' report for each of the other options. The annual cost estimates from those tables for each option are summarized in Table 8–5. The cost estimate for the "current condition" option is based on

TABLE 8–4.   LIFE-CYCLE COST ANALYSIS OF ALL-HYDRAULIC OPTION

(May 1994 Prices, 8% Discount Rate)

Project Year	Type of Expenditure	Projected Cost ($1,000)	Present Value ($1,000)
0	One gate, two machinery trains, plans & specs.	687	687
1	One gate, two machinery trains	552	511
2	One gate, two machinery trains	552	473
10	Rewind motors	11	
10	Replace limit switches	28	18
20	Rewind motors	11	
20	Replace limit switches	28	
20	Replace major control components	21	
20	Replace pumps and tanks	6	14
30	Replace limit switches	28	
30	Replace 1800 rpm motors	12	
30	Replace transformers	5	
30	Replace mag. starters (6)	12	6
40	Rewind motors	11	
40	Replace limit switches	28	
40	Replace major control components	21	
40	Replace pumps and tanks	6	3
50	Rewind motors	11	
50	Replace limit switches	28	1
			1,713

Present value of rehabilitation	$1,671,000
Present value of periodic maintenance	$42,000
Total present value	$1,713,000
Annual cost	
Interest & amortization	
(60 years @ 8.00%)	$138,000
Routine O&M	$8,000
Total annual cost	$146,000

Source: U.S. Army Corps of Engineers, Appendix C.

maintenance and repair costs associated with routine operations of a deteriorating system plus an assumed gate failure every two years.

Clearly, options 2 and 3 are more cost-effective than the other three options. It is not clear whether the difference between option 2 and 3 is sufficient to justify selecting option 2. As it turns out, additional considerations, besides cost, were factored into the decision making process; these are discussed in Section 9.2.

TABLE 8–5.   ANNUAL COST FOR FIVE DESIGN OPTIONS

Option	Annual Cost ($1,000)
1. Current condition	163
2. All-hydraulic system	146
3. Fix existing	148
4. Add variable speed drive	191
5. Add fluid coupling	168

Source: U.S. Army Corps of Engineers, p. 6.

8.6. UNEQUAL LIFETIMES

Our comparative economic evaluations up to this point have been for situations in which each of the options have the same lifetimes. This condition does not always exist, and when it doesn't the question arises of how to deal with unequal lifetimes. Consider the situation we would find ourselves in if the Ajax motors discussed earlier had a six-year lifetime and the Blaylock motors had a four-year lifetime instead of the five-year lifetimes used for both in the original problem statement. To help us focus on the unequal lifetime issue, we will simplify the Ajax/Blaylock motor problem by eliminating the rebuilding and productivity benefit items from our modified analysis.

One way we could proceed is to analyze each option over its lifetime (six years for Ajax; four years for Blaylock). Such a modified present value calculation is summarized in Table 8–6. This somehow doesn't seem fair. For example, even though

TABLE 8–6.   PRESENT VALUE CALCULATION FOR UNEQUAL LIFETIMES

	Ajax ($n = 6$) ($1,000)		Blaylock ($n = 4$) ($1,000)	
	Future/ Annual Value	Present Value	Future/ Annual Value	Present Value
Initial cost	—	30.00	—	20.00
Salvage value ($T_{FP,0.2,6} = 0.335$)	−4.00	−0.34	—	—
Maintenance ($T_{UP,0.2,6} = 3.326$) ($T_{UP,0.2,6} = 2.589$)	1.00 —	3.33 —	— 2.00	— 5.18
Electricity ($T_{GP,0.2,0.05,6} = 3.675$) ($T_{GP,0.2,0.05,4} = 2.759$)	3.00 —	11.02 —	— 3.50	— 9.66
Total present value of costs		43.01		34.84

Ajax's electricity costs are less each year then Blaylock's, Ajax's total electricity costs over its six-year life are greater than Blaylock's total electricity costs over its four-year life. In essence, we are penalizing the Ajax motors for having a longer lifetime.

Another way we could compare the motors is to use four years as the analysis period for both options. If we do so, we can account for all the downstream costs of the Blaylock motors. However, we couldn't give the Ajax motors any credit for their salvage value (which doesn't become available until the sixth year). Neither can we help offset Ajax's higher initial costs with the savings in maintenance and electricity during the fifth and sixth year.

If we try to rectify this imbalance by using a six-year analysis period, other difficulties arise. Specifically, we would have to invest $20,000 at the end of the fourth year in a new set of Blaylock motors. But then the Blaylock motors would still have 50% of their useful life remaining at the end of the analysis period; a capability for which they would get no credit. We need to develop an approach that provides a "level playing field" for all options under consideration. Let's review three different approaches to handling unequal lifetimes in a manner that treats all options fairly.

Common Multiple of Lifetimes

One approach is to select an analysis period that is a common multiple of the lifetimes of all the options. In this case, we can conduct the comparison over a twelve-year analysis period. We summarize the comparative analysis in Table 8–7. After the fourth and eighth years, we would have to purchase new sets of Blaylock motors, and after the sixth year we would have to buy a new set of Ajax motors. But at the end of

TABLE 8–7. PRESENT VALUE CALCULATION FOR COMMON MULTIPLE OF LIFETIMES ($n = 12$)

	Ajax ($1,000)		Blaylock ($1,000)	
	Future/ Annual Value	Present Value	Future/ Annual Value	Present Value
Capital cost (at $n = 0$)	—	30.00	—	20.00
Capital cost (at $n_1 = 4$) ($T_{FP,0.2,4} = 0.482$)	—	—	20.00	9.64
Capital cost (at $n_2 = 6$) ($T_{FP,0.2,6} = 0.335$)	30.00	10.05	—	—
Capital cost (at $n_3 = 8$) ($T_{FP,0.2,8} = 0.233$)	—	—	20.00	4.65
Salvage value (at $n_2 = 6$) ($T_{FP,0.2,6} = 0.335$)	−4.00	−1.34	—	—
Salvage value (at $n_4 = 12$) ($T_{FP,0.2,12} = 0.112$)	−4.00	−0.45	—	—
Maintenance ($T_{UP,0.2,12} = 4.439$)	1.00	4.44	2.00	8.88
Electricity ($T_{GP,0.2,0.05,12} = 5.324$)	3.00	15.97	3.50	18.63
Total present value of costs		58.67		61.80

the twelfth year, there would be no unexpired lifetime for either option, and there would be no residual costs or incomes for either option that are unaccounted for. Based on the total present value of all costs over the twelve-year analysis period, Table 8–7 leads us to conclude that Ajax is the better choice.

Annualized Costs

A second approach is to express all costs as equivalent annual costs instead of present value costs. In many cases, that means using the inverse of the T_{UP}, T_{FP}, and T_{GP} factors used in the present value calculation. The differences in lifetimes between the candidate motors then requires no special treatment and we can use an annualized cost comparison similar to the one we did in Section 8.4 for a five-year analysis period. The results are shown in Table 8–8. While the bottom line numbers are different from the bottom line numbers in Table 8–7, the conclusion is the same—Ajax is the preferred option.

Note that under this framework for dealing with unequal lifetimes, we don't have to explicitly deal with the replacement transactions at the end of each useful life cycle, since those transactions, whenever they occur, are now imbedded in the value of the the uniform annual equivalent costs.

TABLE 8–8.   ANNUALIZED VALUE CALCULATION FOR UNEQUAL LIFETIMES

	Ajax ($n = 6$) ($1,000)		Blaylock ($n = 4$) ($1,000)	
	Future/ Present Value	Annual Value	Future/ Present Value	Annual Value
Initial cost				
$(T_{PU,0.2,6} = 0.301)$	30.00	9.02	—	—
$(T_{PU,0.2,4} = 0.386)$	—	—	20.00	7.73
Salvage value	−4.00	−0.40	—	—
$(T_{FU,0.2,6} = 0.101)$				
Maintenance	—	1.00	—	2.00
Electricity				
$(T_{GU,0.2,0.05,6} = 1.105)$	3.00	3.32	—	—
$(T_{GU,0.2,0.05,4} = 1.066)$	—	—	3.50	3.73
Total equivalent annual costs		12.94		13.46

Conversion of Present Values

A third approach takes one of the present value results listed in Table 8–6, say the present value for Ajax motors calculated on the basis of $n = 6$, and converts it into an equivalent present value for Ajax based on $n = 4$. This adjusted present value for Ajax can then be compared to the present value of the Blaylock option, also calculated on the basis of $n = 4$. The first step in this appoach makes use of the capital recovery factor defined by Equ. (8–16) to transform the present value ($P_6 = 43.01$) associated with $n = 6$ into an equivalent series of uniform annual payments.

$$U = P_6 T_{PU,0.2,6} = 43.01 \cdot 0.301 = 12.93 \tag{8–22}$$

Now this value of U can be converted back into a present value based on any other value of n using the uniform series present worth factor T_{UP}. For $n = 4$, the conversion of the Ajax present value is

$$P_4 = U T_{UP,0.2,4} = 12.93 \cdot 2.589 = 33.48 \tag{8–23}$$

Comparing this to the Blaylock present value of 34.84 (also based on $n = 4$) listed in Table 8–6, we see that Ajax is the preferred choice. This result is consistent with the results of the other two approaches summarized in Tables 8–7 and 8–8. Rather than convert the Ajax present value from $n = 6$ to $n = 4$, we can apply the same procedure to convert the Blaylock present value from $n = 4$ to $n = 6$.

In either case we can combine Equs. (8–22) and (8–23) together with the relationship $T_{UP} = 1/T_{PU}$ to get

$$P_6 T_{PU,0.2,6} = P_4 T_{PU,0.2,4}$$

In general, the conversion formula to convert the present value P_{n_1} for analysis period n_1 into an equivalent present value P_{n_2} for analysis period n_2 is

$$P_{n_1} T_{PU,i,n_1} = P_{n_2} T_{PU,i,n_2} \tag{8–24}$$

8.7. THE EFFECT OF TAXES

Corporations are subject to income taxes just as individuals are. Because income tax rules treat different types of expenditures differently, they can influence the relative economic attractiveness of design options. For example, we know that for U.S. federal individual income tax purposes, a $500 monthly rental payment on an apartment is treated differently than a $500 monthly payment on a home mortgage loan. Specifically, the mortgage interest component of the latter payment is tax deductible, while none of the rental fee is deductible. Thus, the U.S. federal income tax system provides a strong incentive to individuals to purchase their own home rather than to rent. In this section we will examine the key aspects of the U.S. corporate income tax system that may significantly affect design decisions. We start by illustrating in Figure 8–4 the basic elements of income that affect tax obligations.

At the left side of Figure 8–4 we display the total annual income received by the corporation. As with individual income taxes, the corporate income tax system allows companies to take deductions on its total income to come up with the amount of taxable income. There are two main kinds of deductions available to corporations. First, companies can deduct all qualifying operating expenses from their total income. This

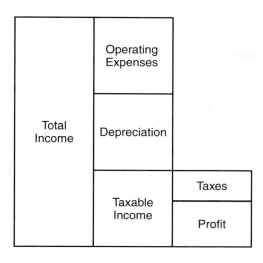

Figure 8–4. Depreciation, Taxable Income, and Taxes. *(Adapted from Dieter p. 386, with permission from McGraw-Hill Companies.)*

includes all regularly recurring expenses such as payroll, electricity and other utility bills, and consumable supplies such as pencils and paper. The second kind of deduction is called depreciation. This represents an allowance for the fact that industrial equipment and property lose value over time and eventually have to be replaced. We will examine the specific formulas for calculating depreciation. For now, it is sufficient to note that the taxable income is determined by subtracting operating expenses and depreciation from total income.

The taxable income is that amount of total annual income that is subject to income taxes. The amount of taxes due on the taxable income depends on the structure of the tax code, but is usually determined by a formula that includes a percentage of taxable income, with the numerical value of the percentage based on the income bracket the company falls into.

After the taxes are subtracted from the taxable income, the remaining funds are the company's profit. Before looking more closely at both the depreciation and tax structures, it is useful to keep in mind that federal tax laws were established by the federal government and are subject to revision at any time by the government. Several major changes in U.S. corporate income tax rules have occurred since 1980. While some of the numerical details described in our discussion may be modified in the future, the basic concepts are likely to remain relevant.

Depreciation

Any depreciation calculation has two components. The first is the period over which the capital investment is depreciated. This may depend on the nature of the investment. For example, according to U.S. federal tax rules, investments in telephone equipment have a depreciation period of fifteen years, while investments in computers are depreciated over five years. The second element of the calculation is the amount of depreciation that can be claimed in each year.

MACRS.   The depreciation formula currently most commonly used is the Modified Accelerated Cost Recovery System (MACRS). The percentage of initial investment that can be depreciated each year under MACRS is listed in Table 8–9 for depreciation periods of 3, 5, 7, 9, 10, and 15 years. Because MACRS assumes that investments are made in the middle of the year, only half a year's worth of depreciation is allowable in the first year. At the end of the depreciation period, another half-year's worth of depreciation is available. Note that the numbers in each column add up to 1, meaning the entire investment is depreciated even though there may be a salvage value in some cases at the end of the depreciation period.

Straight line.   The current tax laws allow depreciation to be calculated on a straight line basis as an alternative to MACRS. This method takes the difference between the initial investment and any salvage value, and divides by the depreciation period to get a uniform depreciation amount for each year of the depreciation period.

Tax Rates

Once we know how much depreciation we can claim in a given year, we next have to apply the tax rates to the allowable depreciation to determine the tax implications of the capital cost of the design alternatives. The U.S. federal tax rates depend on the taxable income of the firm according to Table 8–10.

TABLE 8–9.   MACRS DEPRECIATION SCHEDULE

Year	Depreciation Period (years)				
	3	5	7	10	15
1	0.333	0.200	0.143	0.100	0.050
2	0.445	0.320	0.245	0.180	0.095
3	0.148	0.192	0.175	0.144	0.095
4	0.074	0.115	0.125	0.115	0.077
5	—	0.115	0.089	0.092	0.069
6	—	0.058	0.089	0.074	0.062
7	—	—	0.089	0.066	0.059
8	—	—	0.045	0.066	0.059
9	—	—	—	0.065	0.059
10	—	—	—	0.065	0.059
11	—	—	—	0.033	0.059
12–15	—	—	—	—	0.059
16	—	—	—	—	0.030

Source: Dieter, p. 384

TABLE 8–10. MARGINAL TAX RATES

Taxable Income ($)	Marginal Tax Rates
<50,000	0.15
50,000–75,000	0.25
75,000–10,000	0.34
100,000–350,000	0.36
>350,000	0.34

Source: Dieter, pp. 385–386.

Effect on Design Choice

Let us examine the effects of the federal income tax structure on the choice between the Ajax and Blaylock motors. We will use the present value decision rule with a 20% discount rate. Without tax effects, the economic analysis is summarized in Table 8–2. To incorporate the tax impacts, we will use the MACRS depreciation schedule summarized in Table 8–9 and assume that both motors have a five-year depreciation period. Further we will assume that your company's taxable income this year is expected to be approximately $500,000. According to Table 8–10, the firm is then in the 34% tax bracket.

The top half of Table 8–11 repeats the data from Table 8–2 for convenience. The tax effects displayed in the lower half of Table 8–11 include the tax implications for capital investments and operating costs. First we list the depreciation that can be claimed in each year's tax returns over the next six years, according to MACRS. These are negative quantities because they lead to income tax deductions; they effectively reduce the costs of purchasing the motors. The actual tax deduction in each year is obtained by multiplying that year's depreciation by the tax rate, 34% in this case. Those tax benefits are listed immediately below the corresponding depreciation amounts. Since these benefits are not realized until the end of each year for which they are claimed, they have to be discounted to convert them to equivalent present values.

While the tax benefits of capital investments are spread out over the depreciation period, all operating costs can be deducted during the year in which they occur. For this problem, all maintenance, electricity, and rebuilding expenses are treated as operating costs and are deducted at the 34% tax rate. Since the present value of those costs have already been listed in the upper part of Table 8–11, we apply the 0.34 factor directly to those values. We were not able to use a similar approach for the tax effects of the capital investments because the chronological pattern of the tax deductions was different from the chronological pattern of the expenditures.

Finally, we have to recognize that the productivity benefit and the salvage value terms provide income to the company. We have to pay tax on that income at the 34% rate. That extra tax obligation is treated as an increased cost in the year in which it is received. So the 0.34 factor is applied directly to the corresponding discounted values that appear in the upper part of Table 8–11.

TABLE 8–11.   EFFECT OF INCOME TAXES ON ECONOMIC ANALYSIS

	Ajax ($1,000)		Blaylock ($1,000)	
	Future/ Annual Value	Present Value	Future/ Annual Value	Present Value
Initial cost	30.00	30.00	20.00	20.00
Rebuilding at end of 3rd year ($T_{FP,0.2,3} = 0.577$)	—	—	3.00	1.73
Salvage value ($T_{FP,0.2,5} = 0.403$)	–4.00	–1.61	—	—
Maintenance ($T_{UP,0.2,5} = 2.99$)	1.00	2.99	2.00	5.98
Productivity benefit ($T_{UP,0.2,5} = 2.99$)	–0.50	–1.50	—	—
Electricity ($T_{GP,0.2,0.05,5} = 3.25$)	3.00	9.75	3.50	11.37
Depreciation*				
1st year @ 0.200	–6.00		–4.00	
2nd year @ 0.320	–9.60		–6.40	
3rd year @ 0.192	–5.76		–3.84	
4th year @ 0.115	–3.45		–2.30	
5th year @ 0.115	–3.45		–2.30	
6th year @ 0.058	–1.74		–1.16	
Tax benefits from deprec. @ 0.34				
1st year ($T_{FP,0.2,1} = 0.833$)	–2.04	–1.70	–1.36	–1.13
2nd year ($T_{FP,0.2,2} = 0.694$)	–3.26	–2.26	–2.18	–1.51
3rd year ($T_{FP,0.2,3} = 0.579$)	–1.96	–1.13	–1.31	–0.76
4th year ($T_{FP,0.2,4} = 0.482$)	–1.17	–0.56	–0.78	–0.38
5th year ($T_{FP,0.2,5} = 0.402$)	–1.17	–0.47	–0.78	–0.31
6th year ($T_{FP,0.2,6} = 0.335$)	–0.59	–0.20	–0.39	–0.13
Tax benefits from maint. + rebuild. + elect. @ 0.34		–4.33		–6.49
Tax burdens from product. + salv. @ 0.34		1.06		
Total present value of costs		30.04		28.37

*Depreciation for the Ajax motors is based on neglecting salvage value. Assume no depreciation for rebuilding Blaylock motors.

The net effect of federal income taxes is to reduce the present value of costs of the Ajax motors from the $39,630 listed in Table 8–2 to $30,040, a reduction of $9,590. The present value of the costs of the Blaylock motors is reduced by $10,710 from $39,080 to $28,370 due to tax effects. Thus, the impact of federal income taxes in this case is to increase the slight economic advantage the Blaylock motors already enjoyed over the Ajax motors.

This analysis reveals that the current U.S. federal income tax structure allows immediate deductions for operating expenses, but postpones tax deductions for capital investments into the future according to a depreciation schedule. When future transactions are discounted to obtain their equivalent value, this tax structure bestows a comparative advantage to design options which have a low initial cost but high operating costs over alternatives with high initial costs but low downstream operating costs.

8.8. ACCOUNTING FOR INFLATION

Our economic analyses in this chapter have not explicitly dealt with the effects of inflation. Rather, we have submerged inflation effects within our cost estimates for future transactions, such as the salvage value of the Ajax motors in five years, the rate at which electricity prices will increase, and the discount rate. There are some circumstances under which we may prefer to separate inflationary effects from other sources of changes in price. For example, we may feel confident in estimating, in the absence of inflation, the future costs of certain technologies. Consider computer chips, which have followed a steady downward path for several decades. We may feel quite confident extrapolating that trend over the next five years. In such a situation, we might want to separate that estimate from a more uncertain estimate for the anticipated inflation rate over the same time period. In this section, we will define inflation, describe how to explicitly isolate its effects on future transactions, and how to calculate an inflation-free discount rate for use in present value calculations.

Definition of Inflation

You may remember your parents or grandparents talking about the good old days when an ice-cream cone cost only a nickel, and the price of a movie ticket was only a quarter. What they probably didn't tell you was that back then in the good old days, the minimum wage was only $1.00 an hour. This simultaneous change in both prices and wages is the essence of the concept of inflation, which we define to be the annual average increase in all prices and wages.

To determine whether the real cost of an object or service is changing with time, we must examine not only the change in price, but also the change in wages over the same time period. A good way to do this is to examine how many hours a typical person would have to work in order to earn enough money to purchase the object or service. For example, if the price of a typical textbook in 1998 is $60 and the average wage earner in 1998 makes $10/hour, a typical person has to work six hours in order to earn the money to buy the book. Now, what if ten years earlier the same book cost $30? Has the book gotten more expensive since then? Not if the average wage back then was $5/hour. The real price has not changed since an average person had to work six hours in 1988 to afford the book. What has changed is the value of the dollar. In this example, the 1998 dollar is only worth 50¢ relative to the 1988 dollar.

Units of Monetary Measure

This discussion suggests that there are two units for expressing the monetary value of prices and wages. The first is the widely used measure, which we call nominal, or current, dollars. This term refers to the face value of the bills and coins at the time of the transaction. In terms of nominal dollars, the price of the textbook increased from $30 to $60 over the past ten year period. The second approach is to measure all transactions, regardless of when they occur, in terms of the value of the dollar in a reference year. For the textbook example, we could choose to express the transactions in terms

TABLE 8–12.   UNITS OF MONETARY MEASURE

Entity	Units	1988 Amount	1998 Amount
Textbook	Nominal (current) dollars	$30	$60
Hourly wage	Nominal (current) dollars	$5	$10
Textbook	Real (constant) 1988 dollars	$30	$30
Hourly wage	Real (constant) 1988 dollars	$5	$5
Textbook	Real (constant) 1998 dollars	$60	$60
Hourly wage	Real (constant) 1998 dollars	$10	$10

of either 1988 or 1998 dollars. We refer to this unit of measure as real, or constant, dollars. The three most common ways to describe these hypothetical changes in textbook prices and hourly wages between 1988 and 1998 are summarized in Table 8–12. Note that all three units of monetary measure show that a typical person had to work six hours to earn enough money to buy a textbook either in 1988 or 1998.

More generally, if we choose to use constant dollars to describe the magnitude of future transactions, the most convenient reference year for expressing constant dollars is the year in which the choice among design alternatives is made.

Real Discount Rate

Consider a transaction F that occurs n years in the future. The relationship between the transaction expressed in terms of nominal dollars and that same transaction expressed in terms of real dollars is

$$F_{real} = \frac{F_{nom}}{(1+f)^n} \tag{8–25}$$

where f is the anticipated average annual inflation rate over the period of n years. We earlier expressed the salvage value of the Ajax motors five years from now as $4,000 in terms of nominal dollars. If we choose to express the salvage value in terms of real dollars, and the estimated annual inflation rate over the five years is 3%, then Equ. (8–25) gives the salvage value in terms of real dollars as

$$F_{real} = \frac{4}{(1+0.03)^5} = 3.45 \tag{8–26}$$

Now consider determining the present value of the salvage transaction. If we express the future transaction in terms of nominal dollars, then we convert it to an equivalent present value using the nominal discount rate according to Equ. (8–5) which we rewrite here as

$$P = \frac{F_{nom}}{(1+i_{nom})^n} \tag{8–27}$$

You may recall that we applied this to the salvage value of the Ajax motors in Table 8–2 using a 20% discount rate to get

$$P = \frac{4.00}{(1 + 0.20)^5} = 1.61 \tag{8-28}$$

Starting with the future transaction expressed in terms of real dollars, the conversion to an equivalent present value requires us to modify Equ. (8-27) to read

$$P = \frac{F_{real}}{(1 + i_{real})^n} \tag{8-29}$$

Now we must determine what value to use for the real discount rate. This rate must be based on the fact that the present value of a future transaction should not depend on the particular system of units used to describe that future transaction. Hence, we can equate Equs. (8-27) and (8-29) to get

$$\frac{F_{real}}{(1 + i_{real})^n} = \frac{F_{nom}}{(1 + i_{nom})^n} \tag{8-30}$$

If we substitute Equ. (8-25) into Equ. (8-30) and solve for i_{real} we get

$$i_{real} = \frac{i_{nom} - f}{1 + f} \tag{8-31}$$

Note that if the inflation rate is very small, that is, $f \ll 1$, then

$$i_{real} \cong i_{nom} - f$$

or the real discount rate is approximately the difference between the nominal discount rate and the inflation rate. For the Ajax salvage value, we use Equ. (8-31) to get

$$i_{real} = \frac{0.20 - 0.03}{1 + 0.03} = 0.165$$

Inserting this and the value of F_{real} from Equ. (8-26) into Equ. (8-29) gives

$$P = \frac{3.45}{(1 + 0.165)^5} = 1.61$$

which is the same as that found using Equ. (8-28). This should not be a surprise since we defined i_{real} precisely so that the present value P would be the same regardless of whether we expressed F in terms of nominal dollars or real dollars. Of course, we have to be consistent and use the discount rate (real or nominal) that is associated with the monetary units (real or nominal) used to describe the future transactions.

8.9. ADDITIONAL CONSIDERATIONS

In Sections 8.3 and 8.4 we introduced a set of transformation factors to account for the time value of money when monetary transactions occur at different points in time. We gather those factors together in Table 8-13 and discuss some of the modifications needed for those factors when some of the underlying assumptions are changed.

TABLE 8–13.   SUMMARY OF TRANSFORMATION FACTORS

Factor	Name	Comments
$T_{FP,i,n} = (1 + i)^{-n}$	Present worth factor	
$T_{UP,i,n} = \dfrac{[(1 + i)^n - 1]}{i(1 + i)^n}$	Uniform series present worth factor	The inverse of this is known as the capital recovery factor
$T_{AP,i,n} = \dfrac{[(1 + i)^n - i \cdot n - 1]}{i^2(1 + i)^n}$	Arithmetic series present worth factor	This is multiplied by the annual increment and added to the value of P obtained by using the uniform series present worth factor. See below for example.
$T_{GP,i,e,n} = \dfrac{[(1 + i)^n - (1 + e)^n]}{(1 - e)(1 + i)^n}$	Geometric series present worth factor	$i \neq e$ $i = e$
$T_{GP,i,e,n} = \dfrac{n}{(1 + i)}$		
$T_{FU,i,n} = \dfrac{i}{[(1 + i)^n - 1]}$	Sinking fund factor	The inverse of this is known as the uniform series future worth factor
$T_{FU,i,n,m} = \dfrac{i(1 + i)^{n-m}}{[(1 + i)^n - 1]}$	Generalized sinking fund factor	m is the year in which the transaction F occurs; n is the period over which the annual payments U are made.
$T_{GU,i,e,n} = \dfrac{i[(1 + i)^n - (1 + e)^n]}{[(1 + i)^n - 1](i - e)}$	Geometric series sinking fund factor	

Note that we have included the arithmetic series present worth factor $T_{AP,i,n}$ in Table 8–13 even though we have not encountered this factor elsewhere in this chapter. It is used when an annual series of payments increases by a given amount each year (as opposed to the geometric series factor which handles annual series of payments that increase by the same percentage each year). For example, the annual maintenance costs of the Blaylock motors are $2,000, and the present value of those costs over the five-year lifetime of the motors is $5,980 (see Table 8–2). Suppose we modified the problem by saying that the annual maintenance cost was $2,000 the first year and is expected to increase $200 per year over the five years. The present value of those costs would then be

$$P = \$2,000 \cdot T_{UP,0.20,5} + \$200 \cdot T_{AP,0.20,5}$$

or, from Table 8–13

$$P = \$5,980 + \$200 \cdot \frac{[(1 + 0.20)^5 - 0.20 \cdot 5 - 1]}{0.20^2(1 + 0.20)^5}$$

$$P = \$5,980 + \$200 \cdot 4.91 = \$6,962$$

Note that the arithmetic series factor $T_{AP,0.20,5}$ is multiplied by the annual increment and then added to the present value based on a uniform series calculation. This contrasts with the geometric series factor $T_{GP,0.20,0.05,5}$ (which has the annual escalation rate e imbedded in it) that is multiplied by the value of the first year transaction.

Variations to Downstream Transaction Patterns

We have treated most operating costs as expenses that occur at the end of each year. A more realistic approach would be to treat these as expenses that occur continuously throughout the year. One approach to approximating continuous expenditures is to calculate each downstream transaction as if it occurred at the start of the year. Then average those results with those obtained by the end-of-year calculations.

For shorter term projects, it may be preferable to conduct the analysis in terms of months or weeks rather than years as the unit of time. That way, we can model payroll and other operating expenses more accurately. The only difference in a weekly or monthly analysis period is that an equivalent interest rate associated with the shorter time period should be the basis for the discount rate. The magnitude of this equivalent interest rate depends on the number of periods over which the interest is compounded. For example, if interest on an investment P is compounded monthly at the rate i_{mon} then the future value F of that investment at the end of one year (12 months) is

$$F = P(1 + i_{mon})^{12}$$

The equivalent annual interest rate i_{ann} should lead to the same result, that is,

$$F = P(1 + i_{ann})$$

Equating these two provides

$$i_{ann} = (1 + i_{mon})^{12} - 1$$

We can use the binomial theorem to expand the expression in parentheses to get

$$i_{ann} = (1 + 12 \cdot i_{mon} + 66 \cdot i_{mon}^2 + 220 \cdot i_{mon}^3 + \ldots) - 1$$

$$= 12 \cdot i_{mon} + 66 \cdot i_{mon}^2 + 220 \cdot i_{mon}^3 + \ldots$$

For small interest rates, $i_{mon} \ll 1$, so

$$i_{ann} \cong 12 \cdot i_{mon}$$

A similar argument for interest that is compounded quarterly leads to

$$i_{ann} \cong 4 \cdot i_{quar}$$

Long project times.   When project durations span a generation, say $n > 30$ years, an interesting philosophical question arises over the proper value to assign to the discount rate. Recall that the discount rate reflects the decision maker's time value of money by reducing the value of a future transaction. One, albeit oversimplified, explanation is that the decision maker doesn't value costs and revenues that occur in the distant future as much as those that occur in the present. However, what if the downstream costs are borne by someone other than the decision maker, in particular, by a future generation?

Perhaps the most vexing of contemporary intergenerational equity issues is the disposal of the high-level radioactive waste products from civilian nuclear power plants. Because of the high level of lethal radioactivity and the slow rate of decay, any "permanent" disposal system has to maintain its integrity for tens of thousands of years. We are dealing with making decisions now that affect future generations for a period into the future that is greater than all recorded human civilization. Suppose we are comparing two options for the waste repository: Option A, which costs more to build but less to operate; and Option B, which costs less to build but more to operate. Even with a very low discount rate, the higher operating costs of Option B in future generations will be discounted so heavily as to be negligible. For example, consider a $1,000,000 cost incurred a thousand years from now. For a discount rate of $i = 0.01$, the present value of that $1,000,000 is

$$P = \frac{1,000,000}{(1 + 0.01)^{1,000}} = \frac{1,000,000}{20,960} = 47.71$$

or, less than $50! It's hard to imagine that our descendants who will bear those costs will agree with our choice of discount rate, which essentially shifts the cost of maintaining the waste repository from our generation to theirs. In the mid-1980s when the U.S. Department of Energy compared several options for such a facility, they used a discount rate of zero.

Internal Rate of Return

Finally, it is possible to combine the present value approach with the annual rate of return approach described in Section 8.2. One particular way of doing this is with the internal rate of return (IRR) method. Instead of selecting a discount rate for use in a present value type of analysis, an IRR analysis uses the appropriate transformation factors from Table 8–13 except we treat i as the unknown. We then solve for the value of i that makes the present value equal to zero. We'll illustrate this by modifying our application in Section 8.2 of the average annual rate of return rule. For convenience, we reproduce the key expressions from that earlier discussion.

Recall from Equ. (8–1) that the incremental investment in the Ajax motors is

$$\text{Incremental Investment} = 10.00\ (\$10^3) \tag{8–32}$$

The downstream benefits are listed in Equ. (8–2) as

$$\text{Downstream Benefits} = \text{Avoidance of Rebuilding Cost} + \text{Income from Salvage}$$

$$+ \text{Annual Savings on Maintenance Costs over Five Years}$$

$$+ \text{Increased Revenue from Productivity Enhancement}$$

$$+ \text{Annual Savings on Electricity Costs} \tag{8–33}$$

$$= [3.00 + 4.00 + (10.00 - 5.00) + 2.50 + (3.50 - 3.00)$$

$$+ (3.68 - 3.15) + (3.86 - 3.31) + (4.05 - 3.47)$$

$$+ (4.25 - 3.65)]\ (\$10^3)$$

$$= 17.26\ (\$10^3)$$

We now modify each term in Equ. (8–33) by the appropriate transformation factor from Table 8–13 to get the present value of the downstream benefits as

Present Value of Downstream Benefits =

$$\{3.00\ T_{FP,i,3} + 4.00\ T_{FP,i,5} + [(10.00 - 5.00)/5]\ T_{UP,i,5} + (2.50/5)\ T_{UP,i,5} \quad\quad (8\text{--}34)$$
$$+ (3.50 - 3.00)\ T_{GP,i,5}\}\ (\$10^3)$$

Notice that the $T_{UP,i,5}$ terms are multiplied by the value of the annual transaction and $T_{GP,i,5}$ is multiplied by the value of the transaction in the first year. Writing out the transformation factors turns Equ. (8–34) into

$$\text{Present Value of Downstream Benefits} = \left\{ 3.00\ \frac{1}{(1+i)^3} + 4.00\ \frac{1}{(1+i)^5} \right.$$
$$\left. + 1.00\ \frac{[(1+i)^5 - 1]}{i(1+i)^5} + 0.50\ \frac{[(1+i)^5 - 1]}{i(1+i)^5} + 0.50\ \frac{[(1+i)^5 - (1.05)^5]}{(i-0.05)(1+i)^5} \right\}\ (\$10^3)$$

Equating this with Equ. (8–32) yields the following equation for i:

$$\frac{3.00}{(1+i)^3} + \frac{4.00}{(1+i)^5} + 1.50\ \frac{[(1+i)^5 - 1]}{i(1+i)^5} + 0.50\ \frac{[(1+i)^5 - (1.05)^5]}{(i-0.05)(1+i)^5} = 10.00$$

The solution to this equation is $i = 0.179$. Thus, the internal rate of return for the extra investment in the Ajax motors is 17.9%.

8.10. COST ESTIMATING

Our discussion of engineering economics so far has assumed that the estimates for capital and operating costs are given. In many situations, you as the design engineer have to provide these estimates. We briefly discussed sources of cost information in Section 3.2; here our focus is on methodological considerations. We divide the discussion according to three types of cost information: preliminary equipment and process costs, costs of purchased products, and fabrication costs. We also address the issue of incorporating indirect costs into your estimates.

Preliminary Equipment and Process Costs

Early in the design process you might need a preliminary estimate of the cost of a major piece of equipment, or even of an entire facility. A quick estimate can be obtained using an economy-of-scale rule keyed to a base size of similar equipment/facilities with a known cost. For example, suppose you want to estimate the cost of a 1,000 hp reciprocating compressor. You are in luck, because reciprocating compressors are included in the several dozen categories of equipment covered in a widely used chemical engineering handbook.[4] The general approach to using the data in Perry's Table 25–49 to estimate the capital cost C of a system of size Q is

$$C = C_{ref}\left(\frac{Q}{Q_{ref}}\right)^n \qquad (8\text{--}35)$$

where C_{ref} is the cost of the reference equipment whose size is Q_{ref}. Values for C_{ref}, Q_{ref}, and n are listed in Table 25–49 for each category of equipment. In the particular case of a reciprocating compressor, Perry provides

$$C_{ref} = \$133{,}000$$

$$Q_{ref} = 300 \; hp$$

$$n = 0.84$$

Substituting these values into Equ. (8–35) yields

$$C = \$133{,}000\left(\frac{1{,}000}{300}\right)^{0.84} = \$365{,}700$$

Several notes of caution are appropriate at this point. First, Equ. (8–35) and the associated numerical value of n are valid only in the size range specified by Perry. For the reciprocating compressor, the valid range listed by Perry is $1 < C < 20{,}000$ hp, so we are well within the range of validity. Second, the published values for C_{ref} are for a specified date; to bring them up to date, they have to be multiplied by a factor to account for the effects of inflation.

Cost of Purchased Products

Many systems that you will design will include components that you purchase from a vendor. A good example are the electric motors we have been discussing throughout this chapter. We will concentrate here on the estimates for the capital cost of the motors. You may have obtained the cost data from the manufacturers' catalogues or you may have contacted the vendors directly and requested an estimate. When comparing price quotes from different vendors, you must be sure that the estimates provided are comparable. Some price quotes may be FOB, or the price at the manufacturer's plant or warehouse. Another basis for price quotes includes the cost of delivery. Still another approach, which could be a major item for sales of large equipment (such as a boiler), includes installation charges. If installation costs are not included in the price, you may have to provide your own estimate or obtain an estimate from someone who is experienced with installing that equipment. It could very well be that the differences in installation costs for, say the Ajax and Blaylock motors are significant. And don't forget sales taxes, which could vary depending on the state in which your client is located and the state in which the vendor is located.

---

[4]Green, pp. 25–69.

Another issue to consider is whether or not the advertised price or the price quoted to you is negotiable. Some vendors won't even discuss price over the telephone. Also, the cost may depend on the number of units you are ordering, how quickly you need them, and how and when you plan to pay for them.

Fabrication Costs

For many design projects, we eventually need detailed calculations of fabrication costs. For machining operations, the fabrication costs depend on many factors. Some of these factors are: the material being machined; the equipment used (lathe, drill press, milling machine, etc.); the specific tool used (brazed carbide, throwaway carbide, high speed steel, etc.); the operation that is being performed and the rate at which it is being performed; the desired tolerances and surface finish; and the cost of labor. These calculations can be quite complicated.

As an example, consider the cost of turning a 3.5″ diameter, 19″ long shaft on a lathe. A brazed carbide tool is used and the shaft is made from 4340 steel. The results of the cost breakdown are summarized in Table 8–14. Each term in Table 8–14 is itself the result of a separate calculation. For example, the tool change cost is determined from

$$\text{tool change cost} = \frac{MDLt_d}{3.82f_r vT}$$

where

M = Labor and overhead cost, \$/min f_r = feed per revolution, in.
D = diameter of shaft, in. v = cutting speed, fpm
L = length of shaft, in. T = tool life, total time to dull
t_d = time to change and reset tool   tool, min.

TABLE 8–14.   COST OF TURNING A STEEL SHAFT ON A LATHE

Category	Cost (\$)
Feed	1.48
Rapid traverse	0.11
Load and unload	0.92
Setup	0.43
Tool change	0.49
Tool depreciation	0.13
Tool resharpening	1.48
Rebrazing	0.16
Tip cost	0.10
Grind wheel	0.02
Total	\$5.33

Source: Machining Data Handbook, pp. 21–4 thru 21–27.

Indirect Costs and Profit

If you are designing and/or building the system for an external client, additional considerations beyond those discussed above come into play. You must translate the cost data into a cost estimate for your client, either as part of a proposal, or as part of an invoice (bill) for services rendered and products delivered. If your company just charged all its clients exactly what it cost you to obtain the parts and materials from vendors, your company would not stay in business very long.

Since you spent your time acquiring the cost data, selecting the hardware, and ordering it from the vendor, your invoice should rightfully include a charge for the value of your time in addition to the price for the equipment. If your salary is the equivalent of $20/hour and you spent two hours working on this project, you should charge your client $40 for the time you spent.

Keep in mind that your company not only pays your salary, it contributes to your health insurance, social security, and retirement plan. Those fringe benefits are real costs to your company and need to be recovered in the form of income from its clients. In addition, there are many other costs of operating a business, such as the rent for the office space, the phone, light, and heating bills, cost of office supplies, postage, copy machines, and income taxes. We refer to such expenses as indirect costs (also known as overhead) because they cannot be directly attributed to the specific project. These indirect costs have to be covered somehow, and many firms account for them by adding a fixed percentage onto the direct costs. Your company also is entitled to earn a profit, so a company fee, also in the form of a percentage of either the direct costs or total costs, may be an explicit charge item. In addition, for major projects involving considerable uncertainties, a contingency account may be established to deal with extra costs that cannot be identified at this stage of the project. This acts as a sort of economic "safety factor" so we can handle unforeseen expenses.

Cost Estimates for Chittenden Locks Project

Let us now examine some of the typical cost estimates for the Chittenden Locks project to illustrate the application of some of these concepts. Let's start by examining the estimate in Table 8–4 of $552,000 for the cost of rehabilitating each gate or $1,655,000 for the three gates. We will dissect certain elements of this cost and trace them through several layers of detail.

The $1,655,000 total comes from a base cost of $1,357,000 to which is added a 15% contingency account of $298,000 and $94,000 for construction management costs. The base cost of $1,357,000, in turn, has the five components listed in Table 8–15.

Now let's focus on the direct cost in Table 8–15. It is the sum of the costs associated with the seven major tasks listed in Table 8–16. The level sensor item in Table 8–16 refers to the instruments for determining the water level in the lock. The other items are self-explanatory. We can examine any of these tasks at the next level of detail. Let's look at the concrete work, which can be subdivided into five subtasks as shown in Table 8–17.

TABLE 8–15. COMPONENTS OF BASE COST

Components of Base Cost	Cost ($1,000)
Direct cost	1,039
Overhead	104
Home Office	91
Profit	74
Insurance and Bonding	49
Total	1,357

Source: U.S. Army Corps of Engineers, Appendix B, Summary p. 2.

TABLE 8–16. DIRECT COSTS OF MAJOR TASKS

Task	Cost ($1,000)
Concrete at decks and pits	247
Cylinder and hydraulic installation	480
Level sensor	12
Gate work	94
Electrical	56
Controls	126
Remote controller	24
Total	1,039

Source: U.S. Army Corps of Engineers, Appendix B, Summary p. 2.

TABLE 8–17. DETAILED COST ESTIMATES FOR CONCRETE SUBTASKS

Subtask	Cost ($1,000)
Demolition of slab	12
Concrete demolition	68
Reinstall concrete deck	55
Concrete in pit	100
Tie into existing drainage system	3
Steel cover plate	9
Total	247

Source: U.S. Army Corps of Engineers, Appendix B, Detail p. 1–3.

As shown in the elevation view of Figure 8–3, the gears, motors, and other parts of the current machinery are housed in pits underneath a concrete walkway. The first item in Table 8–17 refers to the demolition of that concrete surface slab. The second item refers to removing concrete walls and other structures within the existing pits. Finally, we'll use our microscope to examine the concrete demolition subtask at the next level of detail (see Table 8–18).

TABLE 8–18.   COST BREAKDOWN
FOR CONCRETE DEMOLITION SUBTASKS

Item	Cost ($1,000)
Equipment	15
Semi-skilled worker	26
Skilled worker	14
Haul and dispose	3
Standby and overtime	10
Total	68

Source: U.S. Army Corps of Engineers, Appendix B, Detail p. 1.

This discussion and the data summarized in Tables 8–15 thru 8–18 illustrates the enormous amount of detail that goes into estimating costs for a design project the size of the Chittenden Locks project. Of course, we examined these costs in the reverse order in which they were estimated during the project itself. Our starting point, the $552,000 cost of rehabilitating each gate, is the end result of a process that started with the cost estimates prepared for each detailed item, such as those listed in Table 8–18.

8.11. CLOSURE

We have developed a set of techniques for comparing the relative economic effectiveness of alternative designs. A good design engineer will select the specific approach best suited for a particular situation, including the one which best meets the needs of the client.

One of the most important considerations in the choice of an economic evaluation technique is to establish a level playing field so a fair comparison can be made of all the design options. Sometimes that is not as easy as it sounds because of the subjective considerations that inherently enter into this type of analysis. In particular, choosing the value of the discount rate involves subjective considerations. Also, it is unavoidable in engineering economics to be in the position of estimating the cost of, or the income to be generated from, future activities. Unless your crystal ball gives you a clearer picture of the future than mine does (I dropped mine in the swamp), there are many ways that these kinds of uncertainties enter into an economic comparison of design alternatives. One approach to dealing explicitly with these uncertainties will be discussed in Chapter 9.

Our treatment of the material has been introductory; there are many textbooks devoted entirely to the subject of Engineering Economics and some engineering curricula include a separate course in the topic.

8.12. REFERENCES

DEGARMO, E. P., W. G. SULLIVAN, and J. A. BONTADELLI. 1989. *Engineering Economy, 8th Edition.* New York: Macmillan Publishing Co.

DIETER, G. E. 1991. *Engineering Design: A Materials and Processing Approach, 2nd Edition.* New York: McGraw-Hill, Inc.

GREEN, D. W., ed. 1984. *Perry's Chemical Engineering Handbook, 6th Edition.* New York: McGraw-Hill Book Co.

Machinability Data Center. 1980. *Machining Data Handbook, 3rd Edition.* Volume Two. Cincinnati: Machinability Data Center.

NEWMAN, D. G. 1991. *Engineering Economic Analysis, 4th Edition.* San Jose, CA: Engineering Press Inc.

RIGGS, J. L. and T. M. WEST. 1986. *Engineering Economics, 3rd Edition.* New York: McGraw-Hill Book Co.

SHTUB, A. et. al. 1994. *Project Management: Engineering, Technology, and Implementation.* Englewood Cliffs, NJ: Prentice-Hall, Inc.

U.S. Army Corps of Engineers. May 1994. *Large Lock Gate Machinery Rehabilitation, Hiram M. Chittenden Locks, Lake Washington Ship Canal, Seattle, Washington.* Detailed Project Report. U.S. Army Corps of Engineers, Seattle District.

8.13. EXERCISES

1. Machine *A* costs $8,500 and has an annual operating cost of $4,500. Machine *B* costs $7,000 and has an annual operating cost of $4,800. Each machine has an economic life of 10 years. What is the annual rate of return on the additional investment in machine *A*?
   ◆ *Source:* Dieter, p. 680.

2. You are designing a piece of equipment for a client who ignores the time value of money in making investment decisions. You have narrowed your design choices down to two options: *A* and *B*. You estimate option *A* has an initial cost of $8,000, and annual operating costs of $400. You expect option *B* to have an initial cost of $6,500, and annual operating costs of $800. Each machine has an economic life of 10 years. Estimate the payback period for your client's potential additional investment in design *A*.

3. As chief engineer for a small manufacturing firm, you are considering whether to invest $40,000 in a new piece of equipment that will save $3,000 annually in operating costs. If the equipment has an expected lifetime of 20 years and a salvage value of $5,000, is this an attractive investment at the discount rate of 15% using the present value criterion?

4. Redo the analysis summarized in Table 8–2 using a 10% discount rate.

5. A company is faced with the decision of selecting either machine *A* or *B*, both of which are capable of doing the same kind of job. Machine *A* costs $1,250 with annual maintenance

and operating costs of $150 for the first 10 years and $180 for the following 10 years. Machine B costs $1,500 and annual operating and maintenance expenses are $100 for the 20-year period. Both machines have zero salvage value at the end of their economic lives. By how much should the manufacturer of machine A increase or decrease the purchase price of the machine so that it is competitive with machine B at a discount rate of 15%?

6. An electric utility company has determined that it needs to bring 600 mW of new generating capacity on-line six years from now to meet anticipated peak loads during the heating season. The utility has narrowed down the choices to a coal power plant or a set of combustion turbines fired by natural gas. The coal plant, which takes six years to build, has a capital cost of $1,200/kW. The combustion turbines can be erected in two years and cost $250/kW. The projected operating costs in six years are 2.1 cents/kWh for the coal plant and 5.7 cents/kWh for the combustion turbine. Each plant is expected to operate at full capacity for 3,000 hrs annually throughout its 20-year lifetime. The above quoted capital and operating costs are expressed in terms of today's dollars. The operating costs of the combustion turbine are expected to keep pace with inflation, which is estimated to average 3% annually over the lifetime of the project. However, the prospects of increasingly stringent regulations dealing with acid rain indicate that coal plant operating costs will probably increase 1% a year over and above inflation. Since the utility in question is municipally owned, it pays no income taxes. Using a real discount rate of 3%, should the utility start building the coal plant now or wait four years to begin building the gas turbine plant? How much does the expected annual plant usage have to change before it affects the choice?

7. Redo Exercise 3 using an equivalent annual cost analysis.

8. The Environmental Protection Agency has ordered your oil refinery to process its waste liquids before discharging them into the local bay. You estimate that it will cost you $30,000 to satisfy this requirement this year. However, by gradually making no-cost adjustments in the refining process, you estimate you can reduce waste processing costs by 5% annually over the next 10 years, which is the remaining life-time of the refinery. On the other hand, an outside company has offered to process your waste for you every year for the next ten years at an annual fee of $15,000. If your company uses a 10% discount rate, which is the preferred course of action?

9. You are considering upgrading some manufacturing facilities by purchasing one of three different machines, each with the same production capacity. Machine A costs $30,000, has a life of 40 years, annual maintenance costs of $1,500, and salvage value of $5,000. Machine B costs $20,000, has a life of 20 years, annual maintenance of $2,000, and salvage value of $3,000. Machine C costs $10,000, has a life of 10 years, annual maintenance of $4,000, and no salvage value. Determine the most economical choice based on minimizing the present value of total costs. Use an annual discount rate of 10%. Assume that initial costs, annual maintenance, and discount rates are constant throughout the analysis period.

10. Your client purchased a workstation computer three years ago for $30,000. Now they are considering expanding their computing capability by either buying a new computer or adding components to the existing one. Additional components can be purchased for the present computer at a cost of $10,000. You estimate that the improved computer will have a life of 5 years with an annual operations, maintenance, and repair (OMR) cost of $2,500. It is not expected to have any salvage value at the end of the period. A second option is to buy a new ABC computer at a cost of $35,000. This unit will have an expected life of 15 years, an annual OMR cost of $1,500, and a salvage value of $5,000. Another option is to purchase a new XYZ computer for $27,000. For this computer, the expected life is 10 years, with an OMR cost of $2,000, and a salvage value of $3,000. The existing computer has a

trade-in value of $5,000 if either of the new computers is purchased. All three computer systems have the same computational capability. Assuming an interest rate of 10%, which alternative should your client choose based on a present value decision rule?

11. A steam boiler costs $25,000 installed and has an estimated life of 8 years. By the addition of certain auxiliary equipment, an annual saving of $400 in operating costs can be obtained, and the estimated life of the installation can be increased by 50%. Assuming that interest on capital is 12%, what is the maximum expenditure justifiable for the auxiliary equipment? The salvage value for either alternative is estimated at $5,000.

12. Redo the problem summarized in Table 8–11 using straight line depreciation over a five-year depreciation period and a tax rate of 0.46.

13. Redo Exercise 4 to include the effect of federal income taxes. Use the MACRS depreciation schedule for $n = 7$, and assume that the taxable income of the company in each year is between $200,000 and $300,000.

14. The annual operating cost of a proposed ventilation system in a supermarket is $20,000 for the first year. This cost is expected to increase at the rate of 10% a year.
    a) What is the annual operating cost in the fifth year of operation expressed in terms of current (nominal) dollars?
    b) If the estimated average annual inflation rate over the next five years is 6%, express the fifth-year operating cost in terms of constant (real) dollars.
    c) If the nominal discount rate is 12%, use the results of part (a) to calculate the present value of the fifth-year operating cost.
    d) If the nominal discount rate is 12%, use the results of part (b) to calculate the present value of the fifth-year operating cost.

15. The annual maintenance cost for a company's computer network is $30,000 for the first year. This cost is expected to increase at the rate of 5% a year.
    a) What is the annual operating cost in the fifth year of operation expressed in terms of current (nominal) dollars?
    b) If the estimated average annual inflation rate over the next five years is 4%, express the fifth-year operating cost in terms of constant (real) dollars.
    c) If the nominal discount rate is 10%, use the results of part (a) to calculate the present value of the fifth-year operating cost.
    d) If the nominal discount rate is 10%, use the results of part (b) to calculate the present value of the fifth-year operating cost.

16. Redo Exercise 3 but calculate the internal rate of return instead of assuming a 15% discount rate.

17. Recalculate the present value of the maintenance costs shown in Table 8–2 assuming that these costs occur at the beginning of each year rather than the end of each year.

18. Consider the $3,000 electricity cost listed in Table 8–1 for the first year of operation of the Ajax motors. Assume that cost was paid in 12 equal monthly installments rather than at the end of the year. What is the difference in the present value using a discount rate of 10%?

19. Use Equ. (8–35) together with data from *Perry's Chemical Engineering Handbook* to estimate the cost of a 200 hp centrifugal pump.

20. Provide a detailed breakdown of your estimates of the cost of manufacturing a wooden pencil (with eraser top).

CHAPTER 9

Decision Making

9.1. OVERVIEW

We frequently describe design as a problem solving activity. From another perspective, we can think of design as the act of choosing among alternatives. In fact, the design process involves both activities, and the nine-step model of design introduced in Chapter 1 explicitly identifies selecting the preferred alternative (another way of saying decision making) as Step 7. In this chapter we will examine decision making processes and introduce several techniques and tools for aiding decision making under complex conditions.

The first aspect of decision making that we examine is the challenge Jane faced in Chapter 1. She had to choose alternative automobile bumper designs on the basis of initial cost, damage control, drivability, and recyclability. This is an example of a multiple-criteria decision making problem, and we examine the topic in considerable detail.

A second aspect of choosing among design options are the assumptions Jane must make about the future operating conditions that her design will encounter. Since her crystal ball fell into the swamp with her and is caked with mud, Jane does not have a totally clear vision of the future. However, these uncertainties and Sandra's willingness to take risks in the face of these uncertainties can be systematically incorporated into the selection process using the techniques discussed later in this chapter. This is all well and good if Sandra is willing to take the time to answer Jane's questions about priorities and risks. As we saw in Chapter 2, Sandra is a very busy person and may not be able to take the time needed to provide Jane with the information she needs.

Finally, we combine our treatment of multiple criteria decision making with our discussion of decision making under risk to examine multiple attribute utility theory as a decision making tool. The time invested in Step 7 has to be in proportion to the size of the project. An additional consideration affecting how much effort to devote to this step is the consequences of making the wrong choice. Welcome back to the swamp!

9.2. MULTIPLE CRITERIA

In Chapter 8 we compared design alternatives using a single criterion—economic value. We learned that there are several ways to define economic value, and in any given situation one of these definitions is selected as the basis for deciding among design alternatives. However, in many engineering design situations, it may not be possible or desirable to reduce all features or functions of design options to a single criterion such as economic value.

In this section we will examine how to choose among design alternatives when multiple criteria are involved. Consider the Chittenden Locks problem discussed in Chapter 8. In addition to the economic analysis, the Corps of Engineers also evaluated each of the four options with respect to reliability and flexibility. Table 9–1 shows the Corps' ranking of each option with respect to each criterion. Since the all-hydraulic alternative is ranked first with respect to each of the criteria, it is clear that it is the preferred design.

TABLE 9–1.   CORPS OF ENGINEERS RANKING OF DESIGN
OPTIONS FOR NAVIGATION LOCKS

Design Options	Criteria		
	Reliability	Cost	Flexibility
All-hydraulic	1	1	1
Fix existing	3	2	2
Add variable speed drive	2	4	2
Add fluid coupling	2	3	2

Source: U.S. Army Corps of Engineers. p. 10.

However, there are many circumstances in which one option does not clearly dominate all the others. We illustrate that situation by revisiting the automobile bumper design problem featured in Chapters 1 and 2. Suppose Jane develops three candidate bumper designs (*A, B,* and *C*). Recall from our discussion in Chapter 1 that the four characteristics that Jane (and by proxy, Sandra) are most interested in are: cost, damage control, recyclability, and drivability. Jane's analysis might reveal that Alternative *A* is the least expensive, alternative *B* provides the best damage control and is most easily recycled, while alternative *C* does the best job of protecting drivability. Her assessment of each design option with respect to each criterion is summarized in Table 9–2. How does Jane use this information to determine which design to select? Jane, good friend, watch out for the alligators!

One approach Jane can take is to choose alternative *B* since it is superior in two categories (damage control and recyclability) while *A* and *C* are superior in one cate-

TABLE 9–2. COMPARISON OF THREE DESIGN OPTIONS WITH RESPECT
TO FOUR DESIGN CRITERIA

Criteria	Design Option A	Design Option B	Design Option C
Cost	excellent	poor	adequate
Damage control	adequate	excellent	poor
Recyclability	poor	excellent	adequate
Drivability	adequate	poor	excellent

gory each. But what if B is only very slightly superior to A and C with respect to damage control and recyclability, but grossly inferior to both A and C in the other two criteria? The information in Table 9–2 is not sufficient to address that question. Similarly, the ranking method used in the Chittenden Locks problem orders the options but does not capture the magnitude of the differences between them. Finally, what if not all of the design objectives are equally important?

There are several approaches you could take if you found yourself in Jane's shoes. First, it is quite possible that Table 9–2 provides adequate information to you (or, preferably, your client) to make a choice that you will feel quite comfortable with. Second, you may wish to have more information, but the pressures of budget and deadlines (constraints frequently faced by design engineers) may not allow you the luxury of obtaining more information. In that case, you will have to suck it up and choose either option A, B, or C although you will have misgivings about all three and wish you weren't forced into this situation. But that's why your client hired you. If the decision were easy to make, your client would have done it himself/herself instead of paying you to do it.

A third possibility is that you may recognize the importance of making the right decision and the adverse consequences of making the wrong decision. You are convinced that it would be prudent to collect more information and use a more sophisticated technique for choosing among options A, B, and C. Most of what we will discuss in the remainder of this chapter is about developing and applying more sophisticated decision-making tools. Even if you don't think a more sophisticated approach is warranted for this particular situation, it is clear that design engineers are frequently confronted with much more complicated choices, and with much more at stake. Consider the complexities of selecting the best design for a new computer chip, a tunnel under the English Channel, an oil refinery, a telecommunications satellite, an automobile transmission, or a de-icing system for airplane wings.

There are several issues that warrant further discussion. First, how do we develop the list of criteria, and how can we be sure that this is an appropriate list? Second, how do we develop the evaluation scale (excellent, adequate, poor)? Third, how do we assign locations along this scale (ratings) to each option for each criterion? Finally, how do we use those ratings to select the best option? We will examine each of these questions in the remainder of this section.

Identifying Criteria

It is not always clear what the criteria should be. One thing we know is that we should include as criteria only those attributes in which at least one of the options is distinguished from the others. For example, if all three design alternatives for the automobile bumper had the same cost, we can eliminate cost as a criterion because it has no value for comparing options. However, in many cases the criteria are selected before we develop the design options. In fact, the criteria serve as the guidelines for developing the options. Under these circumstances, we would retain the cost criterion until we subsequently determine if all options are rated equal with respect to that criterion. At that point the superfluous criterion can be deleted.

Jane's uneasiness about choosing a preferred design alternative based on the information in Table 9–2 may indicate that her list of criteria is not fully capturing the important features of automobile bumper design. Perhaps the inclusion of additional criteria (aesthetics, weight, or ease of installation on existing vehicles) would make it easier for her to decide.

On the other hand, the criteria may not be clearly defined, or not clearly distinguishable from each other. In either case, we can always refine the criteria list after gaining additional insights from other phases of the design process. In particular, criteria may change as new design options are developed and raise questions not previously considered. Or the client may change their mind halfway through the project, realizing that what they now consider to be an important criterion had not previously been properly accounted for. These considerations reinforce the notion of design as an iterative process, with actions in one location of the design swamp having a ripple effect at other locations.

Criteria Metrics

Once we have settled on the criteria by which the design options are going to be evaluated, we need to establish a unit of measure or "metric" for each criterion. For example, a logical metric for the cost of an automobile bumper design is "dollars." The choice of metrics for some of the other criteria may not be so obvious. One possible metric for the recyclable criterion might be, "aluminum content as a percent of total mass." Another candidate metric for the same criterion could be, "the inverse of the number of different materials used." Perhaps we should measure recyclability by a combination of these two metrics. This is a situation where there is no "right" answer. There may be many good answers and some may be better than others. Ultimately it is a matter of client preference and your professional judgment. For a less quantifiable criterion, such as aesthetics, the metric might consist of a set of qualitative descriptors ranging from "very pretty" at one end of the scale to "very ugly" at the other end. As with many other aspects of design, establishing criteria metrics may involve subjective judgments.

Let's say we have overcome these difficulties and have established metrics for each criterion. The problem we now face is how to compare bumper design *A,* which costs $1,000 and contains 15% aluminum, with bumper design *B,* which costs $2,000 but contains 80% aluminum. Because the criteria have different metrics, we are in the awkward position of comparing apples to oranges. We can overcome this situation by establishing a common evaluation scale and mapping each criteria metric onto this scale. We now turn our attention to this task.

Evaluation Scales

In fact, we have already established a common evaluation scale for the bumper design project; it consists of three points: excellent, adequate, and poor. While not explicitly describing how we did it, we arrived at Table 9–2 by transforming each of the criteria metrics to a location on this common scale. This might have consisted of something like the following mapping for the cost criterion:

> $< \$1,000 \Rightarrow$ "excellent"
> $> \$1,000$ but $< \$2,000 \Rightarrow$ "adequate"
> $> \$2,000 \Rightarrow$ "poor"

A similar set of mapping instructions could be developed for each of the other criteria.

Now we are no longer comparing apples and oranges. Each design option is rated with respect to each criterion using this common scale. However, we found this particular common scale to be less than desirable, because it still left us uncertain as to which option should be selected.

One of the shortfalls of the scale used in Table 9–2 is that, in the absence of more detailed mapping instructions, we don't know how much better "excellent" is than "adequate" and how much better "adequate" is than "poor." An alternative approach, is to replace the verbal ratings by symbolic ratings:

> excellent = +
> adequate = o
> poor = –

However, this still does not overcome the limitations of not having a quantifiable scale. Such descriptive scales suffer from the same weakness as the ranking method depicted in Table 9–1.

Numerical scales.   This rating problem can be overcome by establishing a numerical scale and assigning locations on that scale to "excellent," "adequate," and "poor." One such scale is a low of 0 to a high of 10, and one possible assignment of ratings to that scale is excellent = 10, adequate = 5, and poor = 0. However, if we apply this scale to Table 9–2, we see that all three design options achieve a score of 20, and we are not much better off than we were before. Before, we were not sure which option was best; now our rating system tells us that they are all equal.

The fact that the three options all end up with a score of 20 is partially a result of the way we assigned the ratings to the scale. If we had made a slightly different assignment, for example, excellent = 10, adequate = 5, and poor = 1; then the results are $A = 21$, $B = 22$, $C = 21$. While very close, this approach gives a slight edge to option B. Whether or not you are now willing to select B over A and C is another issue. Remember, any such rating system is designed to be an aid to decision-making. You are under no obligation to blindly follow the dictates of an arithmetical result. If you are uncomfortable with selecting B because its superiority over A and C is not substantial enough, that's fine.

One approach to clarifying the situation would be to establish a finer scale, with two intermediate locations such as good = 6.67 and fair = 3.33 as replacements for adequate = 5. This gives us a little more flexibility in assigning ratings to each option. This is the approach we take in the next example using a ten-point scale with nine intermediate locations.[1]

Consider the problem discussed in Chapter 2 of transmitting power between two parallel shafts. The three specific design options being considered are designated as A, B, and C. The fifteen criteria against which each option will be evaluated are indicated in bold in the hierarchical objectives tree shown in Figure 2–12 and repeated here as Figure 9–1. To simplify our discussion, let's focus on just two of the criteria displayed at the bottom of Level 3 of the objectives tree in Figure 9–1 (high torque capability and bearing loads). We will measure high torque capability (abbreviated from here on as torque) in ft./lb. and bearing loads (abbreviated from here on as load) in lb. These are displayed in Table 9–3.

Since these metrics have different units, we need to develop a common scale so that we can surmount the problem of comparing apples to oranges. One such scale, involving numerical scores from 0 (useless) to 10 (perfect) is displayed in Table 9–3. There is nothing magical about the choice of 0 and 10 as the endpoints of the range for this scale. It is primarily a matter of convenience. A scale that ran from 1 to 5 might be just as useful.

You may feel uncomfortable with making the fine distinctions called for with this scale, such as distinguishing between "satisfactory" and "adequate." In those situations, a coarser scale that omits the odd-numbered performance levels in Table 9–3, may be more appropriate. As we mentioned earlier, there is no single "right way" to select a common scale. If your company/client does not specify the nature of the scale, you may pick one that you feel comfortable with.

End points.   The next step is to establish the end points of the criteria metrics that map into the end points of the common scale. This may not be as easy as it sounds, but once the end points have been established, we can assume that equally spaced intermediate values match up with the equally spaced increments on the common scale. Let's assume that in a specific design context, we've concluded that a load of 5,000 lb. should be considered "perfect" and a load of 500 lb. should be considered

---

[1]Cross, p. 128.

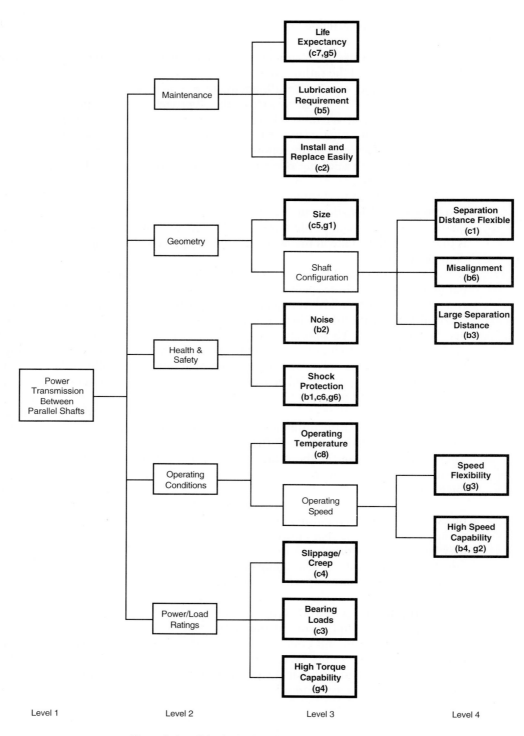

Figure 9–1. Criteria Arranged into Hierarchical Groups

TABLE 9–3. COMMON SCALE AND CRITERIA METRICS

Common Scale		Criteria Metrics	
Performance Level	Value	Torque (ft.-lb.)	Load (lb.)
Perfect	10	50,000	5,000
Excellent	9	45,150	4,550
Very Good	8	40,300	4,100
Good	7	35,450	3,650
Satisfactory	6	30,600	3,200
Adequate	5	25,750	2,750
Tolerable	4	20,900	2,300
Poor	3	16,050	1,850
Very Poor	2	11,200	1,400
Inadequate	1	6,350	950
Useless	0	1,500	500

Source: Adapted from Cross, p. 128, with permission from John Wiley & Sons, Ltd.

"useless." For the same problem, we decide that the appropriate end points for the torque criterion are 50,000 ft.-lb. and 1,500 ft.-lb. These end points and the equally spaced intermediate values are listed in Table 9–3.

We can now begin the process of selecting the best of the three options. Let's say that the three design options we are examining have the load and torque capabilities listed as the "raw scores" in Table 9–4. Using the mapping instructions contained in Table 9–3, we can identify the corresponding values on the common scale and insert them as shown in the corresponding columns in Table 9–4. The sum of the common scale values provides the total score for each option. In this case, with the comparison covering only two criteria, option *A* has a slight advantage over option *B,* and both *A* and *B* are substantially better than option *C.* However, the comparative evaluation process needs to cover all fifteen criteria before we can decide which is the best option.

In many situations it is helpful to normalize the total score by dividing the total for each option by the sum of the total scores for all options. These normalized totals are shown in the last row of Table 9–4. Note that the sum of the normalized totals is unity.

Relative Importance of Criteria

Up until now, we have treated all criteria as being equally important. Of course, there is no requirement for such treatment and in many situations, the criteria are clearly not equally important. In this subsection, we want to examine ways to deal with criteria that have unequal significance.

The simplest approach is to assign numerical weights to each criterion to indicate its importance relative to all other criteria. Ideally these weights will reflect your client's judgment of relative importance. The weights can be normalized in the same

TABLE 9–4. PARTIAL COMPARATIVE EVALUATION OF THREE DESIGN OPTIONS

		Design Option					
		A		B		C	
Criteria	Units	Raw Score	Value on Std. Scale	Raw Score	Value on Std. Scale	Raw Score	Value on Std. Scale
Load	lb.	4,200	8	4,600	9	1,100	1
Torque	ft.-lb.	35,000	7	26,000	5	21,000	4
Total			15		14		5
Normalized total			0.441		0.412		0.147

way the ratings of design options were normalized in Table 9–4. The disadvantage of this approach is the absence of an explicit procedure for developing the numerical values of the weights. You or your client may be uncomfortable relying on nothing but "feel" and "judgment" as the basis for the numerical weights. Section 9.3 and much of the remainder of this section are devoted to examining more rigorous approaches for developing these weighting factors. Generally these techniques involve making pairwise comparisons between criteria rather than assigning weights by examining the list of criteria in its entirety.

Pairwise comparisons. We start by pairing up each criterion with every other criterion one at a time and judging which of the items in each pair is more important than the other. When the number of criteria is small, this task is straightforward. Consider the four criteria listed in Table 9–2 for the bumper design problem. We use these to construct Table 9–5 as an aid for conducting pairwise comparisons. A common approach is to select a criterion listed in the first column and move across that row, entering a 1 in the corresponding column when the selected criterion is judged to be more important, and a 0 when the selected criterion is judged to be less important, than each of the other criteria. For example, the row labeled "Cost" in Table 9–5 reflects the judgment that cost is less important than damage control and drivability but more important than recyclability. The overall rankings of importance

TABLE 9–5. RANKING MATRIX AND RANKING OF CRITERIA

Criteria	Cost	Damage Control	Recyclability	Drivability	Row Total	Normalized Weights
Cost	n.a.	0	1	0	1	0.167
Damage Control	1	n.a.	1	0	2	0.333
Recyclability	0	0	n.a.	0	0	0
Drivability	1	1	1	n.a.	3	0.500

is found by adding the entries in each row. In this case, drivability (with a row total value of 3) is the most important criterion, followed by damage control, cost, and recyclability in that order. Clearly, different people could make different assignments of 1 and 0 and thereby arrive at a different ordering of the criteria. Ideally, the client should make these assignments so that our decision analysis properly incorporates the client's priorities. However, in many cases, it is not possible or feasible to extract this information from the client. Under these circumstances you may have to infer from your knowledge of your client what choices you think your client would make.

The 4×4 ranking matrix is indicated by the enclosed box in Table 9–5. Not all sixteen elements in this matrix are independent. Clearly, we don't compare the relative importance of a criterion to itself so no pairwise comparisons are needed for the four diagonal elements of the matrix. Also, not all the remaining twelve pairwise comparisons are independent. For example, if cost is deemed less important than damage control (leading to a 0 in the first row and second column of the ranking matrix), that implies that damage control is more important than cost (hence the 1 in the second row and first column). Note that in the case of four criteria, six pairwise comparisons had to be made. In general, if there are N criteria, then the number of pairwise comparisons that have to be made are

$$PC = \frac{N(N-1)}{2} \tag{9-1}$$

This approach is satisfactory as long as N is relatively small. However, consider the shaft power transfer design problem where $N = 15$. Equ. (9–1) gives

$$PC = \frac{N(N-1)}{2} = \frac{15(14)}{2} = 105$$

Conducting 105 pairwise comparisons would not only be tedious, but runs the very real risk of self contradiction.[2] A more efficient approach (and one that has a better chance of being consistent) would be to organize the criteria into logical subgroups, and construct a hierarchical structure such as the objectives tree shown in Figure 9–1. With the aid of the tree, we only have to form the pairwise comparisons within each group at each level of the hierarchy. The number of pairwise comparisons that are needed is determined by applying Equ. (9–1) separately to each group depicted in Figure 9–1. The result is

$$PC = \overbrace{\frac{3(2)}{2} + \frac{2(1)}{2}}^{\text{level 4}} + \overbrace{\frac{2(1)}{2} + \frac{2(1)}{2} + \frac{3(2)}{2} + \frac{2(1)}{2} + \frac{3(2)}{2}}^{\text{level 3}} + \overbrace{\frac{5(4)}{2}}^{\text{level 2}}$$

$$PC = 3 + 1 + 1 + 1 + 3 + 1 + 3 + 10 = 23$$

[2]Research suggests that individuals have difficulty making consistent pairwise comparisons among items in a list when the list has more than 5 to 9 items (Saaty, 1980 p. 17).

This is considerably less than the 105 comparisons required if we attempted to deal with the fifteen criteria without introducing these groupings. So an objectives tree has substantial benefits in decision-making as well in problem formulation (as originally discussed in Chapter 2).

Weighting Factors

Depending on how sophisticated your analysis needs to be in this phase of decision making, there are two ways to interpret the values listed in the row totals column of Table 9–5. One approach is to use the row totals in Table 9–5 not only as an indication of the ranking of the criteria in order of importance, but as a basis for assigning numerical weights to each of them. It is traditional to normalize weights so that they add up to 1. This is done by dividing each row total by the sum of all row totals. The normalized weights are shown in the last column of Table 9–5.

A more sophisticated view of the weighting process recognizes the limitations of the approach just described. In particular, our approach to ranking the criteria in Table 9–5 involved assigning values of either 1 or 0 to the outcome of each pairwise comparison. There was no room for intermediate measures of relative importance. If you decide that a more elaborate process for determining weights is justified, one approach is the analytical hierarchy process described in Section 9.3. We could also use the technique of multiple attribute utility theory described in Section 9.8. For now we focus on the mechanics of using objectives trees to establish weights for problems involving many objectives.

Using objectives trees to assign weights.   Regardless of the specific approach used to establish the weights, the objectives tree format is a convenient way to organize your thinking regarding weighting factors and to display the numerical values of the weighting factors. It is also a very efficient approach. As we just demonstrated, organizing the fifteen objectives for the shaft power transmission problem in the form of the objectives tree shown in Figure 9–1 allows us to determine the weighting factors using only twenty-three pairwise comparisons as opposed to the 105 pairwise comparisons that would be required if we had not subdivided the fifteen objectives into hierarchical groups.

We will call each box in Figure 9–1 a category. A category may be an objective, such as each of the five categories in Level 4 of Figure 9–1 (recall that the objectives in Figure 9–1 are identified by the boxes with the thick borders). A category may also be just a convenient way of grouping other categories; incorporating several objectives at lower levels of the hierarchy. This is the case with the "operating speed" category in Level 3 of Figure 9–1 and the "geometry" category in Level 2.

The objectives tree format calls for developing the weighting factors in two stages. First, we determine the importance of each category relative to other categories within the same group (categories that spring from a common category at the next highest level in the tree); second, we account for the importance of each group relative to other groups.

We define k as the importance of each category relative to other categories within the same group. We will normalize the k values within each group so that they add to unity:

$$\sum_{\substack{\text{within} \\ \text{each group}}} k = 1 \qquad (9\text{--}2)$$

and display the value of k in the lower left corner of the category box. We start at Level 1 in Figure 9–1; since there is only one category in Level 1 (this is the case for all objectives trees), the relative importance factor for that category is $k = 1$. This value is displayed in Figure 9–2. We then move to Level 2, where we have one group of five categories. We need ten pairwise comparisons to establish values of k for each category. Assume we have done that and displayed the resulting normalized k values in the lower left corner of each category box in Level 2 of Figure 9–2. Note that Equ. (9–2) is satisfied. Similarly, we obtain the k values for the five groups in Level 3 and the two groups in Level 4; these are also shown in Figure 9–2.

Now we need to establish the relative weight (w) of a category by modifying the k value to account for the importance of the group within which the category resides. We start at Level 1 where we define the relative weight of the single category to be unity. This value is displayed in the lower right corner of the category box.

For successive levels, we define the relative weight of a category to be the relative importance of that category within its own group (k) multiplied by the relative weight (w) of the category in the next highest level from which the group springs. At Level 2, the weight of the category in Level 1, from which this group of five categories arises, was unity ($w = 1.0$). Therefore, the relative weight of each category in Level 2 of Figure 9–2 is the same as its relative importance.

Now let's move to Level 3. Consider the "high torque capability" category as an example. As indicated in Figure 9–2, its relative importance within its own group is $k = 0.30$. The group belongs to the "power/load ratings" category at Level 2, which has a relative weight of 0.25. Hence the relative weight of the high torque capability is $(0.30)(0.25) = 0.075$ as shown in the lower right corner of the category box in Figure 9–2.

We now refine the comparative evaluation matrix shown in Table 9–4 to include the relative weights associated with each criteria from Figure 9–2. The modified version is shown in Table 9–6. We obtain the entries in Table 9–6 for the raw score and value on standard scale columns from Table 9–4. To obtain the entries in the weighted value columns, we multiply the scores in the value on standard scale column by the corresponding relative weights. The sum of the weighted values numbers for these two criteria yield the totals and normalized totals shown in the last two rows of Table 9–6. Comparing the normalized totals in Table 9–6 with those in Table 9–4 demonstrates the effect of using the weighting factors in this example. Using the weighting factors, Option B fares somewhat less favorably relative to Options A and C than in the unweighted analysis. Of course, determining which is the preferred design option (A, B, or C) requires us to conduct the operations described above for all fifteen criteria.

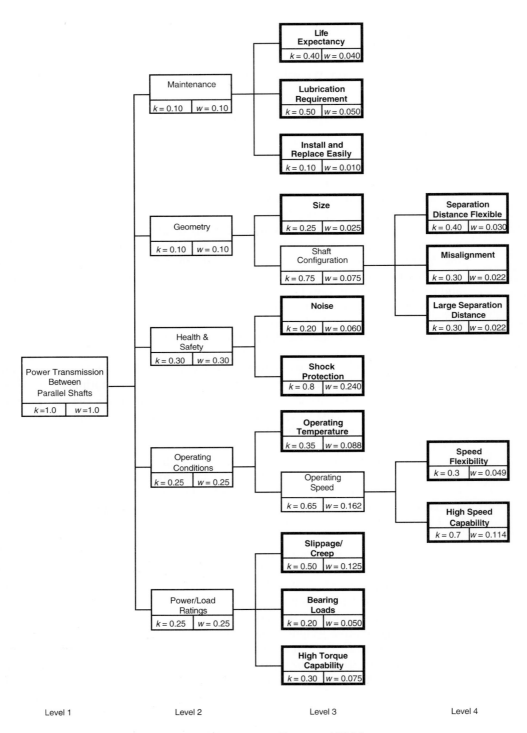

Figure 9–2. Relative Importance Factors and Weights

TABLE 9-6. PARTIAL COMPARATIVE EVALUATION WITH WEIGHTING FACTORS OF THREE DESIGN OPTIONS

Criteria	Units	Relative Weight	Design Option A Raw Score	A Value on Std. Scale	A Weighted Value	B Raw Score	B Value on Std. Scale	B Weighted Value	C Raw Score	C Value on Std. Scale	C Weighted Value
Life expectancy	years	0.0400									
Lubrication		0.0500									
Installation		0.0100									
Size	ft.^{3}	0.0250									
Separation flexible		0.0300									
Misalignment	degrees	0.0225									
Large separation	ft.	0.0225									
Noise	dB	0.0600									
Shock protection		0.2400									
Operating temp.	°F	0.0875									
Speed flexible		0.0488									
High speed	rpm	0.1138									
Slippage/creep		0.1250									
Bearing loads	lb.	0.0500	4,200	8	0.400	4,600	9	0.450	1,100	1	0.050
High torque	ft.-lb.	0.0750	35,000	7	0.525	26,000	5	0.375	21,000	4	0.300
Total		1.0000			0.925			0.825			0.350
Normalized Total					0.440			0.393			0.167

9.3. ANALYTICAL HIERARCHY PROCESS

Let us examine some of the weaknesses of the approach described in Section 9.2. In Table 9–3, we established performance metrics for each design criterion and mapped those values to a common scale. We can establish those metrics and mappings before knowing what specific design options are being considered. But the a priori judgment regarding the endpoints of the metrics may turn out to be faulty. At one extreme, the scale may be too coarse to differentiate among the design options. At the other extreme, one or more options may have a value off one end of the scale. In either case, we may have to recalibrate the scale once we have identified the design options. In addition, it may not be possible to quantify the performance levels of criteria for insertion into Table 9–3. Even if that were possible, it may be difficult to associate any given performance level with a value on the common scale.

In Figure 9–2 we assigned weights to each of the design criteria. Suppose you (more accurately, your client) decide that criterion x is twice as important as criterion y and that criterion y is three times as important as criterion z. That implies that criterion x is six times as important as criterion z. However, there is no guarantee that a head-to-head comparison of x and z would lead to that result. To the contrary, it is conceivable that a head-to-head comparison between x and z will lead to the conclusion that z is more important than $x;$ a direct contradiction to the result obtained indirectly. Hence, the weights that you assign to the criteria may depend on how you make the comparisons.

In many situations, you cannot justify going any farther. You make your best effort to develop the numbers using the approach described, and recognize that you cannot justify further fine-tuning given all the subjective considerations and gaps in performance data. On the other hand, there are situations where the stakes may be quite high. In a $10,000,000 project, it may be worth spending $10,000 of engineering time to use a more sophisticated approach. Such decisions are difficult, but that's what you get paid for. A cookbook method is no substitute for experience and sound professional judgment in what is inherently a subjective process. However, if you and your client deem that a more sophisticated approach is warranted, the Analytical Hierarchy Process (AHP) method described in this section is worth considering. You can use AHP to compare design options with respect to each criterion and also to compare the relative importance of criteria for the purpose of assigning weights to them. One of AHP's advantages is that it provides a diagnostic tool for assessing the consistency of the preference and importance ratings.

For the sake of illustrating the approach, let's use AHP to rate the three design options (*A, B, C*) for shaft power transmission with regard to torque. Recall that when we rated the options in Table 9–4, we first established a relationship between the magnitude of the torque and a value on the 0–10 common scale. Then we estimated the magnitude of the torque associated with each option. Thus, we based our ratings for each option with respect to the torque criterion by assessing the performance of each option with respect to a fixed, *absolute* scale. In contrast, AHP is based on judgments of pairwise relative preferences between options. AHP uses a *relative* scale that requires sufficient information be available to compare the design options to each other.

TABLE 9–7.   PAIRWISE COMPARISON MATRIX OF RELATIVE PREFERENCES FOR THREE DESIGN OPTIONS WITH REGARD TO TORQUE

	A	B	C
A	1	5	7
B	1/5	1	4
C	1/7	1/4	1

To do this, we first set up a 3×3 matrix **R** shown in Table 9–7. Each element r_{ij} represents our preference for the option in row i relative to the option in column j with respect to torque capability. To quantify our responses, we use the nine-point scale shown in Table 9–8 for pairwise comparison of relative preferences. The range of numerical values on this nine-point scale has been tested and found to effectively reflect the differences in preference levels used by decision makers.[3]

Starting with the first row of the relative preference matrix in Table 9–7, we have $r_{11} = 1$ since we are comparing Option A to itself. Moving to the second column, we judge that we strongly prefer Option A to Option B in terms of the torque imposed on the shafts, so $r_{12} = 5$. We also judge that Option A is very strongly preferred to Option C with regard to shaft torque, so $r_{13} = 7$.

We now move to the second row in Table 9–7 and compare Option B to each of the others. We have already indicated a strong preference for A over B and accordingly set $r_{12} = 5$. That means that the reciprocal relationship must hold; we have a strong preference against B relative to A in terms of torque capability, so $r_{21} = 1/5$. Moving across the second row, we have $r_{22} = 1$ and assign $r_{23} = 4$ as a measure of our preference for B over C with regard to torque. Finally, we use the reciprocals to complete the third row of the matrix.

TABLE 9–8.   RELATIVE SCALE FOR PREFERENCES IN PAIRWISE COMPARISONS

Judgment of Preference	Numerical Rating
Extremely preferred	9
	8
Very strongly preferred	7
	6
Strongly preferred	5
	4
Moderately preferred	3
	2
Equally preferred	1

[3]Andersen, p. 148.

Consistency of Preferences

Let's examine the consistency of the pairwise judgments we just made. Our expressed preference for A over B led to $r_{12} = 5$, and our expressed preference for B over C led to $r_{23} = 4$. If we were perfectly consistent, these preferences would be transitive, that is, our preference for A over C should be

$$r_{13} = r_{12}r_{23} = 20$$

Since our directly expressed preference for A over C was $r_{13} = 7$, our preference judgments are inconsistent. We can see from this exercise how difficult it would be to maintain consistency in a general $n \times n$ relative preference matrix for which the requirement for consistency is

$$r_{ik} = r_{ij}r_{jk} \tag{9–3}$$

for all i, j, and k. We postpone further discussion of consistency until later in this section, where we discuss a technique for quantifying the degree of inconsistency in any AHP matrix and for assessing the need to reduce the magnitude of the inconsistency.

Normalized Preferences and Overall Preferences

We divide each r_{ij} in Table 9–7 by the sum of the values in that column to arrive at the normalized preferences

$$n_{ij} = \frac{r_{ij}}{\sum\limits_{i=1}^{i=n} r_{ij}} \tag{9–4}$$

shown in Table 9–9.

A useful measure of overall preference for each option is obtained by calculating the average of the preferences in each row:

$$O_i = \frac{1}{n} \sum_{i=1}^{i=n} n_{ij} \tag{9–5}$$

We indicate those averages to the right of the matrix in Table 9–9. Thus, for example, the overall preference value for Option A is 0.709. Note that the sum of the overall preferences is unity.

TABLE 9–9.   NORMALIZED PREFERENCES AND OVERALL PREFERENCES FOR THREE DESIGN OPTIONS WITH REGARD TO TORQUE

	A	B	C	Overall Preferences
A	0.745	0.800	0.583	0.709
B	0.149	0.160	0.333	0.214
C	0.106	0.040	0.084	0.077

We can compare the overall preference values in Table 9–9 with the values listed for each option in the "Value on Std. Scale" columns and the "Torque" row of Table 9–6 (after normalization). Although the relative ranking of the three options is the same in both cases, the numerical values are different. This illustrates the sensitivity of any decision making exercise to the specific algorithm used.

We have to construct a normalized preference matrix similar to the one shown in Table 9–9 for each of the other fourteen criteria in the shaft power transmission problem. To keep from being overburdened by the arithmetic, we confine ourselves to the three criteria (loads, torque, and slippage) that form one of the groups in Level 3 of the hierarchy depicted in Figure 9–1. Assume that we establish preferences among design options *A, B,* and *C* with respect to bearing loads and slippage following the same procedure that we used with regard to torque that led to Table 9–9. In this case, we'll skip the details and display just the overall preferences in Table 9–10.

Relative Importance of Criteria

Now let's use AHP to assign relative importance factors to the three criteria in the bottom group of Level 3 in Figure 9–1. In Section 9.2 we quickly asserted that k (slippage) = 0.50, k (bearing) = 0.20, and k (torque) = 0.30. Now we use AHP to construct pairwise comparison matrices to develop these factors systematically and test the consistency of the results. We start with a nine-point scale shown in Table 9–11 for pairwise comparison of the relative importance of criteria.

The similarity between Table 9–11 and Table 9–8 are obvious, but let's reflect a moment on the differences between them and how they are used. We use Table 9–11 to compare the relative importance of the criteria by which we will judge all design options. On the other hand, Table 9–8 serves as the mechanism to compare design options with respect to each criterion. This distinction will be clarified as we proceed to use Table 9–11.

We now construct the 3×3 pairwise relative importance matrix shown in Table 9–12 for assessing the relative importance of the three criteria (slippage, bearing loads, torque). This matrix is the counterpart to the relative preference matrix in Table 9–7. We use Table 9–11 to fill in the cells of this matrix; let's assume that process leads to the values shown in Table 9–12. For example, the numbers reflect our

TABLE 9–10.   OVERALL PREFERENCES FOR THREE DESIGN OPTIONS WITH REGARD TO BEARING LOADS AND SLIPPAGE

	Overall Preferences	
	Bearing Loads	Slippage
A	0.319	0.379
B	0.074	0.425
C	0.607	0.196

TABLE 9–11.   NINE-POINT RELATIVE SCALE
FOR IMPORTANCE IN PAIRWISE COMPARISONS

Judgment of Preference	Numerical Rating
Extremely more important	9
	8
Very strongly more important	7
	6
Strongly more important	5
	4
Moderately more important	3
	2
Equally important	1

Source: Saaty and Vargas, p. 18.

judgment that slippage is moderately less important than bearing loads while bearing loads are marginally more important (less than moderately more important) than torque.

As we did with Table 9–7, we normalize the importance ratings in Table 9–12. We divide each cell entry by the column total and then calculate the overall importance ratings for each criterion as the average of the normalized values in that row. The results are shown in Table 9–13.

We can treat these overall importance ratings as weights and combine them with the overall preference values in Tables 9–9 and 9–10 and determine a weighted sum similar to what we did in Table 9–6. We end up with the following scores for the three design options with respect to the slippage, bearing loads, and torque criteria:

$$A = (0.211)(0.379) + (0.549)(0.319) + (0.241)(0.709) = 0.426$$
$$B = (0.211)(0.425) + (0.549)(0.074) + (0.241)(0.214) = 0.182$$
$$C = (0.211)(0.196) + (0.549)(0.607) + (0.241)(0.077) = 0.393$$

This approach to developing weighting factors is more sophisticated (and time consuming) than that used in Table 9–5, where we only assigned values of 1 or 0 to each element of the matrix.

TABLE 9–12.   PAIRWISE COMPARISON MATRIX OF RELATIVE
IMPORTANCE RATINGS FOR THREE DESIGN CRITERIA

	Slippage	Bearing Loads	Torque
Slippage	1	1/3	1
Bearing Loads	3	1	2
Torque	1	1/2	1

TABLE 9–13. NORMALIZED IMPORTANCE AND OVERALL IMPORTANCE FOR THREE DESIGN CRITERIA

	Slippage	Bearing Loads	Torque	Overall Importance
Slippage	0.200	0.182	0.250	0.211
Bearing Loads	0.600	0.546	0.500	0.549
Torque	0.200	0.273	0.250	0.241

Measure of Consistency

The relative preference and importance matrices (Tables 9–7 and 9–12) represent individual or collective judgments regarding pairwise comparisons. As such, there are opportunities for inconsistencies to arise as we have already seen. AHP provides the consistency ratio CR as a measure of the extent of those inconsistencies. CR is the ratio of a consistency index CI for the given pairwise comparison matrix to the value of the same consistency index for a randomly generated pairwise comparison matrix:

$$CR = \frac{CI}{CI_{random}} \tag{9–6}$$

where the consistency index CI is defined as[4]

$$CI = \frac{\lambda_{max} - n}{n - 1} \tag{9–7}$$

The λ_{max} term in Equ. (9–7) is the largest positive eigenvalue of the pairwise comparison matrix. CI has been calculated for many randomly generated pairwise comparison matrices of order n. The average results for CI_{random} are listed in Table 9–14.

TABLE 9–14. AVERAGE CONSISTENCY INDEX FOR RANDOMLY GENERATED PAIRWISE COMPARISON MATRICES

n	CI_{random}
2	0.00
3	0.52
4	0.90
5	1.12
6	1.24
7	1.32
8	1.41

Source: Saaty, p. 24.

[4]See the Appendix to this section for a detailed derivation of this expression.

The suggested rule-of-thumb is that if the consistency ratio CR exceeds 0.10, you should seriously consider modifying the elements of the pairwise comparison matrix with the aim of improving the consistency.

Consistency ratios for shaft transmission problem. Here we determine CR for the relative preference matrix included in Table 9–7:

$$\mathbf{R} = \begin{pmatrix} 1 & 5 & 7 \\ 1/5 & 1 & 4 \\ 1/7 & 1/4 & 1 \end{pmatrix}$$

The eigenvalues of this matrix are the roots of the characteristic equation

$$|\mathbf{R} - \lambda \mathbf{I}| = 0$$

or

$$\begin{vmatrix} (1 - \lambda) & 5 & 7 \\ 1/5 & (1 - \lambda) & 4 \\ 1/7 & 1/4 & (1 - \lambda) \end{vmatrix} = 0$$

After expanding and collecting powers of λ we get

$$\lambda^3 + \lambda^2 + 0.857\lambda - 1.207 = 0$$

the largest root of which is

$$\lambda_{\max} = 3.124$$

Then Equ. (9–7) gives

$$CI = \frac{3.124 - 3}{3 - 1} = 0.062$$

Using this together with the CI_{random} value associated with $n = 3$ in Table 9–14, Equ. (9–6) yields

$$CR = \frac{0.062}{0.52} = 0.119$$

This does not meet our consistency criterion that $CR \leq 0.10$, so we should reevaluate the judgments we used to construct Equ. (9–3).

We can perform a similar test for the relative importance matrix in Table 9–12:

$$\mathbf{R} = \begin{pmatrix} 1 & 1/3 & 1 \\ 3 & 1 & 2 \\ 1 & 1/2 & 1 \end{pmatrix}$$

Its maximum eigenvalue is found from

$$\begin{vmatrix} (1-\lambda) & 1/3 & 1 \\ 3 & (1-\lambda) & 2 \\ 1 & 1/2 & (1-\lambda) \end{vmatrix} = 0$$

to be

$$\lambda_{max} = 3.018$$

This leads to

$$CR = \frac{0.018}{0.52} = 0.035$$

which does meet the consistency criterion.

Appendix 9A: Theoretical Basis for AHP Consistency Measure

In this appendix we trace the source of Equ. (9–7). The development is lengthy and subtle, but it clearly reveals the justification for using Equ. (9–7) as the basis for a consistency measurement. Our approach is to consider a matrix that is known to be consistent and which has certain properties in common with the relative preference and importance matrices. The measure of inconsistency is determined by comparing other properties of the relative preference and importance matrices with those of the known consistent matrix. We will develop the approach using a 3×3 matrix and indicate its generalization to an $n \times n$ matrix. Our discussion will be in terms of a relative preference matrix; it is also valid for relative importance matrices. We will continue to use the shaft power transmission problem as the context for our discussion, and focus specifically on the torque criterion.

A consistent relative preference matrix.   Suppose we had a complete and incontrovertible source of information and insight that allowed us to specify overall preferences for the three design options A, B, and C with respect to the torque criterion. To simplify the notation in the subsequent algebra, let's relabel the design options as A_1, A_2, and A_3 for the 3×3 case and A_i ($i = 1, 2, \ldots, n$) for the $n \times n$ case.

Let the overall preference for option A_1 with respect to this criterion be p_1, with similar notation for the absolute preferences of the A_2 and A_3 options with respect to the same criterion. Starting with the identity for the absolute preference for p_1

$$p_1 = p_1 \tag{9–8}$$

we can transform the left side into

$$\frac{1}{3}\left[\left(\frac{p_1}{p_1}\right)p_1 + \left(\frac{p_1}{p_2}\right)p_2 + \left(\frac{p_1}{p_3}\right)p_3\right] = p_1 \tag{9–9}$$

or

$$\left(\frac{p_1}{p_1}\right)p_1 + \left(\frac{p_1}{p_2}\right)p_2 + \left(\frac{p_1}{p_3}\right)p_3 - 3p_1 = 0 \tag{9–10}$$

You may wonder why we do such an apparently pointless manipulation. The answer is that experience tells us that it will lead to something useful. If you can be patient as the point unfolds, your patience will be amply rewarded. The person who developed this technique (Saaty) probably tried many other approaches that led to dead ends before he recognized the value of this approach. We are benefiting from his experience, so we can start immediately with an approach that we know will be successful.

Similarly, we can start with the overall preference identities for p_2 and p_3 and develop equations that are counterparts to Equ. (9–10):

$$\left(\frac{p_2}{p_1}\right)p_1 + \left(\frac{p_2}{p_2}\right)p_2 + \left(\frac{p_2}{p_3}\right)p_3 - 3p_2 = 0 \qquad (9\text{–}11)$$

$$\left(\frac{p_3}{p_1}\right)p_1 + \left(\frac{p_3}{p_2}\right)p_2 + \left(\frac{p_3}{p_3}\right)p_3 - 3p_3 = 0 \qquad (9\text{–}12)$$

In matrix form, Equs. (9–10), (9–11), and (9–12) are

$$\begin{bmatrix} \dfrac{p_1}{p_1} & \dfrac{p_1}{p_2} & \dfrac{p_1}{p_3} \\[2ex] \dfrac{p_2}{p_1} & \dfrac{p_2}{p_2} & \dfrac{p_2}{p_3} \\[2ex] \dfrac{p_3}{p_1} & \dfrac{p_3}{p_2} & \dfrac{p_3}{p_3} \end{bmatrix} \begin{bmatrix} p_1 \\ p_2 \\ p_3 \end{bmatrix} = 3 \begin{bmatrix} p_1 \\ p_2 \\ p_3 \end{bmatrix} \qquad (9\text{–}13)$$

We'll rewrite Equ. (9–13) in more compact form as

$$\mathbf{Rp} = 3\mathbf{p} \qquad (9\text{–}14)$$

where $\mathbf{R}$ is the matrix whose elements are

$$r_{ij} = \frac{p_i}{p_j}$$

and $\mathbf{p}$ is a column vector whose rows are p_i.

In the $n \times n$ case, the matrix form is

$$\begin{bmatrix} \dfrac{p_1}{p_1} & \dfrac{p_1}{p_2} & \cdots & \cdots & \dfrac{p_1}{p_n} \\[2ex] \dfrac{p_2}{p_1} & \dfrac{p_2}{p_2} & \cdots & \cdots & \dfrac{p_2}{p_n} \\[2ex] \vdots & & \ddots & & \vdots \\[1ex] \vdots & & & \ddots & \vdots \\[1ex] \dfrac{p_n}{p_1} & \dfrac{p_n}{p_2} & \cdots & \cdots & \dfrac{p_n}{p_n} \end{bmatrix} \begin{bmatrix} p_1 \\ p_2 \\ \vdots \\ \vdots \\ p_n \end{bmatrix} = n \begin{bmatrix} p_1 \\ p_2 \\ \vdots \\ \vdots \\ p_n \end{bmatrix} \qquad (9\text{–}15)$$

or

$$\mathbf{Rp} = n\mathbf{p} \qquad (9\text{--}16)$$

We can refer to the 3×3 matrix in Equ. (9–14) as the relative preference matrix for the design options A_1, A_2, and A_3 [with an obvious generalization for Equ. (9–16)]. This label is appropriate since each row in the matrix consists of ratios of the preference of one of the design options to the preferences of all the other design options for this criterion. These are the relative preferences. Another important similarity between $\mathbf{R}$ and the relative preference matrix included in Table 9–7 is that they are both reciprocal matrices:

$$r_{ij} = \frac{1}{r_{ji}}$$

We will take advantage of the properties of reciprocal matrices to develop our measure of consistency.

λ_{max} for consistent relative preference matrix.   We arrived at Equs. (9–9), (9–10), and (9–11) by starting with the assumption that the p_i are known. Now let us be more realistic and recognize that we will not start with the p_i. On the contrary, our starting point will be the relative preference matrix (as it was when we built Table 9–9), and we will use it to determine the overall preferences p_i. So, now we treat the p_i as the unknowns. From this perspective, Equs. (9–10) through (9–12) are linear homogeneous (their right hand sides are 0) equations in the p_i. One (trivial) solution is $p_1 = p_2 = p_3 = 0$. This solution of course, is of no interest to us since we are seeking nontrivial preference measures. The only way there can be a nontrivial solution is if the determinant of the coefficient matrix vanishes. In matrix form, this condition is expressed as

$$|\mathbf{R} - 3\mathbf{I}| = 0 \qquad (9\text{--}17)$$

This is the classic eigenvalue problem. The eigenvalues are the roots of Equ. (9–17). In expanded form we have

$$\begin{bmatrix} \left(\dfrac{p_1}{p_1} - 3\right) & \dfrac{p_1}{p_2} & \dfrac{p_1}{p_3} \\[2ex] \dfrac{p_2}{p_1} & \left(\dfrac{p_2}{p_2} - 3\right) & \dfrac{p_2}{p_3} \\[2ex] \dfrac{p_3}{p_1} & \dfrac{p_3}{p_2} & \left(\dfrac{p_3}{p_3} - 3\right) \end{bmatrix} = 0 \qquad (9\text{--}18)$$

Oh, one more minor adjustment. If our starting point is the relative preference matrix $\mathbf{R}$, then not only is the $\mathbf{p}$ vector unknown in Equ. (9–16), but the coefficient n is also unknown. So let us replace the known value $n = 3$ in Equ. (9–18) by the unknown eigenvalues λ. Thus, Equ. (9–18) becomes the following equation for finding λ:

$$
\left[
\begin{array}{ccc}
\left(\dfrac{p_1}{p_1} - \lambda\right) & \dfrac{p_1}{p_2} & \dfrac{p_1}{p_3} \\[2ex]
\dfrac{p_2}{p_1} & \left(\dfrac{p_2}{p_2} - \lambda\right) & \dfrac{p_2}{p_3} \\[2ex]
\dfrac{p_3}{p_1} & \dfrac{p_3}{p_2} & \left(\dfrac{p_3}{p_3} - \lambda\right)
\end{array}
\right] = 0
\qquad (9\text{–}19)
$$

This is a cubic equation, the three roots of which are the eigenvalues $\lambda_1, \lambda_2, \lambda_3$.

We now invoke some results from matrix theory. First, the trace of a matrix (the sum of the elements along the main diagonal) is equal to the sum of the eigenvalues:

$$
\text{tr } \mathbf{R} = \lambda_1 + \lambda_2 + \lambda_3 \qquad (9\text{–}20)
$$

But since $\mathbf{R}$ is a reciprocal matrix, its diagonal elements are all unity, that is, $r_{ij} = 1$, for all $i = j$. Therefore,

$$
\text{tr } \mathbf{R} = 3 \qquad (9\text{–}21)
$$

and Equ. (9–20) gives

$$
\lambda_1 + \lambda_2 + \lambda_3 = 3 \qquad (9\text{–}22)
$$

Hold on, we are almost finished making our point. Note from Equ. (9–13) that the second and third rows in the $\mathbf{R}$ matrix are multiples of the first row. Since $\mathbf{R}$ has only one independent row, in the language of matrix theory, $\mathbf{R}$ has a rank of one. That, in turn, means that $\mathbf{R}$ has only one non-zero eigenvalue. So we can set $\lambda_2 = \lambda_3 = 0$ and Equ. (9–22) can be rewritten as

$$
\lambda_{\max} = 3
$$

Clearly, the result for a consistent relative preference matrix of order n is

$$
\lambda_{\max} = n \qquad (9\text{–}23)
$$

Consistency index.   Finally, we can see the light at the end of the tunnel! We've traveled an admittedly tortuous route to establish that the maximum eigenvalue of a consistent relative preference matrix is equal to the order of the matrix. This property serves as the basis for quantifying the inconsistency of any relative preference matrix. In particular, we define the consistency index for any relative preference matrix of order n as

$$
CI = \frac{\lambda_{\max} - n}{n - 1}
$$

Thus, if the relative preference matrix that we construct is consistent, then its $\lambda_{\max}$ will be equal to n and $CI = 0$. Then the CR for that matrix (see Equ. 9–6) will also be 0 and our consistency test that $CR < 0.10$ will be satisfied. To the extent that our relative preference matrix is inconsistent, we will find $\lambda_{\max} > n$, leading to a non-zero CI and CR. To prevent the magnitude of the inconsistency from being unduly influenced by the size of the matrix, we introduce the $(n - 1)$ term as a normalizer of the $\lambda_{\max} - n$ difference.

9.4. DECISION MAKING UNDER UNCERTAINTY

In our earlier discussion about choosing among alternatives, we assumed that we knew all future costs (in the Chapter 8 economic analysis) or performance characteristics (in the Sec. 9.2 multiple attribute decision analysis). Of course, there are many uncertainties for systems that are still in the design stage regarding future costs, events, and performance characteristics. A sophisticated decision rule should attempt to incorporate these uncertainties into the decision making process.

Phases of Decision Making

We now describe the five major phases in selecting among alternatives when faced with uncertainties. You may find it helpful in any such decision making exercise to identify each phase explicitly to ensure that you have laid the proper groundwork for making the decision.

Phase 1: Specify all the alternatives to be included in the exercise.

In Chapter 6 we talked about the value of generating many design concepts before choosing the best one. However, the resources required to include all alternatives in the decision making process may not be available. In that situation, you need to give considerable thought to reduce the alternatives to a manageable number, by eliminating ahead of time the least promising options, or by combining and consolidating them. This phase requires careful consideration, because the subsequent decision making procedure can only help you select from the alternatives that you specify in this phase. It can be very frustrating to complete a decision analysis only to discover that you wanted to include additional options, or to modify the existing options. It is a good idea to invest time during this phase to ensure that the alternatives being examined are the ones that really interest you, and that you have properly defined them.

Phase 2: Specify all relevant events that might occur subsequent to making the decision, which could affect the outcome of the decision, but over which the decision maker has no control.

These events are the source of uncertainty in the analysis. Will OPEC double the price of oil five years from now? Will the government tighten food safety regulations next year? Will the design be exposed to a 100 mph wind during its lifetime? As in Phase 1, you want to keep the number of relevant events to be included at a manageable level. You want to specify only those events that may significantly affect the outcome. Remember, the decision analysis procedure (like many engineering analysis techniques) involves using a relatively simple model to help improve your understanding of a very complicated real life situation. The simpler the model, the easier it will be to utilize it; you might not have the resources needed to use a more sophisticated model. But don't leave something out that you feel is crucial. Striking the appropriate balance between simplicity and completeness is generally very difficult.

Phase 3: Estimate the probability of occurrence of every relevant event specified in Phase 2.

This is the place where we have to use our crystal ball to quantify the likely occurrence of the relevant events. While no one has a flawless crystal ball, there are a variety of techniques available for developing such probability estimates in a reasonably systematic and unbiased manner. It is imperative that the set of relevant events is complete; that is, the sum of their probabilities of occurring is unity.

Phase 4: Quantify the outcome of every possible combination of decision alternatives and relevant events.

We use the term outcome to mean the value of the design parameter or performance characteristic that is the basis for choosing among the design options. In the case of multiple attributes, the outcome would be the final scores obtained via a weighted sum of the individual scores, as discussed in Section 9.2. If we have successfully completed Phases 1, 2, and 3, Phase 4 is a relatively straightforward step.

Phase 5: Use a predetermined decision rule to select the design alternative that yields the most desirable result.

We must select a decision rule to conduct this phase of the analysis. The most widely used decision rule is to choose the alternative that either maximizes the expected monetary value or minimizes the expected cost of the outcome. We will examine other decision rules later on in this chapter.

Case Study of Designing a Strategy for Field Testing Under Uncertain Conditions

Suppose you are a test engineer for an equipment manufacturer. You want to use an old portable electric generator to supply power for a test being conducted at a remote field test facility. You estimate the cost of conducting the test to be $10,000 if the generator works. If the generator fails during the test, it will cost $25,000 for repairs and added test costs. This constitutes *Phase 2 : Specify the relevant events.* Either the generator works or it doesn't. Once you begin the test, you have no control over whether the generator works. You further conclude that no other uncertainties are relevant to this problem.

Your immediate responsibility is to decide whether to spend $15,000 to overhaul the generator before testing begins. This is *Phase 1: Specify the alternatives.* Your choices are to start the test with the generator in its current condition or overhaul the generator before using it. No other options are available to you. We introduce the symbols U and O to represent the use and overhaul options respectively.

Suppose your best estimate is that there is a 70% chance the generator will work if used in its current condition. That is, $\Pr(WU) = 0.70$ and $\Pr(FU) = 0.30$ where we use W and F to represent "working" and "failing" respectively. You also are convinced that the generator will not fail during the test if it is overhauled [$\Pr(WO) = 1.00$, $\Pr(FO) = 0.00$]. This constitutes *Phase 3: Estimate the probability of each relevant event occurring.* Note that $\Pr(WU) + \Pr(FU) = 1$ and $\Pr(WO) + \Pr(FO) = 1$. This ensures that we have a complete set of events associated with each design option and that all other events are irrelevant to this formulation of the problem.

The information provided in the above problem statement is sufficient to quantify the cost of conducting the test under each combination of design choice and event. We summarize the costs in Table 9–15. This constitutes *Phase 4: Quantifying the outcome of every combination of decision alternatives and relevant events.*

TABLE 9–15.   COST OF FIELD TESTING UNDER DIFFERENT OPTIONS AND EVENTS

	Events	
Design Options	Generator Works (W)	Generator Fails (F)
Use generator in current condition (U)	$10,000	$35,000
Overhaul generator before using (O)	$25,000	n.a.

We come now to *Phase 5: Application of a decision rule.* Let's say that we will choose the option that minimizes the expected cost of testing. Let $C(WU)$ be the cost of the test if the generator works without it first being overhauled, with a similar notation for the overhaul option. The expected cost (EC) of using the generator in its current condition is

$$EC(U) = C(WU)\Pr(WU) + C(FU)\Pr(FU) = (10)(0.70) + (35)(0.30) = 17.5$$

The expected cost of overhauling the generator before using it is

$$EC(O) = C(WO)\Pr(WO) + C(FO)\Pr(FO) = (25)(1.00) = 25.0$$

Since using the generator in its current condition provides the lowest expected cost of testing, that is, $EC(U) < EC(O)$, that is the preferred option. Clearly, variations in the decision rule, or set of relevant events and associated probabilities could lead to a different choice.

Handling Additional Uncertainties

The format used in Table 9–15 for displaying information and facilitating decision making becomes awkward if we wish to incorporate additional uncertainties into the analysis. For example, in the original formulation of the problem, we lumped all failure possibilities into a single event F. Suppose now that we want our decision analysis to be more realistic (and hence more complicated) by reflecting different failure possibilities. In particular, we want to consider the possibility that the generator could fail at three different stages in the test program, leading to different amounts of damage (D_1, D_2, D_3) to other equipment, and hence differences in overall test costs. Note that what we are doing here is modifying Phase 2 of our analysis to include more failure events. We also need to modify Phase 3 to specify the probability of those failure

TABLE 9–16. EFFECT OF DAMAGE ON TESTING
COSTS

Probability of damage (D_i) occurring as a result
of generator failure

$\Pr(D_1|F) = 0.10$	$\Pr(D_2|F) = 0.30$	$\Pr(D_3|F) = 0.60$

Cost of testing as a result of generator failure

D_1	D_2	D_3
$20,000	$40,000	$90,000

events occurring and Phase 4 to quantify the effect that those failures have on the test-ing costs. Assume that we have collected the additional data needed for this more so-phisticated analysis. We summarize this information in Table 9–16, where the nota-tion $\Pr(x|y)$ means the probability of x occurring given that y has occurred. We call this the conditional probability of x, given y. Note that the set of "damage" events is complete, that is, $\Pr(D_1|F) + \Pr(D_2|F) + \Pr(D_3|F) = 1.00$.

The tabular format is not conducive to combining the information in Tables 9–15 and 9–16 to perform the calculations needed to select the preferred option. In the next section we introduce a new graphical tool for displaying this kind of informa-tion and providing a convenient format for selecting the best option from the avail-able choices.

9.5. DECISION TREES

When decisions involve more than one set of uncertainties (such as the modified problem presented at the end of Section 9.4) or more than one decision point, a tree-type format is a more useful decision aid than a tabular format. We will use both ver-sions of the field test problem from Section 9.4 to illustrate the process of constructing and solving decision trees.

Constructing Decision Trees

Decision trees are graphical representations that consist of four elements:

- Branches: straight lines (———) that terminate at each end with one of three types of nodes
- Decision nodes: depicted as squares (□)
- Event, or chance nodes: depicted as circles (○)
- Payoff nodes: depicted as price tags (◁)

Every decision tree is a collection of branches connected to each other by nodes. We now outline the four stages in constructing decision trees.

Step 1: Initial Node

All decision trees are constructed from the left end starting with a decision node that represents the first decision point.

Step 2: Initial Branches

Construct one branch emanating from that decision node to represent each option. Label each branch to identify the option that it represents. For our example problem, we need two branches, one to represent the option of using the generator in its current condition, and the other to represent the option of overhauling the generator before using it. We show the beginning of our decision tree in Figure 9–3.

Step 3: Terminate Each Branch in a Node

The right end of each branch must terminate in one of the three types of nodes (decision, chance, or payoff). To determine which type is needed, we ask, "If we proceed along this branch, are we next confronted with another decision, or an event over which we have no control?" If we are confronted with another decision, we terminate that branch with a decision node and go to Step 4a. If we are confronted with an event over which we have no control, we terminate that branch with a chance node and go to Step 4b. If we are not confronted with a decision or an event, we terminate that branch with a payoff node and go to Step 4c.

For this example problem, if we decide to use the generator in its current condition, we are next confronted with the event of a possible generator failure. Hence, according to Step 4b, that branch in Figure 9–3 terminates in a chance node. On the other hand, if we choose to overhaul the generator before using it, there are no further decisions to be made and no further uncertainties to encounter. According to Step 4c, that branch in Figure 9–3 terminates in a payoff node. The current form of the decision tree is shown in Figure 9–4.

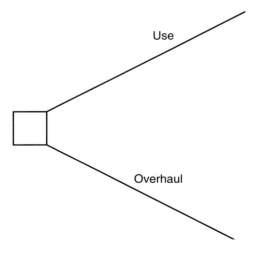

Figure 9–3.   Stage 1 of Decision Tree Construction

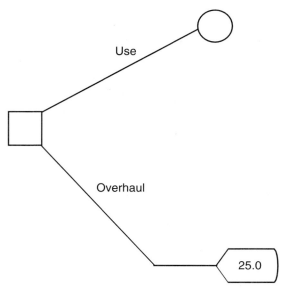

Figure 9–4.   Stage 2 of Decision Tree Construction

Step 4: Repeat Cycle until Completion

Step 4a: If a branch terminates in a decision node, return to Step 2 to construct branches emanating from this node to represent the options being considered at this decision point.

Step 4b: If a branch terminates in a chance node, construct one branch emanating from the chance node to represent each possible outcome of the event. Each branch emanating from a chance node should be labeled with the probability of that outcome occurring. After constructing and labeling these branches, return to Step 3.

    In this example problem, we draw two branches emanating from the chance node to represent the probabilities of the generator working or failing. See Figure 9–5 for the current form of the decision tree.

Step 4c: If a branch terminates with a payoff node, no further branches are constructed. Every combination of branches and nodes that begin with the originating decision node and end with a payoff node is called a path. Each payoff node is labeled with the value of the outcome associated with the path.

    To continue with the example problem, we return to Step 3. We note that in the original formulation of the problem, once the generator either works or fails, there are no further decisions or uncertainties.[5] That leads us to Step 4c and completion of the decision tree. The completed decision tree is shown in Figure 9–6.

---

[5]The revised formulation of the problem involving the additional uncertainties over the level of damage caused by generator failure will be analyzed later in this section.

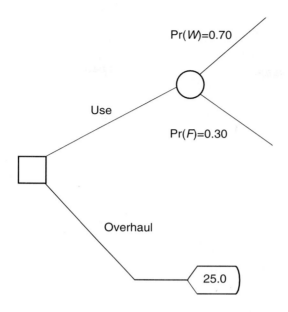

Figure 9–5.   Stage 3 of Decision Tree Construction

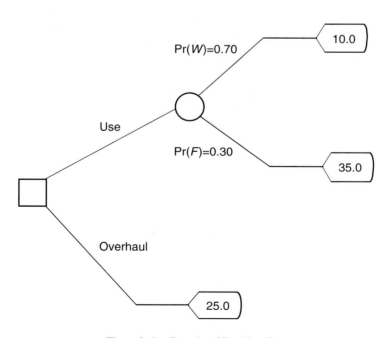

Figure 9–6.   Completed Decision Tree

The completed decision tree in Figure 9–6 is a graphical representation of the information contained in Table 9–15. The tree incorporates all decision alternatives and uncertainties into a convenient format. Even if the decision tree is not used any further, it is a valuable tool for clarifying the decision options being considered, the uncertainties being accounted for, the probabilities of those uncertainties occurring, and the magnitude of the outcomes. In principle, there is no limit on the number of decision alternatives and uncertainties that can be incorporated into a decision tree.

Solving Decision Trees

The general approach to solving a decision tree is to move backwards through the tree (from right to left) until we reach the originating decision node (recall from Step 1 in constructing trees, all decision trees start at the left end with a decision node). The rules for solving decision trees are summarized in the following steps.

Step 1: Trace Back From Payoff Node
Select any payoff node and move to the left to trace the branch from that node back to the next node.

Step 2a: Calculate Expected Value at Chance Node
If the next node encountered is a chance node, calculate the expected value of all nodes (chance, decision, or payoff) connected immediately to the right of the encountered node. Insert that expected value into the encountered node.

For the decision tree in Figure 9–6, let us start with the payoff node at the top of the tree (its value is 10.0). When we trace its branch back to the left, the first node we encounter is a chance node. That chance node has one other payoff node connected to it (its value is 35.0). The expected value of those two payoffs is

$$EC = (0.70)(10.0) + (0.30)(35.0) = 17.5$$

We insert this value into the encountered node. That corner of the decision tree is now depicted in Figure 9–7.

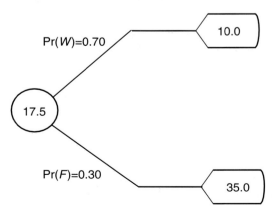

Figure 9–7. Stage 1 of Decision Tree Solution

Step 2b: Select Best Option at Decision Node

If the next node encountered is a decision node, select the branch that leads to the best value among all nodes (chance, decision, or payoff) connected immediately to the right of the encountered decision node. Transfer the best value to the decision node and "prune" the rejected branches (symbolized by a double hash mark across the branch).

In the example problem, if we continue moving to the left, we next encounter the decision node. Of the two branches emanating from the decision node, the top one has an expected cost of 17.5 (shown in Figure 9–7) and the bottom one (shown in Figure 9–6) has an expected cost of 25.0. Since our decision rule is to minimize the expected costs, we select the top branch, transfer its expected value to the decision node, and prune the bottom branch. The results are shown in Figure 9–8.

Step 3: Repeat Steps 2a and 2b Until All Nodes Have Been Accounted For

The numerical value assigned to each appropriate node in Steps 2a and 2b captures all relevant information to the right of that node. Further stages in solving the tree do not utilize portions of the tree to the right of any node to which a

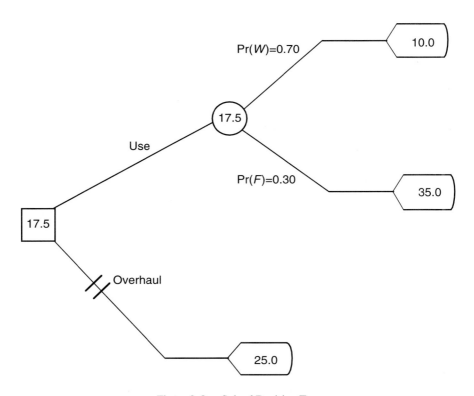

Figure 9–8.   Solved Decision Tree

value was assigned in Steps 2a and 2b. For the example problem, there are no other nodes so we have completed the solution.

Documentary Value of Decision Trees

This solution (do not overhaul) is identical to the one we reached in Section 9.4 without using a decision tree. It could be argued that, for this particular problem, there is no advantage of a decision tree over the tables and equations used in Section 9.4. The power of the tree format will soon become apparent as we consider more complicated combinations of options and uncertainties.

However, even for the simple problem just solved, decision trees such as the one displayed in Figure 9–8 provide a complete compact history of the decision making process. All decisions, all uncertainties and their assumed probabilities of occurring, and all payoffs are explicitly identified in an easy-to-read format. You can present it to your client or use it several months later, after you've moved on to other projects, to help you recall the details.

For this problem, the tree allows us to recapitulate our options, and recall the reasons for choosing not to overhaul the generator. It is clear from the tree that if we choose not to overhaul the generator, and if the generator works (there is a 0.70 probability that it will work), then the test will cost us 10.0. If however, the generator fails, the cost of testing escalates to 35.0. However, since the expected cost of 17.5 is less than the 25.0 cost of the overhaul, the option of not overhauling is preferred.

Sensitivity Analysis

One of the difficulties of dealing with uncertainties is coming up with the probabilities of the uncertain events occurring. In this case, how certain are we that the probability of the generator failing is 0.30? What if it is 0.40? We don't need to agonize too much over the accuracy of the estimate because we can easily test to see if the decision to use the generator without overhauling it is sensitive to changes in the probability estimate. We know that if the probability of failure is 0.30, our best option is to use the generator without overhauling it. We can now ask, "How much will the probability of failure have to be before we would choose to overhaul the generator before using it?" This question can be answered easily by modifying the original decision tree so that $Pr(F) = x$. That means, of course, that $Pr(W) = 1 - x$. The modified tree is shown in Figure 9–9.

We calculate the expected cost at the chance node

$$EC(U) = (1 - x)(10.0) + (x)(35.0) = 10.0 + 25.0(x)$$

The brink of switching options occurs when the expected costs of the options emanating from the decision node are the same, that is, when $EC(U) = EC(O)$. This leads to

$$10.0 + 25.0(x) = 25.0$$

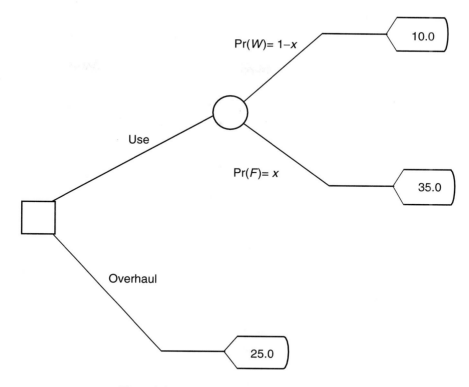

Figure 9–9.   Decision Tree for Sensitivity Analysis

Solving for x yields

$$x = 0.60$$

That is, overhauling the generator is not justified unless the probability of generator failure is at least 0.60 [$Pr(F) > 0.60$]. Hence, we do not have to know precisely what the probability of generator failure is. All we need to be able to estimate is whether it is greater than or less than 0.60. The tree format is conducive to carrying out these kinds of sensitivity studies.

Common Costs

Notice that all payoff nodes included the 10.0 cost of conducting the test. Payoff nodes added the cost of failures or overhaul to the base testing cost. This raises the question of whether we could have subtracted this common cost from all payoffs and only included the incremental costs associated with failure or overhaul. If we go back to Figure 9–6 and subtract 10.0 from each payoff node, we find that the relative expected

costs of the option branches are not affected by such a move and the same option will
be chosen. Similarly, regarding Figure 9–9, reducing the cost of each payoff node by
10.0 does not affect the value of $x = 0.60$ that defines the case where both options have
equal expected costs. However, we will see situations later in the chapter where the
exclusion of common costs can affect the outcome. For this reason it is a good idea to
always include common costs.

Conditional Uncertainties

Let's now look at the more complicated formulation introduced in Section 9.4, involving different levels of damage occurring depending on when the generator failed. That
is, let us replace the cost of failure by the three costs of failure and their associated
probabilities summarized in Table 9–16. We do this by inserting the additional
branches and probabilities to the left of the probability node in Figure 9–6. This yields
the tree depicted in Figure 9–10. Note that the damage probabilities are conditional
on the prior event that the generator failed.

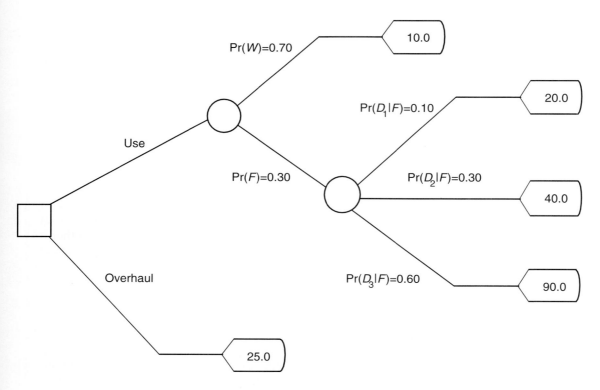

Figure 9–10.   Decision Tree with Conditional Uncertainties

We solve the tree using the same steps outlined before, starting at the payoff nodes at the right end of each branch and moving to the left until we reach the original decision node at the extreme left end of the tree. The solution to the tree shown in Figure 9–10 is depicted in Figure 9–11. The expected cost of failure has risen to 68.0 from 35.0. This is enough to increase the expected costs of not overhauling the generator to 27.4. Since this is now higher than the cost of overhauling, the best choice under these circumstances is to overhaul the generator.

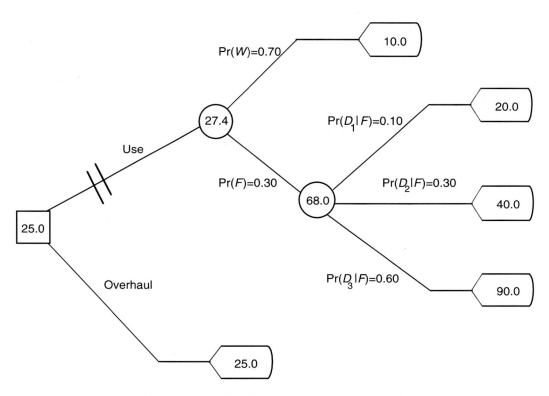

Figure 9–11.   Solved Decision Tree with Conditional Uncertainties

Sequential Design Decisions

Many decisions must be made over the course of a design project. Some decisions must be made before others, and those earlier decisions usually affect the choices that are available down the road. Let's modify the field test problem to illustrate how decision trees can help identify the best sequence of decisions (only one decision was required in the above problem; whether or not to overhaul the generator).

Let's add another option to the field test problem—investing $5,000 in a 100% accurate diagnosis of the generator. We don't know what the result of the diagnosis in this case will be, but if it is infallible, the probability that it will predict the generator to be good must be 0.70 (the same probability that the generator is good). If the diagnosis predicts the generator to be good more than 70% of the time or less than 70% of the time, it cannot be infallible because there would be some instances in which the diagnosis and actual performance would be in conflict. We will address the issue of an imperfect diagnosis later on in this section.

Constructing the tree.   We begin by constructing a new branch emanating from the original decision node representing this diagnosis option as shown in Figure 9–12. This branch terminates in a chance node that represents the result of the diagnosis. Emanating from the chance node are the two possible outcomes—the generator is good (G) or the generator is bad (B). Then come the chance nodes representing the uncertainties of generator failure and damage level.

For this case of an infallible diagnosis, the probability of the generator failing is $\Pr(F) = 0.0$ if the diagnosis says it is good, and $\Pr(F) = 1.0$ if the diagnosis says that the generator is bad. Thus, the main effect of the diagnosis option is to modify the probabilities associated with the uncertain events.

For the infallible diagnosis, there is no further decision dilemma: If the diagnosis says the generator is good, we will obviously use it without overhauling; if the diagnosis says the generator is bad, we will obviously overhaul it before using it. Hence, strictly speaking, we could eliminate those decision nodes and the obviously nonviable branches emanating from them to simplify the tree. However, we include them here in preparation for the discussion in the next subsection of a fallible diagnosis. The only adverse effect of including them is to clutter up the tree with branches that clearly will be pruned. (We've also explicitly added the $5,000 cost of the diagnosis to each of the affected payoff nodes in Figure 9–12.)

Solving the tree.   The tree is solved by working backwards from the payoff nodes, employing the expected cost calculation at each chance node encountered, and selecting the best option and deleting all others at each decision node encountered. As mentioned above, the probabilities associated with the infallible diagnosis are such as to make some of the expected cost calculations unnecessary, but we will go through the motions anyway as preparation for our subsequent relaxation of the infallibility assumption. The top part of the tree is identical to the tree in Figure 9–11, so we can copy those results. Let's focus on the bottom part of the tree, the part that contains the branches and paths associated with the diagnosis option. As we work back to the first decision node encountered, we prune the undesired branches. The next node encountered is another chance node that has an expected cost of 19.5. Then we arrive at the original decision node at the far left end of the tree. We see that the expected cost of the diagnosis option (19.5) is less than the ex-

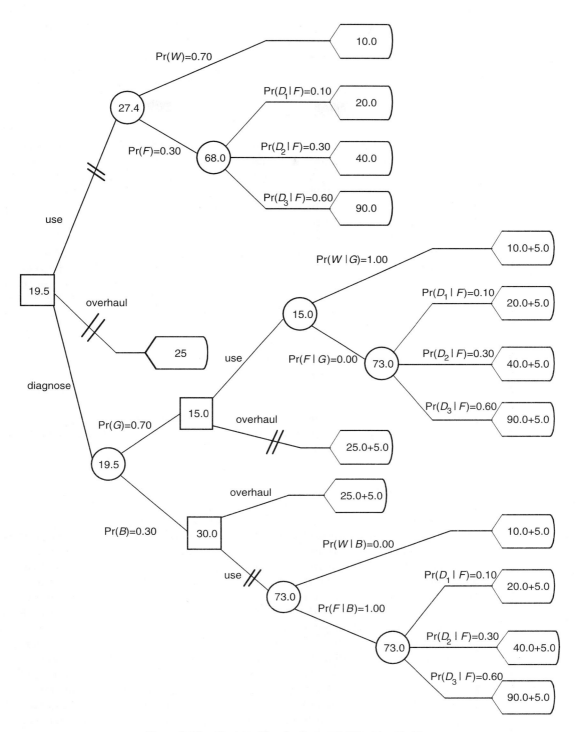

Figure 9–12. Decision Tree for Sequential Decision Problem

pected cost of the other two options. Hence, our best option is to pay the $5,000 for the diagnosis.

Value of Additional Information

For this situation, we concluded that it was worth it to invest $5,000 to obtain the additional information from the diagnosis. We can approach this issue of paying for additional information from a different perspective. Suppose we didn't know ahead of time how much this particular diagnostic procedure would cost. Instead, we pose the question, "How much should we be willing to pay for the diagnosis?" To answer this question we modify Figure 9–12 by replacing the $5,000 known diagnosis cost in the payoff nodes by the unknown cost x.

Then we can proceed to solve the bottom part of the tree by carrying x along as part of our arithmetic manipulations until we get back to the original decision node. Without going through the details, we end up with the expected cost of the diagnosis option:

$$EC(D) = 14.5 + x$$

We see that the least cost option from among the others is

$$EC(O) = 25.0$$

We should be willing to pay any amount for the diagnosis that makes $EC(D) < EC(O)$. Thus, we conclude that the diagnosis is worth $11,500. If it costs any less, we should select it. If it costs more, we should reject it.

Imperfect Information

Now let us take a more realistic view of this diagnostic tool. While a good diagnostic tool should be highly reliable, it is unlikely that it is infallible. Let us assume that this particular tool will predict that the generator is good with a probability of $\Pr(G) = 0.70$. In this sense, it is the same as the infallible tool. This is an overall performance indicator and does not necessarily mean that every diagnosis is correct. To illustrate, Table 9–17 shows a situation where the diagnosis predicts the generator is good in seven out of ten trials and the generator works in seven out of ten trials. However, both trial 8 and trial 10 involve incorrect diagnoses.

For this example, let's say that of those times when the tool diagnoses the generator as good, the probability that the generator will work is $\Pr(W|G) = 0.98$ (that of course also means that $\Pr(F|G) = 0.02$). We will also assume that $\Pr(W|B) = 0.10$ and $\Pr(F|B) = 0.90$. With these probability estimates, we transform the decision tree in Figure 9–12 into the one shown in Figure 9–13. After solving the tree, we conclude that the expected cost of the test with this imperfect diagnosis is 20.3 (greater than the 19.5 expected cost of the test with the perfect diagnosis). Since the expected cost of

TABLE 9–17. EXAMPLE PERFORMANCE OF DIAGNOSTIC TOOL

Trial No.	Diagnosis	Actual Performance
1	good	works
2	bad	fails
3	good	works
4	good	works
5	good	works
6	bad	fails
7	good	works
8	bad	works
9	good	works
10	good	fails
Total	7 good—3 bad	7 works—3 fails

conducting the test after using the imperfect diagnosis option is still lower than that of any other option, the diagnosis remains the preferred option.

9.6. BAYESIAN DECISION MAKING

In Section 9.5 we looked at a decision making situation involving a sequence of uncertainties, in which the probability of the second event (generator failure) occurring depended on whether the first event (diagnosis of a bad generator) occurred. A legitimate question is the source of such probability estimates. In many cases, data useful for making probability estimates is not available in the desired form, and it has to be manipulated in order to be useable.

Let us assume that is the case here. We will focus on the sequences of two uncertain events associated with the diagnosis branch in Figure 9–13. We can delete the intervening decision nodes to allow us to focus on the chance nodes. They are redrawn in Figure 9–14. Note that Figure 9–14 is not a decision tree (there are no decision nodes or payoff nodes). It is just a way of keeping track of key probability relationships, and we call it a probability tree.

In setting up the diagnosis option in Section 9.5, we assumed that the values for $Pr(G)$ and $Pr(B)$ were known. In this section, we want to relax that assumption since it is unlikely that we will know these probabilities ahead of time. Similarly, we are unlikely to have values for $Pr(W|G)$, $Pr(F|G)$, $Pr(W|B)$, and $Pr(F|B)$. In fact, each of the six probability estimates shown in Figure 9–14 frequently have to be calculated from other available data. We will develop the procedure for conducting those calculations in this section.

We started the problem with an estimate for the probability of generator failure, that is,

$$Pr(W) = 0.70, Pr(F) = 0.30 \tag{9–24}$$

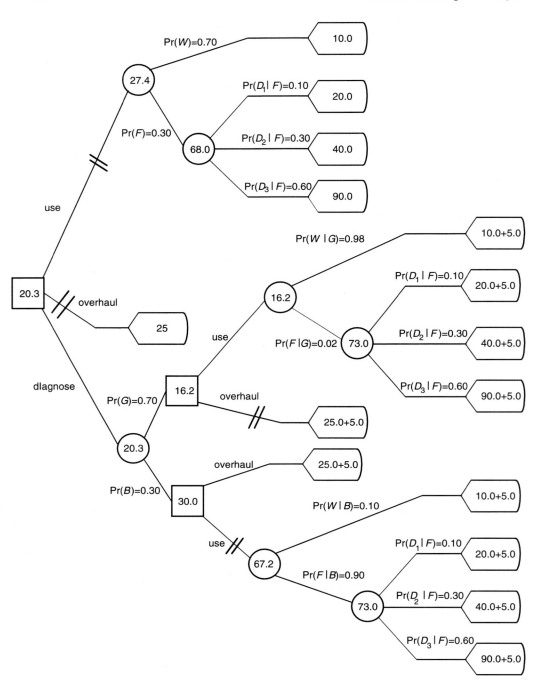

Figure 9–13.   Decision Tree for Sequential Decision Problem with Imperfect Information

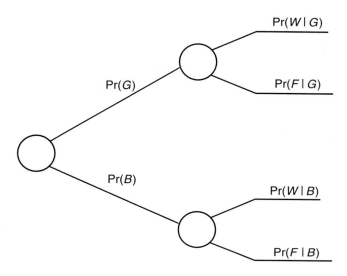

Figure 9–14.   Probability Tree for Generator Failure Data

The results of the diagnosis allow us to revise the probability of failure estimates, that is, to obtain values for $\Pr(W|G)$, $\Pr(F|G)$, $\Pr(W|B)$, and $\Pr(F|B)$. To do this, we need some data on the track record of the particular diagnostic tool we are using. This consists of the prior diagnostic record of the tool on generators known to be working and on generators known to have failed. This data will be in the form of values for $\Pr(G|W)$, $\Pr(B|W)$, $\Pr(G|F)$, and $\Pr(B|F)$. Let us assume that we have collected the following data:

$$\Pr(G|W) = 0.95$$
$$\Pr(B|W) = 0.05$$
$$\Pr(G|F) = 0.15$$
$$\Pr(B|F) = 0.85$$

(9–25)

We can present the data from Equs. (9–24) and (9–25) in the form of a probability tree as shown in Figure 9–15.

What we want to do now is to use the data in Figure 9–15 to get values for the probabilities listed in Figure 9–14. We start by tracing each path and multiplying the probabilities of the associated events to obtain the probabilities of both events in each path occurring, as shown in Figure 9–16.

The notation

$$\Pr(x) \cap \Pr(y)$$

represents the probability that both x and y occur and is called the path probability. Note that the path probability is clearly commutative, that is,

$$\Pr(x) \cap \Pr(y) = \Pr(y) \cap \Pr(x)$$

(9–26)

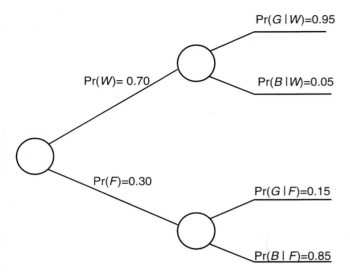

Figure 9–15.   Probability Tree for Diagnostic Tool Data

We can obtain values for $\Pr(G)$ and $\Pr(B)$ by recognizing that

$$\Pr(G) = \Pr(G)\cap\Pr(W) + \Pr(G)\cap\Pr(F)$$
$$\Pr(B) = \Pr(B)\cap\Pr(W) + \Pr(B)\cap\Pr(F)$$

$$(9\text{--}27)$$

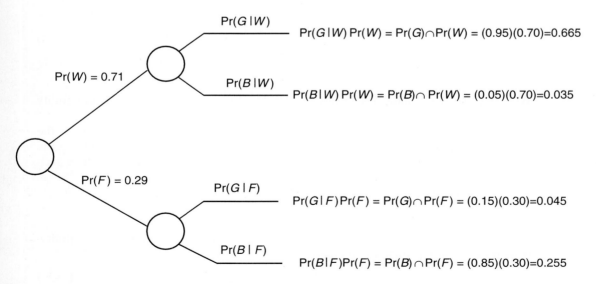

Figure 9–16.   Path Probabilities for Diagnostic Tool Data

Substituting the values for the path probabilities from Figure 9–16 into Equs. (9–27) leads to

$$\text{Pr}(G) = 0.665 + 0.045 = 0.710$$

$$\text{Pr}(B) = 0.035 + 0.255 = 0.290$$

(9–28)

We can now solve for the desired probabilities with the aid of the tree format. We have redone Figure 9–14 as Figure 9–17 by adding the path probabilities from Figure 9–16 [taking advantage of Equ. (9–26)] and inserting the numerical values for $\text{Pr}(G)$ and $\text{Pr}(B)$ from Equ. (9–28).

We can now use Figure 9–17 to solve for each of the conditional probabilities:

$$\text{Pr}(W|G) = \frac{\text{Pr}(W \cap G)}{\text{Pr}(G)} = \frac{0.665}{0.71} = 0.937$$

$$\text{Pr}(F|G) = \frac{\text{Pr}(F \cap G)}{\text{Pr}(G)} = \frac{0.045}{0.71} = 0.063$$

(9–29)

$$\text{Pr}(W|B) = \frac{\text{Pr}(W \cap B)}{\text{Pr}(B)} = \frac{0.035}{0.29} = 0.121$$

$$\text{Pr}(F|G) = \frac{\text{Pr}(F \cap B)}{\text{Pr}(B)} = \frac{0.255}{0.29} = 0.879$$

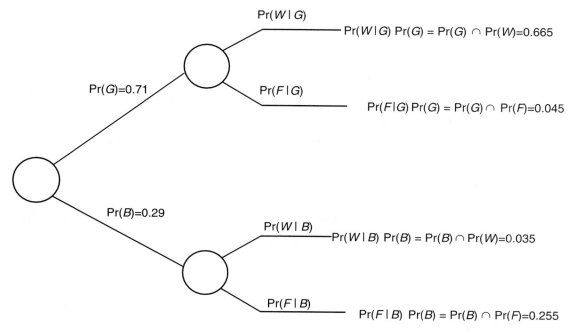

Figure 9–17.   Reversed Probability Tree for Generator Failure

The revised numerical values for probabilities from Figure 9–17 and Equ. (9–29) can be inserted into Figure 9–13 and the problem resolved. See Figure 9–18 for the revised solution.

Bayes' Theorem

We will reorganize the calculations just completed so we can develop a general equation for using supplementary information to update prior probabilities. The resulting equation is known as Bayes' theorem. Let us start with the first expression in Equ. (9–29) and reproduce it here as Equ. (9–30):

$$\Pr(W|G) = \frac{\Pr(W \cap G)}{\Pr(G)} \tag{9-30}$$

We want to rewrite the right hand side in terms of the conditional probabilities $\Pr(G|W)$. We begin by using the result in Figure 9–16 that the numerator in Equ. (9–30) is the path probability and can be expressed as

$$\Pr(W \cap G) = \Pr(G|W)\Pr(W) \tag{9-31}$$

We now focus on the denominator of Equ. (9–30). From Equ. (9–27) we can write

$$\Pr(G) = \Pr(G) \cap \Pr(W) + \Pr(G) \cap \Pr(F)$$

which can be transformed with the aid of Equ. (9–31) into

$$\Pr(G) = \Pr(G|W)\Pr(W) + \Pr(G|F)\Pr(F) \tag{9-32}$$

where the last term on the right hand side of Equ. (9–32) is obtained from Equ. (9–31) by substituting F for W. Equations (9–31) and (9–32) can be substituted into Equ. (9–30) to get

$$\Pr(W|G) = \frac{\Pr(G|W)\Pr(W)}{\Pr(G|W)\Pr(W) + \Pr(G|F)\Pr(F)} \tag{9-33}$$

Following this same procedure, we can obtain a similar expression for $\Pr(F|G)$ in the form

$$\Pr(F|G) = \frac{\Pr(G|F)\Pr(F)}{\Pr(G|W)\Pr(W) + \Pr(G|F)\Pr(F)} \tag{9-34}$$

Now instead of having only two possible conditions of the generator, W and F, suppose we had n possibilities, each one designated by W_i. Then Equs. (9–33) and (9–34) can be combined to provide the general expression

$$\Pr(W_i|G) = \frac{\Pr(G|W_i)\Pr(W_i)}{\sum_{i=1}^{n}\Pr(G|W_i)\Pr(W_i)} \tag{9-35}$$

Following the same procedure, we can derive a similar expression for $\Pr(W_i|B)$.

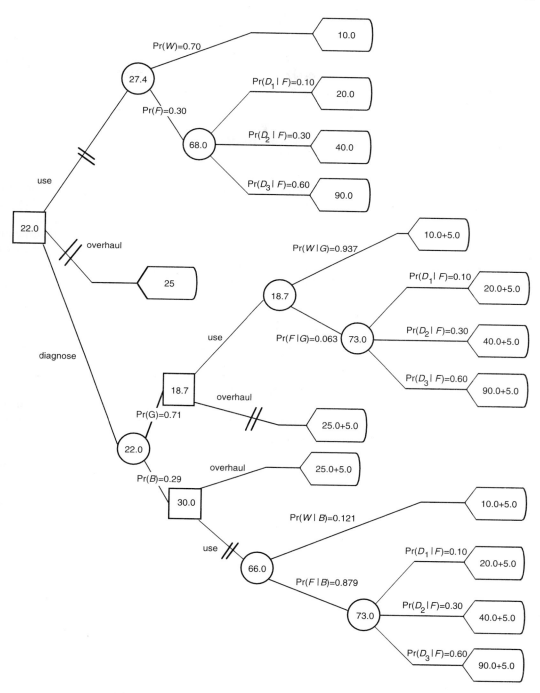

Figure 9–18.   Decision Tree with Revised Probability Estimates

We can now state Bayes' theorem in its most general form in which the conditional probabilities of n events x_i, given a prior event y, can be expressed as

$$\Pr(x_i|y) = \frac{\Pr(y|x_i)\Pr(x_i)}{\sum_{i=1}^{n}\Pr(y|x_i)\Pr(x_i)} \qquad (9\text{--}36)$$

9.7. INCORPORATING ATTITUDE TOWARD RISK

In Sections 9.3 through 9.5, we used an expected value decision rule to select the best design option from among the alternatives. In many situations this rule does not adequately model the choices most people actually make. Consider the situation depicted in Figure 9–19 of choosing between Option A which has a guaranteed income of $1,000, and Option B, which provides a 50% chance of obtaining a $2,000 income and a 50% chance of a zero income. In Figure 9–19, $p = 0.5$ represents the probability of the best of the uncertain outcomes occurring. In this case, since there are only two uncertain outcomes, the probability of the other one occurring must be $1 - p = 0.5$.

If we apply a maximum expected value decision rule to this problem, we get

$$EV(A) = \$1,000$$

$$EV(B) = 0.5(\$2,000) + 0.5(\$0) = \$1,000$$

Thus, an expected value decision maker would be indifferent regarding A and B. Both options are equally attractive since they both have an expected value of $1,000. However, when presented with these two options, most people express a clear preference for Option A. That is, most people would prefer the certainty of receiving the $1,000 rather than the risky option. The maximum expected value decision rule just does not do a good job of capturing people's preferences in this case. In this section we will develop another decision rule, one that explicitly incorporates a quantitative representation of the decision maker's attitude toward risk.

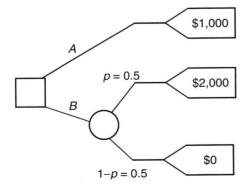

Figure 9–19.   Decision Tree for Options A and B

Utility

Since monetary measures of outcomes in the example depicted in Figure 9–19 did not lead to a satisfactory decision rule, we look for an alternate measure of outcomes. We define the term "utility" as the measure of the true worth of an outcome. If we can establish a relationship between monetary outcomes and their utility, we have the basis for formulating a "maximum expected utility" decision rule. This relationship will be in the form of constructing a utility function $u(\$)$ that assigns a numerical value of utility to each monetary outcome. Such a relationship is depicted in Figure 9–20.

In the following subsection, we illustrate how to construct such a function. Before doing so, we establish some key characteristics of utility. To do this we generalize the decision problem illustrated in Figure 9–19 by not specifying a numerical value for p or for the outcome of Option A. The modified decision tree is shown in Figure 9–21.

Now, if for a given set of values for x and p, we conclude that options A and B are equally attractive, that must mean that their utilities are equal, that is,

$$u(A) = u(B)$$

or, in terms of p and x,

$$u(\$x) = pu(\$2,000) + (1 - p)u(\$0) \tag{9–37}$$

In order to construct the utility function for this problem, we now examine a series of three subsidiary decision problems. In each of these problems, we consider a different value of x in Figure 9–21 and ask what value of p is needed to make Options A and B equally attractive.

Three Subsidiary Problems.   For the first subsidiary problem, we will set $x = \$2,000$. Note that this is the largest of all possible payoffs for the original problem depicted in Figure 9–19. The decision tree for this subsidiary problem is shown in Figure 9–22. We now pose the question: For what value of p are Options A and B in Figure

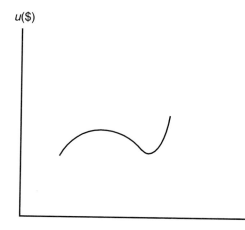

$u(\$)$

$\$$ **Figure 9–20.**   Graph of a Utility Function

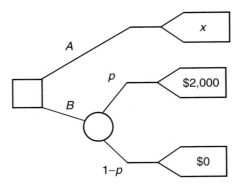

Figure 9–21. Modified Decision Tree

9–22 equally attractive? Clearly, regardless of an individual's attitude toward risk, the only rational response is $p = 1$. With $x = \$2000$ and $p = 1$, Equ. (9–37) becomes

$$u(\$2,000) = u(\$2,000)$$

which is an identity. This means that we can assign any value we want to $u(\$2,000)$ without violating the property of $u(\$)$ that it represents the true attractiveness of an option. For convenience, we choose

$$u(\$2,000) = 1 \qquad (9\text{–}38)$$

This gives us a point through which the graph of our utility function must pass. Note that the value we assigned to u in Equ. (9–38), $u = 1$, was equal to the value of p ($p = 1$) which made Options A and B in Figure 9–22 equally attractive.

For the second subsidiary problem, set $x = \$0$. Note that this is the smallest of all possible payoffs for the original problem depicted in Figure 9–19. The decision tree for this subsidiary problem is shown in Figure 9–23.

We now pose the same question as we did in Subsidiary Problem 1: For what value of p in Figure 9–23 are Options A and B equally attractive? Clearly, regardless

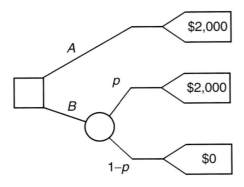

Figure 9–22. Decision Tree for Subsidiary Problem 1

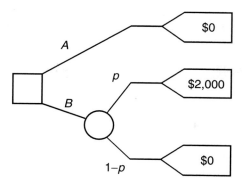

Figure 9–23.   Decision Tree for Subsidiary Problem 2

of an individual's attitude toward risk, the only rational response is $p = 0$. With $x = \$0$ and $p = 0$, Equ. (9–37) becomes

$$u(\$0) = u(\$0)$$

which is an identity. This means that we can assign any value we want to $u(\$0)$. For convenience, we choose

$$u(\$0) = 0 \qquad (9\text{–}39)$$

This gives us another point through which the graph of our utility function must pass. Note that the value we assigned to u in Equ. (9–39), $u = 0$, was equal to the value of p ($p = 0$) which made Options A and B in Figure 9–23 equally attractive.

Thus, regardless of an individual's preference for Option A or Option B in the original problem (Figure 9–19), their utility function for this problem will pass through the two points defined by Equs. (9–38) and (9–39), as is shown in Figure 9–24 (one of those points is the origin).

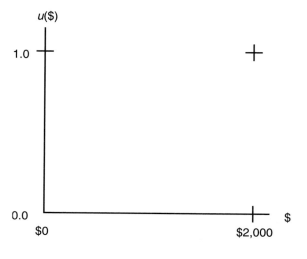

Figure 9–24.   End Points of All Utility Functions For Example Problem

For the third subsidiary problem, we will set $x = \$1,000$. Note that this is an intermediate payoff for the original problem depicted in Figure 9–19. The decision tree for this subsidiary problem is shown in Figure 9–25.

We now pose the same question we posed for Subsidiary Problems 1 and 2: For what value of p are Options A and B equally attractive? Recall that for the first two subsidiary problems, we expect all individuals to give the same answer to this question. Now, for the third subsidiary problem, each individual is likely to give a different response. We can get some clue as to the nature of those responses by observing that the payoffs in Figure 9–25 are identical to those in the original problem presented in Figure 9–19. Hence an individual who preferred Option A in the original problem (for which $p = 0.5$) would require $p > 0.5$ in Figure 9–25 in order for Option B to be equally attractive as Option A. Conversely, an individual who preferred Option B in the original problem would require $p < 0.5$ in Figure 9–25 in order for Option A to be equally attractive as Option B. And an individual who was indifferent toward Options A and B in the original problem would require $p = 0.5$ in Figure 9–25.

Calibrating the utility function.   How can we determine a numerical value for p in the cases where $p \neq 0$ and $p \neq 1$? Suppose you preferred Option A in Figure 9–25 when $p = 0.5$ so we know that $p > 0.5$ in order for A and B to be equally attractive. But if $p = 1.0$ in Figure 9–25, B would clearly be the preferred choice. So we know that p is bounded by $0.5 < p < 1.0$. We now can ask a series of questions designed to narrow this range. Would you still prefer A if $p = 0.6$? Would you still prefer B if $p = 0.9$? We can continue this process until we've narrowed the range as much as needed to select a value for p. Regardless of which numerical value of p we finally settle on, we can substitute it and Equs. (9–38) and (9–39) into Equ. (9–37) to get

$$u(\$1,000) = p \qquad (9\text{–}40)$$

Hence we have a numerical value for $u(\$1,000)$. Three representative locations of this point ($p < 0.5$, $p = 0.5$, $p > 0.5$) are shown in Figure 9–26. This gives us another point through which the graph of our utility function must pass. Note that the value assigned to u in Equ. (9–40) is equal to the value of p in Figure 9–25 that made Options A and B equally attractive.

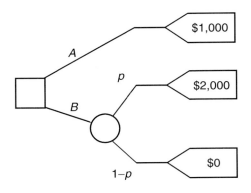

Figure 9–25.   Decision Tree for Subsidiary Problem 3

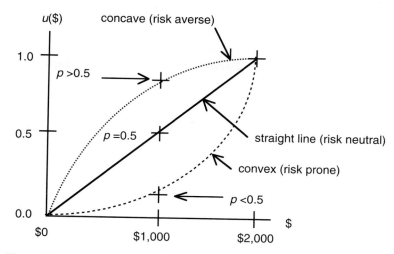

Figure 9–26.   General Form of Risk Averse, Risk Neutral, and Risk Prone Utility Functions

Constructing a utility function.   We now have three points through which our utility function must pass [$u(\$0) = 0$, $u(\$1,000) = p$, and $u(\$2,000) = 1$]. With these three points, we might feel comfortable drawing the utility function. We can also collect more data by selecting another intermediate payoff value; for example, set $x = \$500$ in Figure 9–21 and find an associated value of p and hence a value for u. The intermediate payoff value doesn't have to be one of the actual possible outcomes; any hypothetical outcome between the best possible and worst possible outcomes will work.

We can repeat this exercise with as many intermediate payoff values as necessary to give us enough data to construct the utility function (each intermediate payoff value and the associated value of u gives us a point on the utility curve). Most utility functions will look like one of the curves drawn in Figure 9–26: a concave curve, a straight line, or a convex curve.

A utility function in the form of a straight line indicates that the utility of any outcome is proportional to the dollar value of that outcome. Hence, a maximum expected utility decision rule will lead to the exact same choices as a maximum expected value rule. That is, a decision maker whose utility function is a straight line, is an expected value decision maker, willing to make choices on the basis of the monetary values of the outcomes. We also call such a decision maker a risk neutral decision maker.

A utility function in the form of a concave curve indicates that a unit increase in monetary outcomes is more valuable for small monetary outcomes than for large monetary outcomes. This represents the attitude of a risk averse decision maker, who would choose a small, but certain, benefit over a large but uncertain benefit. This is the decision making framework of a cautious, conservative individual.

A utility function in the form of a convex curve indicates that a unit increase in monetary outcomes is more valuable for large monetary outcomes than for small monetary outcomes. This represents the attitude of a risk prone decision maker, who would choose a large but uncertain benefit over a small, but certain, benefit.

Characteristics of a Utility Function

Let us review the main features of utility functions. In general, a utility function for any decision problem can be constructed by examining a set of subsidiary decision trees of the form shown in Figure 9–27 where L is the least desirable outcome, M is the most desirable outcome, and X is any intermediate outcome. The following properties hold for all utility functions:

1. The least desirable of all outcomes in a given decision analysis problem has a utility of 0; $U(L) = 0$.
2. The most desirable of all outcomes in a given decision analysis problem has a utility of one; $U(M) = 1$.
3. The utility of any intermediate outcome X is equal to the value of p for which the decision-maker is indifferent between Options A and B in the subsidiary problem shown in Figure 9–27.

Once we have constructed a utility function for a given decision analysis problem, we can convert all outcomes to corresponding utility values. We can then invoke a maximum expected utility decision rule and solve the decision tree for the problem.

The concept of utility is not limited to outcomes measured in monetary terms. We can construct utility functions for current u(amps) or pressure u(psi) or whatever units the key design parameters or performance parameters are expressed in. The utility concept is not even limited to outcomes that can be quantified. For example, suppose the characteristic of the design options that is to serve as the basis for a decision is the aesthetic appeal of the design. If we can rank the aesthetic quality of each option, we can identify the most appealing option M [by definition, $u(M) = 1.0$] and

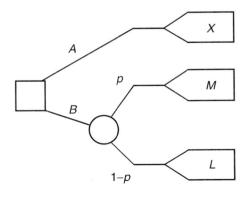

Figure 9–27.   Generic Subsidiary Problem for Constructing a Utility Function

the least appealing option L [by definition, $u(L) = 0.0$]. Then we can use Figure 9–27 to assign a utility value to any intermediate aesthetic outcome X.

An individual's attitude toward risk can vary from one problem to another. The same person can be risk averse in one situation, and risk prone in another. For example, we mentioned earlier that most people presented with the decision problem Figure 9–19 are risk averse. However, if we change the outcomes in Figure 9–19 to $M = \$2$, $L = 0$, $X = \$1$, many of the risk averse people become risk prone. The reason is not hard to understand. In the original problem, the opportunity to receive a payoff of the order of magnitude of $1000 is so attractive that people are not willing to risk the chance of getting nothing. On the other hand, in the modified problem, the $1 payoff is not that attractive. It isn't that much better than getting nothing, and many people would just as soon gamble on getting the $2.

All the utility functions depicted in Figure 9–26 pass through the origin because the least attractive outcome option had a monetary value of 0, so $u(0) = 0$. In general, this isn't so. The least attractive outcome for a given decision analysis can have either a positive or negative value so $u(L) = 0$ where $L \neq 0$.

Also, the utility functions depicted in Figure 9–26 are monotonically increasing; that is, the slope of the utility curve is always positive. In principle, a utility function for a given problem could be nonmonotonic, such as the one shown in Figure 9–20.

Modeling Utility Functions

In the above discussion, we described how a utility function can be constructed from client responses to a series of questions. In many situations it may not be practical to obtain the data needed to do this. The client may not be available, or even if available, unwilling to spend the time with you in an exercise they might not fully understand or appreciate. In addition, the client responses may be inconsistent, and it may not be easy to draw a smooth curve through the data points so obtained. Another approach to constructing utility functions, which is not data intensive, is to build a mathematical model for the utility function by prescribing the parameter in a generic family of curves.

One such family is given by

$$u(s) = \frac{1 - e^{-rs}}{1 - e^{-r}} \tag{9–41}$$

where s is a normalized form of the outcome x, defined as

$$s = \frac{x - x_{worst}}{x_{best} - x_{worst}} \tag{9–42}$$

so that $s = 0$ when $x = x_{worst}$, and $s = 1$ when $x = x_{best}$.

We see from Equ. (9–41) that $u = 0$ when $s = 0$, and $u = 1$ when $s = 1$. Thus, Equs. (9–41) and (9–42) satisfy the requirement that the utility for the best and worst outcomes are 0 and 1 respectively. Each choice for the parameter r specifies a different curve in the family. Figure 9–28 shows several members of this family of curves. When

$$u(s) = \frac{(1-e^{-rs})}{(1-e^{-r})}$$

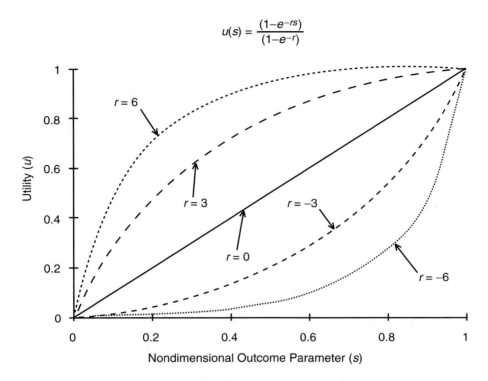

Figure 9–28. Mathematical Model for Family of Utility Functions

$r > 0$, the utility function represents risk averse behavior; when $r < 0$, the utility function represents risk prone behavior. As shown in Appendix 9B, the straight line utility function occurs when $r = 0$. Hence, once you obtain a sense of the nature and strength of the client's risk attitude for a particular situation, you may be able to construct a utility function from Equ. (9–41) that faithfully represents that attitude.

      Equation (9–41) is just one of many possible models for utility functions. The only requirements are that the utility function pass through the points $(L, 0)$ and $(M, 1)$ and that the shape of the curve that passes through those points represents the decision makers preferences for the different possible outcomes.

Field Test Strategy For Risk Prone Decision Maker

Let's return now to the decision problem summarized in Figure 9–11. For convenience, we repeat that diagram here as Figure 9–29. Recall that the outcomes represent the cost of conducting a field test at a remote facility. The tree as displayed reflects a risk neutral attitude, and the application of a decision rule to minimize the

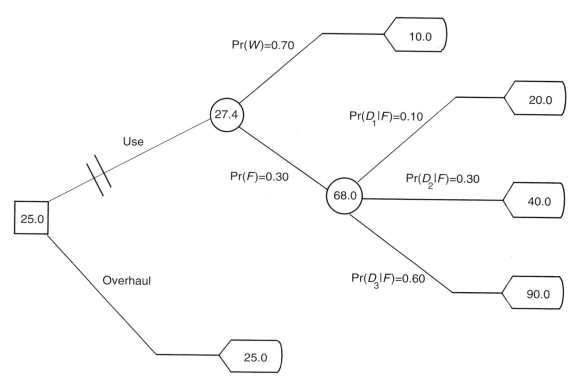

Figure 9–29.   Decision Tree for Risk Neutral Decision Maker

expected cost of conducting the test. The option with the least expected cost is to overhaul the generator before it is used in the test. We want to re-solve the tree to represent a risk prone decision maker and see if we are led to make the same choice.

Let us model this decision maker's willingness to take risks by using Equ. (9–41) with $r = -3$. From the tree, we see that the most attractive outcome is the one with the least cost:

$$x_{best} = 10.0 \qquad\qquad (9\text{–}43)$$

The least attractive outcome is the one with the highest cost:

$$x_{worst} = 90.0 \qquad\qquad (9\text{–}44)$$

We can then set up Table 9–18 to calculate the utility of each outcome. The values in the x column are the outcomes displayed in Figure 9–29. The values in the s column are the nondimensionalized outcomes, obtained from Equ. (9–42). The utilities (u) are then calculated from Equ. (9–41) with $r = -3$. For example, for $x = \$20,000$, we

TABLE 9–18.   UTILITIES FOR FIELD TEST PROBLEM ($r = -3$)

$x(\$1,000)$	s	u
10.0	1.000	1.000
20.0	0.875	0.671
25.0	0.812	0.547
40.0	0.625	0.289
90.0	0.000	0.000

substitute Equs. (9–43) and (9–44) into Equ. (9–42) to get $s = 0.875$. Then Equ. (9–41) yields

$$u(s) = \frac{1 - e^{-(-3)(0.875)}}{1 - e^{-(-3)}} = \frac{1 - e^{2.625}}{1 - e^3} = 0.671$$

The decision tree in Figure 9–29 is redrawn as Figure 9–30 by replacing the monetary outcomes by their utility equivalents listed in Table 9–18. After invoking the maximum expected utility rule, we see that the "use" option is preferred by this decision maker.

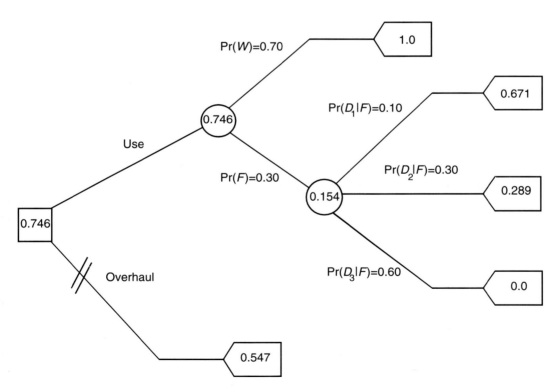

Figure 9–30.   Decision Tree Reflecting Risk Attitude of Decision Maker

The only difference between the decision problems represented by Figs. 9–29 and 9–30 is the attitude toward risk of the decision maker. It is not surprising that they make different choices. We don't know whether the same choices would have been made without the aid of utility theory. Obviously if a decision maker knows ahead of time what option they prefer, there is no need to even conduct the analysis. Like any formal decision making methodology, it should be invoked only when the decision maker does not feel comfortable making the decision without the additional clarification and insight that can be provided by the methodology.

Discounting Utilities

In decision trees where the outcomes are expressed in monetary terms (such as Figure 9–29), the outcomes and expected monetary values at event nodes and decision nodes may need to be discounted using the techniques described in Chapter 8 to account for the fact that not all the outcomes, events, and decisions occur at the same time. When monetary outcomes are converted to utilities using either a risk prone or risk averse utility function, "future" utilities should be discounted in order to appropriately account for the time-value of the monetary outcomes. This discounting should be done at the payoff, decision, or event node as appropriate to account for the time differential. The three-step procedure illustrated in Figure 9–31 should be used:

1. Use the utility function to convert a utility u_a to its corresponding monetary value x_a.
2. Discount the monetary value x_a to its discounted value $x_a(\text{disc})$.
3. Use the utility function to convert the discounted monetary value $x_a(\text{disc})$ to its corresponding "discounted" utility $u_a(\text{disc})$.

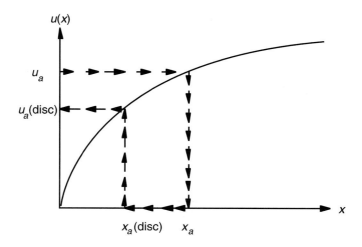

Figure 9–31.   Procedure for Discounting Utilities

Appendix 9B: Simplification of Utility Function Model for $r = 0$

The equation of one family of utility functions was given by Equ. (9–41) as

$$u(s) = \frac{1 - e^{-rs}}{1 - e^{-r}}$$

We want to show that $u(s)$ is a straight line when $r = 0$. When we set $r = 0$ in the above equation, we get the indeterminate form

$$u(s) = \frac{0}{0}$$

This can be resolved by applying L'Hôpital's rule to get

$$u(s) = \lim_{r \to 0} \left[\frac{\dfrac{d}{dr}(1 - e^{-rs})}{\dfrac{d}{dr}(1 - e^{-r})} \right]$$

which yields

$$u(s) = \lim_{r \to 0} \left[\frac{(se^{-rs})}{(e^{-r})} \right] = s$$

which, of course, is the equation of a straight line.

9.8. MULTIPLE ATTRIBUTE UTILITY THEORY (MAUT)

We will use utility theory to revisit the problem discussed in Section 9.2 of choosing between design options *A, B,* and *C* for transmitting power between parallel shafts. In Section 9.2 we used fifteen criteria, unevenly weighted, as the basis for making the design choice. To simplify the following discussion, let's restrict ourselves to using only four of the original fifteen criteria. From Table 9–5, we see that the four most important criteria (the ones with the highest weighting factors) are shock protection, high speed, slippage, and operating temperature.

Table 9–19 displays a hypothetical set of raw scores for these four attributes for design options *A, B,* and *C*. For purposes of subsequent manipulations, we represent each attribute by the index number i. The raw scores are indicated by x_i, the normalized scores by s_i, and the utilities by u_i.

Utility Functions for Each Attribute

Our first step in using multiple attribute utility theory (MAUT) to select a design option is to construct utility functions for each attribute $[u_1(x_1), u_2(x_2), u_3(x_3)$ and $u_4(x_4)]$. Let's assume that the individual utility functions have the form given by Equ. (9–41).

TABLE 9–19. RAW SCORES AND UTILITIES OF FOUR ATTRIBUTES FOR THREE DESIGN OPTIONS

			Design Options										
			A			B			C				
Attributes	i	r_i	Raw Score (x_i)	Normalized Score (s_i)	Utility (u_i)	Raw Score (x_i)	Normalized Score (s_i)	Utility (u_i)	Raw Score (x_i)	Normalized Score (s_i)	Utility (u_i)
Shock protection	1	0	300	1.000	1.000	300	1.000	1.000	50	0.000	0.000
High speed	2	−1	100	0.000	0.000	400	1.000	1.000	100	0.000	0.000
Slippage/Creep	3	0	50	0.000	0.000	100	0.111	0.111	500	1.000	1.000
Operating temperature	4	2	200	0.200	0.381	50	0.000	0.000	800	1.000	1.000

In particular, let's say we have established $r_1 = 0$, $r_2 = -1$, $r_3 = 0$, and $r_4 = 2$. These values are listed in Table 9–19. The corresponding versions of Equ. (9–41) take the form

$$u_1(s_1) = s_1$$

$$u_2(s_2) = \frac{1 - e^{s_2}}{1 - e}$$

$$u_3(s_3) = s_3 \tag{9-45}$$

$$u_4(s_4) = \frac{1 - e^{-2s_4}}{1 - e^{-2}}$$

In order to use Equ. (9–45), we first have to introduce the normalized values s_1, s_2, s_3, and s_4 in accordance with Equ. (9–42). Using the raw scores from Table 9–19, we get

$$s_1 = \frac{x_1 - x_{1_{worst}}}{x_{1_{best}} - x_{1_{worst}}} = \frac{x_1 - 50}{300 - 50} = \frac{x_1 - 50}{250}$$

$$s_2 = \frac{x_2 - x_{2_{worst}}}{x_{2_{best}} - x_{2_{worst}}} = \frac{x_2 - 100}{400 - 100} = \frac{x_2 - 100}{300}$$

$$s_3 = \frac{x_3 - x_{3_{worst}}}{x_{3_{best}} - x_{3_{worst}}} = \frac{x_3 - 50}{500 - 50} = \frac{x_3 - 50}{450} \tag{9-46}$$

$$s_4 = \frac{x_4 - x_{4_{worst}}}{x_{4_{best}} - x_{4_{worst}}} = \frac{x_4 - 50}{800 - 50} = \frac{x_4 - 50}{750}$$

The values of s_i as determined from the above expressions for each option are listed in Table 9–19 in the "Normalized Score" columns. Inserting those s_i values into Equ. (9–45) yields the utility values for each option; they are also listed in Table 9–19.

Composite Utility Function

Next we will combine the individual utility functions given by Equ. (9–45) into a composite utility function for each design option, i.e., $U_A(x_1, x_2, x_3, x_4)$, $U_B(x_1, x_2, x_3, x_4)$, and $U_C(x_1, x_2, x_3, x_4)$. The design option that has the highest composite utility will be the preferred choice.

Before we carry out these steps, let's take a moment to reflect on the role of uncertainty and risk in this type of decision making. We introduced the concept of utility functions in the last section as a technique for embodying attitude toward risk in choosing among design options involving uncertain outcomes. In fact, utility functions are constructed from the decision maker's responses to a series of questions explicitly dealing with their preferences between alternatives under conditions of risk.

In Section 9.6 we introduced decision trees as tools for making choices under conditions of risk. And we showed in Section 9.7 that single attribute utility functions can provide results consistent with the decision maker's risk attitude. In decision making involving risk, where each outcome has multiple attributes, we can use this same technique, except we express each outcome in terms of the composite utility function.

However, MAUT can also be used in multiple attribute decision problems that do not involve risk. We can think of MAUT as an alternative to the weighted objectives and the AHP methods discussed in Sections 9.2 and 9.3. This is the context in which we will develop MAUT in this section. Let us examine two approaches to combine individual utility functions into a composite utility function.

Additive form of multi-attribute utility function.   The first approach to constructing a composite utility function from individual utility functions for each attribute is similar to the technique of weighted sums used in Section 9.2. That is, we assume that the composite utility function is a linear combination of the individual utility functions. For this problem where we have four attributes and a utility function for each one, the composite utility function is given by

$$U(x_1, x_2, x_3, x_4) = \sum_{i=1}^{4} k_i u_i(x_i) \tag{9-47}$$

where the k_i are called scaling constants. In the method of weighted objectives and in the AHP method, we evaluated the k_i by considering the relative importance of the various criteria. The MAUT approach determines the k_i by examining the willingness of the client to make tradeoffs between different attributes. As we will see shortly, this MAUT approach leads to a fundamentally different interpretation of the k_i than the other two approaches.

Since $U(x_1, x_2, x_3, x_4)$ a utility function, we want it to have the same properties of single attribute utility functions. In particular, we want $U = 1$ for the best possible design option. But the best possible design option is one that has the highest possible utility for each of its attributes, that is, $u_1(x_1) = 1$, $u_2(x_2) = 1$, $u_3(x_3) = 1$, and $u_4(x_4) = 1$. Then Equ. (9-47) reduces to

$$1 = k_1 + k_2 + k_3 + k_4 \tag{9-48}$$

Thus, the scaling constants for an additive composite utility function must add up to unity. The worst possible design option is one that has the lowest possible utility for each of its attributes; that is, $u_1(x_1) = 0$, $u_2(x_2) = 0$, $u_3(x_3) = 0$, and $u_4(x_4) = 0$. We want the composite utility of such an option to be 0. We see that Equ. (9-47) guarantees that result.

Determining the scaling constants.   The first step in determining numerical values for the k_i is to establish their relative order. We do this by considering four hypothetical designs, E through H, whose attribute values are drawn from the attribute values associated with the real design options A, B, and C. In particular, each hypothetical design has one of its attributes at the best value from among the real options, and the other three attributes at their worst values from among the real options. So, for example, using Equ. (9-47) and Table 9-19, the hypothetical option E has the form

$$U_E(x_{1t}, x_{2t}, x_3, x_4) = U_E(x_{1_{best}}, x_{2_{worst}}, x_{3_{worst}}, x_{4_{worst}}) = U_E(200, 100, 50\ 50)$$

The four hypothetical options together with the utilities for each of their attributes are listed in Table 9–20. Since the best and worst values of each attribute have utilities of 1.0 and 0.0 by definition, we do not need to list their numerical values in Table 9–20.

We find the composite utility for hypothetical design E from Equ. (9–47) and Table 9–20 to be

$$U_E = k_1 u_1 + k_2 u_2 + k_3 u_3 + k_4 u_4 = k_1(1) + k_2(0) + k_4(0) = k_1$$

A similar process for options F, G, and H leads to

$$U_E = k_1$$
$$U_F = k_2$$
$$U_G = k_3 \qquad\qquad (9\text{--}49)$$
$$U_H = k_4$$

Now we ask the client to indicate their preference among those four hypothetical options. These preferences establish the relative order of the scaling constants. For example, if the client prefers design option E to design F, and in turn prefers F to option G, and G to H, then we conclude from Equ. (9–49) that

$$k_1 > k_2 > k_3 > k_4 \qquad\qquad (9\text{--}50)$$

If the client indicates an ordered preference other than $E > F > G > H$, we can always relabel the options so that Equ. (9–50) is valid under any ordering sequence.

Now let's focus on the relationship between k_1 and k_2. The first two expressions in Equ. (9–49) together with Equ. (9–50) tell us that

$$U_E(x_{1_{best}}, x_{2_{worst}}, x_{3_{worst}}, x_{4_{worst}}) > U_F(x_{1_{best}}, x_{2_{worst}}, x_{3_{worst}}, x_{4_{worst}}) \qquad (9\text{--}51)$$

or numerically,

$$U_E(200, 100, 50, 50) > U_F(50, 400, 50, 50)$$

If we modify design E by reducing the value of x_1 from $x_{1_{best}}$, there will be some point at which we will no longer prefer option E over option F.[6] Let the value of x_1 on which this transition occurs be x_1' and let E' represent the modified version of design option E. Then

$$U_{E'}(x_1', x_{2_{worst}}, x_{3_{worst}}, x_{4_{worst}}) > U_F(x_{1_{worst}}, x_{2_{best}}, x_{3_{worst}}, x_{4_{worst}}) \qquad (9\text{--}52)$$

We assume that this transition occurs for this example when $x_1' = 75$ so that

$$U_E(75, 100, 50, 50) = U_F(50, 400, 50, 50)$$

Using Equ. (9–47), Equ. (9–54) then reduces to

$$k_1 u_1(x_1') = k_2 \qquad\qquad (9\text{--}53)$$

---

[6]The existence of this transition point is guaranteed since if x_1 is reduced all the way to $x_{1_{worst}}$, then $U_E = 0$. Since $U_F > 0$, then $U_F > U_E$ when $x_1 = x_{1_{worst}}$.

TABLE 9-20. HYPOTHETICAL DESIGN OPTIONS AND THEIR ATTRIBUTE UTILITIES

Attributes	i	Hypothetical Design Options							
		E		F		G		H	
		Raw Score (x_i)	Utility (u_i)	Raw Score (x_i)	Utility (u_i)	Raw Score (x_i)	Utility (u_i)	Raw Score (x_i)	Utility (u_i)
Shock protection	1	best	1	worst	0	worst	0	worst	0
High speed	2	worst	0	best	1	worst	0	worst	0
Slippage/Creep	3	worst	0	worst	0	best	1	worst	0
Operating temperature	4	worst	0	worst	0	worst	0	best	1

Since we now know the value of x_1' and we have the form of u_1 given in Equ. (9–45), Equ. (9–53) gives us a relationship between k_1 and k_2. Specifically, with $x_1 = 75$ the first of Equs. (9–45) and (9–46) transform Equ. (9–53) into

$$k_1\left(\frac{75-50}{250}\right) = k_2$$

or

$$k_1 = 10k_2$$

Following the same line of reasoning, we can establish the following relationships between k_2 and k_3:

$$k_2 u_2(x_2') = k_3 \tag{9–54}$$

where x_2' is the value of x_2 for which the client is indifferent between hypothetical design options F and G. Repeating the process with options G and H in Equ. (9–49), we can come up with

$$k_3 u_3(x_3') = k_4 \tag{9–55}$$

Equs. (9–53), (9–54), (9–55), and (9–48) allow us to solve for k_1, k_2, k_3 and k_4. Once we have determined the k_i, the composite utility function can be evaluated for each of the design options.

Ah, if life was so simple. There are some implications in the approach just described that deserve closer scrutiny. In particular, there are alternative approaches to using Equ. (9–48) along with Equs. (9–53), (9–54), and (9–55) for determining the k_i. For example, we could examine the willingness to make tradeoffs between designs E and G in Equ. (9–49). That will lead us to

$$k_1 u_1(x_1') = k_3 \tag{9–56}$$

Now Equ. (9–56) can be used with Equs. (9–53), (9–54), and (9–55) to determine the k_i. This will likely lead to different values for the k_i than the ones we got by using Equ. (9–48) with Equs. (9–53), (9–54), and (9–55). Further, these k_i would not satisfy Equ. (9–48). We could also look at design options E and H or F and H. The point is that there are a variety of tradeoff combinations we can examine and each combination could lead to a different set of k_i. For some of the combinations, the k_i would not satisfy Equ. (9–48).

There are several approaches to deal with this problem. First, we can ignore it since all it means is that the best possible option has a composite utility other than unity, that is,

$$U_E(x_{1_{best}}, x_{2_{best}}, x_{3_{best}}, x_{4_{best}}) \neq 1$$

This may not have any practical consequences since if there was one option for which all design parameters had their best value, that option would clearly dominate all others, and we would not need utility theory to help decide which option to select. Second, we can make adjustments in the values of k_i so that Equ. (9–48) is satisfied. Since the tradeoff transition points that provide the equations for determining the k_i cannot be located precisely, small adjustments in the k_i so that Equ. (9–48) is satisfied, may be acceptable to all interested parties. Third, we can interpret the failure to satisfy Equ. (9–48) as indicating that the composite utility function is not of the form given by Equ. (9–47). Let's now explore a more sophisticated approach to constructing the composite utility function.

Multiplicative Form of Multi-Attribute Utility Function

An alternative to the additive form of the composite utility function discussed above is a composite utility function that involves cross-product terms of the individual utility functions. For the case of four attributes, Appendix 9C shows that we can express the composite utility function $U(x_1, x_2, x_3, x_4)$ as

$$1 + KU(x_1, x_2, x_3, x_4) = \prod_{i=1}^{4} [1 + Kk_i u_i(x_i)] \tag{9–57}$$

where K is a new parameter that we call the composite scaling constant. To explore the significance of K, recall that we defined the multi-attribute utility function $U(x_1, x_2, x_3, x_4)$ and the scaling constants k_i so that $U(x_1, x_2, x_3, x_4) = 1$ when each of the individual utility functions $u_i = 1$. Substituting these into Equ. (9–57) gives

$$1 + K = \prod_{i=1}^{4} [1 + Kk_i] \tag{9–58}$$

as the equation for determining K. Equ. (9–58) expands to

$$\begin{aligned} 1 + K = 1 &+ Kk_1 + Kk_2 + Kk_3 + Kk_4 \\ &+ K^2k_1k_2 + K^2k_1k_3 + K^2k_1k_4 + K^2k_2k_3 + K^2k_2k_4 + K^2k_3k_4 \\ &+ K^3k_1k_2k_3 + K^3k_1k_2k_4 + K^3k_1k_3k_4 + K^3k_2k_3k_4 \\ &+ K^4k_1k_2k_3k_4 \end{aligned} \tag{9–59}$$

which simplifies to

$$\begin{aligned} 1 = k_1 + k_2 + k_3 + k_4 &+ K(k_1k_2 + k_1k_3 + k_1k_4 + k_2k_3 + k_2k_4 + k_3k_4) \\ &+ K^2(k_1k_2k_3 + k_1k_2k_4 + k_1k_3k_4 + k_2k_3k_4) + K^3k_1k_2k_3k_4 \end{aligned} \tag{9–60}$$

Now, if

$$k_1 + k_2 + k_3 + k_4 = 1$$

Equ. (9–60) reduces to

$$\begin{aligned} K[k_1k_2 + k_1k_3 &+ k_1k_4 + k_2k_3 + k_2k_4 + k_3k_4 \\ &+ K(k_1k_2k_3 + k_1k_2k_4 + k_1k_3k_4 + k_2k_3k_4) + K^2k_1k_2k_3k_4] = 0 \end{aligned} \tag{9–61}$$

Since all the $k_i > 0$ the only way Equ. (9–61) can be satisfied is if $K = 0$. We conclude that if the individual scaling constants k_i add up to unity, the composite scaling constant K must be 0. When $K = 0$, the multiplicative composite utility function reduces to the additive form [see Equ. (9–62)].

If the k_i add up to unity, the additive form of the utility function is appropriate to use; otherwise the multiplicative form is appropriate. We determine the k_i for the multiplicative form in exactly the same way described earlier for the additive form. Once the k_i are found, K can be determined from Equ. (9–60). It can be shown that if $\Sigma_{i=1}^{n} k_i > 1$ then $-1 < K < 0$ and if $\Sigma_{i=1}^{n} k_i < 1$ then $K > 0$.[7]

Appendix 9C: Deriving The Multiplicative Form of the Composite Utility Function

Consider the situation where the composite utility function involves cross-product terms of the individual utility functions. To prevent the algebra from becoming too unwieldy, we will derive the multiplicative form for the case of three attributes. We will assume that every pair of two attributes interact, and all three attributes interact with each other in a triplet. Assuming that all these interactions are via multiplication, the composite utility function can be expressed as

$$U(x_1, x_2, x_3) = k_1 u_1(x_1) + k_2 u_2(x_2) + k_3 u_3(x_3)$$
$$+ Kk_1 k_2 u_1(x_1)u_2(x_2) + Kk_2 k_3 u_2(x_2)u_3(x_3) + Kk_1 k_3 u_1(x_1)u_3(x_3) \quad (9\text{–}62)$$
$$+ K^2 k_1 k_2 k_3 u_1(x_1)u_2(x_2)u_3(x_3)$$

We have introduced the new composite scaling constant K so that when $K = 0$, Equ. (9–62) reduces to the additive form. Since we want a compact form of the composite utility function for any number of attributes, we'll do a little rearranging. First we will multiply each term in Equ. (9–62) by K and then add 1 to both sides. We obtain

$$1 + KU(x_1, x_2, x_3) = 1 + Kk_1 u_1(x_1) + Kk_2 u_2(x_2) + Kk_3 u_3(x_3)$$
$$+ K^2 k_1 k_2 u_1(x_1)u_2(x_2) + K^2 k_2 k_3 u_2(x_2)u_3(x_3) + K^2 k_1 k_3 u_1(x_1)u_3(x_3) \quad (9\text{–}63)$$
$$+ K^3 k_1 k_2 k_3 u_1(x_1)u_2(x_2)u_3(x_3)$$

We factor the right hand side of Equ. (9–63) to get

$$1 + KU(x_1, x_2, x_3) = [1 + Kk_1 u_1(x_1)][1 + Kk_2 u_2(x_2)][1 + Kk_3 u_3(x_3)]$$

which can be written in more compact form as

$$1 + KU(x_1, x_2, x_3) = \prod_{i=1}^{3}[1 + Kk_i u_i(x_i)] \quad (9\text{–}64)$$

It can be shown that for the general case of n attributes, Equ. (9–64) takes the form

$$1 + KU(x_1, x_2, x_3, \ldots, x_n) = \prod_{i=1}^{n}[1 + Kk_i u_i(x_i)]$$

[7]Shtub et al., p. 169.

9.9. CLOSURE

We have developed several aids to making design decisions when faced with complicated choices. These complications generally are of two forms. First, we may be faced with choosing among several design alternatives when each option is characterized by multiple criteria. Second, we may have to choose among options when we are uncertain about the consequences of those decisions. We dealt with both issues by examining techniques for systematically incorporating attitudes toward risk and subjective judgments regarding the relative importance of each criterion. The methodologies presented in this chapter (some of which use sophisticated mathematical reasoning) help us to systematically account for subjective considerations. But we should never lose sight of the fact that the underlying judgments that are being modeled are inherently subjective. No amount of matrix manipulation or construction of utility functions can change that. You should never feel obligated to accept the results of a decision making exercise just because it was obtained using a computer. Remember, these are decision aids; you, your boss, or your client is still the decision maker.

The choice of which decision analysis technique to use in a given situation is itself a subjective judgment. Think of these methods as tools in a toolbox. Pick the one that is most appropriate for the task at hand. If your task is to hang a picture on a wall, use some tape or a hammer and nail, not a chainsaw. Further, never impose a particular decision making methodology upon a reluctant client.

9.10. REFERENCES

ANDERSON, DAVID R. et. al. 1986. *Quantitative Methods for Business, 3rd Edition.* St. Paul, MN: West Publishing Co.

CROSS, NIGEL. 1994. *Engineering Design Methods, 2nd Edition.* New York: John Wiley & Sons.

DIETER, G. E. 1991. *Engineering Design: A Materials and Processing Approach, 2nd Edition.* New York: McGraw-Hill, Inc.

GOICOECHA, AMBROSE et. al. 1982. *Multiobjective Decision Analysis with Engineering and Business Applications.* New York, NY: John Wiley & Sons.

KEENEY, RALPH L, and HOWARD RAIFFA. 1976. *Decisions with Multiple Objectives: Preferences and Value Tradeoffs.* New York: John Wiley & Sons.

MEREDITH, DALE D., et. al. 1985. *Design and Planning of Systems, 2nd Edition.* Englewood Cliffs, NJ: Prentice Hall, Inc.

SAATY, THOMAS. 1980. *The Analytical Hierarchy Process: Planning, Priority Setting, Resource Allocation.* New York: McGraw-Hill Book Co.

SAATY T. L. and L. G. VARGAS. 1991. *Prediction, Projection, and Forecasting: Applications of the Analytical Hierarchy Process in Economics, Finance, Politics, Games, and Sports.* Boston: Kluwer Academic Publishers.

SHTUB, AVRAHAM, JONATHAN F. BARD, and SHLOMO GLOBERSON. 1994. *Project Management: Engineering, Technology, and Implementation.* Engelwood Cliffs, NJ: Prentice Hall, Inc.

U.S. Army Corps of Engineers. May 1994. *Detailed Project Report: Large Lock Gate Machinery Rehabilitation.* Seattle WA.

9.11. EXERCISES

1. As vehicle fleet manager for the engineering department of your local city, you are respon-sible for purchasing sedans for use by departmental personnel who visit project sites throughout the city. The following ten attributes may be useful in deciding which brand of automobile to purchase: interior trim, exterior design, workmanship, initial cost, fuel econ-omy, maintenance costs, handling and steering, braking, ride, and comfort. Use an objec-tives tree to group the attributes into at least two but no more than four categories. Then establish relative weighting factors for each attribute and indicate the reasoning you used in developing those factors.
   ◆ *Source:* Dieter, p. 670.

2. Select any group of at least three criteria from the objectives tree constructed for Exercise 13 in Chapter 2. Construct a relative importance matrix for this group and calculate the consistency ratio for your matrix.

3. You are considering welding, adhesives, bolts, and rivets as four options for joining the components of a mechanical system. In the table below, are the initial cost, the probability of failure, and the cost of repairs if failure occurs for each option. Which option has the least expected cost?

	Initial Cost ($)	Probability of Failure	Repair Cost ($)
Welding	5,000	0.075	15,000
Adhesives	4,000	0.100	10,000
Bolts	7,000	0.125	5,000
Rivets	6,000	0.050	7,500

4. Which is the preferred option for Problem 3 if your attitude toward risk is reflected by a utility function of the following shape?

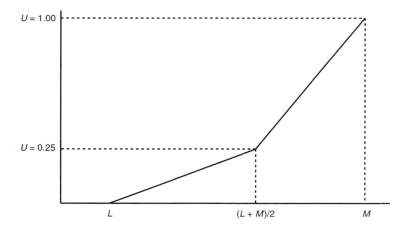

5. A mechanical engineering firm is considering submitting a bid to supply and install the ventilation system for a new transit tunnel. As chief engineer of the company, you estimate that a proposal can be prepared at a cost of $5,000, but that such a bid will have only a 20% chance of being accepted. As an alternative, the company can invest $20,000 on an extensive research study of the project before preparing the proposal. Such a proposal will have a 30% chance of being accepted. If the company does get the job, it can either expand its personnel and staff or subcontract work to other companies. If a major portion of the work is to be performed by subcontractors, it is estimated that there is a probability of 70% and 30% for making a profit of $1.0 million and $2.0 million, respectively. If the work is to be accomplished mostly by expanding staff, there is a probability of 60% and 40% for profits of $0.5 million and $3.0 million, respectively. Based on maximizing the expected value, what should be the optimum strategy for your company?

6. A plant engineer is about to commence an urgent above normal production run. To ensure meeting the desired production rate, she has decided to transfer and incorporate into the production line a piece of equipment from another line at a cost of $5,000. The engineer is considering whether to overhaul this equipment before placing it in the new production line. The piece of equipment costs $800 to overhaul, whereas if she incorporates the item into the process line and it then breaks down it will cost $1,500 to cover the cost of repair and lost time. The engineer estimates that there is a 66% chance that the equipment motor is reliable but is assured that it will be reliable if it is overhauled. A dynamometer test of the motor costs $100 but will only indicate whether the motor is in good or bad condition with a 10% chance that the test results prove invalid. The engineer estimates that there is a 70% chance that the dynamometer test will indicate a reliable motor. If the plant manager wishes to base her decision on expected monetary value, what should be her optimum strategy?
♦ *Source:* Meredith, p. 250.

7. You are a design engineer for an international engineering consulting firm that has been hired by the People's Republic of China to design and build a 500 Mw coal-fired power plant. The government wants the power plant to be completed and generating electricity in six years in order to meet the new electricity requirements associated with its industrialization program. One of the options is to proceed immediately to design and build a conventional coal plant which can be brought on-line in four years. You must decide within the next week whether to recommend to your client to go that route or wait four years in the hope that fluidized bed coal burning technology will be mature enough that you can proceed to design and build the power plant incorporating this new technology. The advantage of this new technology is that the power plant will be significantly more efficient while emitting fewer pollutants without having to use scrubbers to clean up the exhaust gases. However, if your client chooses to wait for the new technology to become available, the plant will not be ready for eight years so the government will have to import power from Russia for two years (costs = 0.5×10^9) to meet the power requirements. Your best estimate of the chances of fluidized bed technology being available after the four year waiting period is 75%. The costs of bringing such a unit on line are estimated at 1.5×10^9. If conventional technology is chosen, the government must decide to go ahead with a plant design that includes scrubbers to reduce air pollution (costs = 2.2×10^9) or a more complicated design (cost = 1.8×10^9) that could either omit scrubbers or have them added on at a later date (added cost of 0.7×10^9). The reason for considering the more complex design option is the uncertainty over whether international pressure over global atmospheric pollution will lead to tougher environmental regulations within the next five years to require scrubbers. Your best estimate at this time is that there is a 60% chance that scrubbers will be required

on all conventional technology (including complex design) coal plants operating beyond six years from now. All costs described above are in terms of net present value. What strategy should you recommend to your client based on minimizing the expected costs.

8. As chief cargo-door engineer for a major aircraft manufacturer, you are responsible for re-designing the door to prevent further accidents such as the one which occurred several years ago on United Airlines flight 811 between Hawaii and New Zealand. In order to have the new cargo doors installed as soon as possible, you are considering whether to redesign the door prior to knowing the official findings of the accident investigation team. You do know that the possible causes of failure have been reduced to either a failure in the latch or a failure in the electric motors that close the door. You estimate that there is a 65% proba-bility that a latch failure will be identified as the cause of the accident and a 35% probabil-ity that the motors will be blamed. The four options you are considering are (a) redesign the latch now at a cost of $25,000; (b) redesign the motor now at a cost of $20,000; (c) re-design both now at a cost of $45,000; and (d) wait for the results of the accident investiga-tion before deciding which component to redesign. If you choose either (a) or (b) and the component which you *did not* redesign is determined to be the cause of failure, it will cost you an additional $45,000 for an accelerated redesign of the latch and an additional $50,000 for an accelerated redesign of the motor. Also, if you choose (d), your design costs are thus associated with an accelerated schedule. Which option has the lowest expected cost?

9. Your engineering design firm is considering bidding on a major project that will yield a profit for the firm of $100,000. You estimate that there is a 20% chance of receiving the contract if you prepare the proposal in-house. An outside consultant could be hired at a cost of $20,000 to prepare the proposal, and this will increase the probability of being awarded the project to about 50%. If you are a risk neutral decision maker, should you hire the consultant? What if your attitude toward risk is reflected by your willingness to trade a 30% chance at making the $100,000 profit under this contract for a guaranteed contract worth $80,000?

10. Redo Exercise 5 assuming that the decision maker's utility function is of the form $u = (1 - e^{-rx})/(1 - e^{-r})$. Determine the preferred option when $r = 4$ and when $r = -3$. Character-ize the risk taking nature of the decision maker for these two cases. Note: The independent variable x has to be non-dimensionalized so that $0 \le u \le 1$.

11. An oil exploration company holds a lease that must either be sold now, held for a year and then sold, or exercised now by drilling for oil. The cost of drilling a well is $200,000 and there is a 50% chance that the well will be dry. It is also estimated that there is a 40% chance of the well producing $400,000 worth of oil, and a 10% chance of producing $1,500,000 worth of oil. If the company decides instead to sell their lease now, they can get $125,000 for it. If they decide to hold the lease for a year and then sell it, there is a 70% chance that oil prices will fall and the value of the lease will decrease to $100,000 and a 30% chance that the lease will increase in value to $500,000. Which option should they choose if their attitude toward risk is reflected by a utility function of the form $u = (1 - e^{-ax})/(1 - e^{-a})$ where $a = -4$ and x is an appropriately scaled nondimensional cost?

12. Your local electric utility company wants to demonstrate the effectiveness of a new tech-nology for using electricity to heat small commercial buildings and is trying to decide what kind of building to use for the demonstration project. Buildings for which the heating re-quirement is a large percentage of total energy use within the building are considered good candidates for this technology. Also, building types whose heating requirements are signifi-cant on a regional basis are likely prospects. Finally, since the particular technology being

demonstrated is only applicable to small buildings, the percentage of buildings within a given category that are smaller than 10,000 ft.^{2} is another good indication of the potential regional impact of the technology. The following data is available on each of these three attributes for four building types in the region served by the utility. Assign weights to each of the three attributes based on the ratio of the attribute range to the attribute mean. Which building type should be selected for the demonstration project? Use a utility function which reflects risk neutral behavior.

	Percent Heating Load of Total Load	Annual Electricity Consumption for Heating (GWh)	Percent Bldgs Under 10,000 ft^2
Small office bldgs	49.6	752	80
Restaurants	20.6	159	92
Retail stores	45.6	654	80
Elementary schools	58.9	483	42

13. A contractor working on an outdoor construction project in a coastal area is reviewing progress on August 1. If normal progress is maintained and no time is lost due to hurricanes, the job will be completed on August 31. However, due to poor weather conditions in the area after August 16, there will be only a 40% chance of finishing on time. It is estimated that there is a 50% chance of a minor hurricane, which will cause a delay of 5 days, and a 10% chance of a major hurricane, which will cause a delay of 10 days. It must be decided now whether to start a crash program on August 2 at an additional cost of $75 per day and finish the project on August 16. As an alternative, the normal schedule can be maintained and progress reviewed on August 31. At that time, if a hurricane has occurred and the project is delayed, there will be a choice of accepting the delay at a certain penalty cost or trying to crash the program then. The penalty cost for delay of completion will be $400 per day for the first 5 days and $600 per day for the second 5 days. The additional cost of a crash program after the hurricane will be $200 per day. The total additional cost is computed as the sum of delay penalty cost and crash cost. It is also estimated that the possible results (outcomes) of a crash program after a minor hurricane causing a 5-day delay will be as follows:

Crash Program Result	Probability	Total Additional Cost
Save 1 day	0.5	$1600 + 800 = $2400
Save 2 days	0.3	$1200 + 600 = $1800
Save 3 days	0.2	$800 + 400 = $1200

The possible results of a crash program after a major hurricane causing a 10-day delay is estimated as follows:

Crash Program Result	Probability	Total Additional Cost
Save 2 days	0.7	$(2000 + 1800) + 1600 = $5400
Save 3 days	0.2	$(2000 + 1200) + 1400 = $4600
Save 4 days	0.1	$(2000 + 600) + 1200 = $3800

Using a utility function of the form $u = (1 - e^{-rx})/(1 - e^{-r})$, determine the preferred option when $r = 5$ and when $r = 4$. Characterize the risk taking nature of the decision maker for these two cases. Note: The independent variable x has to be nondimensionalized so that $0 \le u \le 1$.

♦ *Source:* Meredith, p. 227, with permission from Prentice-Hall, Inc.

14. You are considering using either hydraulic cylinders, pneumatic cylinders, or electric motors to power a mechanical system. Shown below for each option are the capital cost, the probability of failure, and the cost of repairs if failure occurs. Which option has the least expected cost?

	Capital Cost ($)	Probability of Failure	Repair Cost ($)
Hydraulic	5,000	0.001	44,000
Pneumatic	4,000	0.005	16,000
Electric	6,000	0.010	4,000

15. Using AHP, your pairwise comparisons of four design alternatives has yielded the following matrix of relative preferences:

$$\begin{bmatrix} 1 & 6 & 5 & 8 \\ 1/6 & 1 & 3 & 4 \\ 1/5 & 1/3 & 1 & 2 \\ 1/8 & 1/4 & 1/2 & 1 \end{bmatrix}$$

Do these preferences meet the AHP test for consistency?

CHAPTER 10

Optimum Design

10.1 OVERVIEW

In Chapter 9 we examined techniques for selecting the best choice from among a small number of design options. As the number of options increases, the task of evaluating each one of them in detail and comparing it to all other options becomes daunting. Further, there is no guarantee that there isn't a better option out there that we haven't thought of yet. We examine techniques in this chapter that can be useful in overcoming both of these obstacles and permit us to efficiently identify the optimum design from among a large number of options, or even an infinite set of options. (Note from Chapter 1 that the concept of optimum design is part of the ABET definition of engineering design.) Of course, these techniques come at a price. As with most such good deals, there are lots of strings attached. It is just as important to understand the limitations of these techniques as it is to be able to apply them. In fact, with many optimization techniques embedded into mass-marketed software packages, they can easily be applied. Therein lies the danger that they will be misapplied because of the user's inability to appreciate their limitations. Hence, we will take pains in this chapter to explore the fundamental assumptions, to understand the basic strategies underlying the searches for the optimum, and to interpret the results.

We begin by considering an approach to finding the best design for a multi-stage system in which either the number of stages or the number of design options available at each stage is large, but finite. The technique is known as dynamic programming. We will consider applications of dynamic programming to both unconstrained and constrained optimal design problems.

We then turn our attention to constrained optimum design problems when there are several continuous design variables. The fact that the design variables are continuous means that there are an infinite number of design options that can satisfy the constraints. The trick is to use an efficient strategy for locating the optimum design from among this infinite set. The first class of problems we treat are those in which the key relationships involving the design variables are linear. We examine the formulation and characteristics of this class of problems, referred to as linear programming, and we explore the classical solution technique in considerable detail. Computational complexities associated with nonlinearities are examined, and an approximate approach is presented that allows certain types of nonlinearities to be handled using linear programming techniques. Finally, we explore a calculus-based procedure known as Lagrange multipliers that can, in principle, handle general constrained nonlinear optimization problems.

10.2 DYNAMIC PROGRAMMING

In this section we develop an approach to optimizing the design of systems configured in stages and whose design can be characterized as a sequence of design decisions made in each stage. The method, known as dynamic programming, is most useful in situations involving at least four stages with several design options available at each stage. We will explore two design problems that illustrate several variations of the dynamic programming technique. First, we'll discuss a relatively simple problem that allows us to focus on the key dynamic programming concepts. Then we will examine a more challenging problem that dramatically illustrates the power of dynamic programming.

Routing an Electric Power Transmission Line

Electric power transmission lines are familiar sights on the American landscape. These lines frequently deliver power from generating plants to customers hundreds of miles away. The cables, supporting towers, and the corridors in which they are located pose interesting engineering design problems (see Figure 10–1).

Consider a simplified version of the transmission tower location problem. Suppose an electric utility company wants to minimize the cost of building a power line from a generating station at A to customers at E (see Figure 10–2). The construction cost listed in Figure 10–2 (expressed in millions of dollars) for each segment option reflects such things as topography, land acquisition cost, and population densities. The four-stage sequence of design decisions begins at A where we can choose from among two options—route the line through $B1$ or $B2$. At B we have two more options ($C1$ or $C2$). Once we get to C, two options are available to get to D. After we arrive at one of the D options, there is only one way to get to E.

List of design options.   One approach to finding the optimum design is to use "brute force"; that is, calculate the cost of all possible alternatives and select the one that has the lowest cost. There are only eight options in this case, so calculating the cost of each one is not a major burden. Table 10–1 displays the results. Clearly the optimum (least cost) design is Option 6 which costs $52 million.

We performed three arithmetic operations for each of the eight options, or twenty-four operations. We now examine whether the dynamic programming ap-

Figure 10–1.   Electric Power Transmission Corridor *(David Parker/Science Photo Library/Photo Researchers, Inc.)*

proach can identify the optimum design with less calculation effort. As we move through each stage, we will keep track of the number of computations required.

Stage 1 analysis.   We can solve many dynamic programming problems by moving either forward or backward through the stages. To illustrate the backward approach, let us start at point E and work our way back to point A one stage at a time.

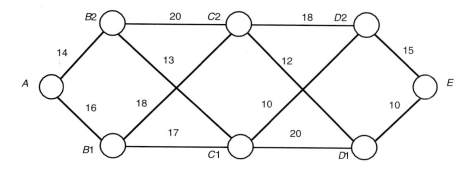

Figure 10–2.   Segment Options for Locating Transmission Line

TABLE 10–1.   DESIGN OPTIONS AND COSTS FOR ROUTING A TRANSMISSION LINE

Design Option	Segments	Segment Costs	Total Cost
1	A-B1-C1-D1-E	16 + 17 + 20 + 10	63
2	A-B1-C1-D2-E	16 + 17 + 10 + 15	58
3	A-B1-C2-D1-E	16 + 18 + 12 + 10	56
4	A-B1-C2-D2-E	16 + 18 + 18 + 15	67
5	A-B2-C1-D1-E	14 + 13 + 20 + 10	57
6	A-B2-C1-D2-E	14 + 13 + 10 + 15	52
7	A-B2-C2-D1-E	14 + 20 + 12 + 10	56
8	A-B2-C2-D2-E	14 + 20 + 18 + 15	67

The two segments that end at point E start at $D1$ and $D2$. For each of these starting points, there is only one way to get to point E. The costs of the segments are listed in Table 10–2. This Stage 1 analysis doesn't require any calculations. All we've done is copy data from Figure 10–2 to Table 10–2.

TABLE 10–2.   STAGE 1 ANALYSIS FOR ROUTING A TRANSMISSION LINE

Beginning of Segment	End of Segment	Cost of Segment	Cost From Beginning of Segment to E
$D1$	E	10	10
$D2$	E	15	15

Stage 2 analysis.   We have four possible segments in Stage 2, with $C1$ and $C2$ as possible beginning points and $D1$ and $D2$ as possible end points. For each of the end points, we know from Table 10–2 how much it costs to get to E. Table 10–3 lists the costs of each segment from C to D and the remaining costs to point E. We have performed four additions through this stage.

We are now ready to invoke the fundamental concept of dynamic programming. We can get from $C1$ to E via either $D1$ or $D2$. Table 10–3 tells us that the route through $D1$ costs \$30 million while the route through $D2$ costs \$25 million. Thus, if we choose a route that goes through $C1$, then the optimum routing from $C1$ to E is via $D2$, irrespective of how we got to $C1$. We place a checkmark in the last column of Table 10–3 to identify $C1$-$D2$-E as the optimum routing of all the routes that go through $C1$. We make a similar argument for the routes that pass through $C2$ and mark $C2$-$D1$-E as the optimum of these routes.

Stage 3 analysis.   We have four possible segments in Stage 3, with $B1$ and $B2$ as the beginning points and $C1$ and $C2$ as the end points. The checkmarks in Table 10–3 give us the minimum cost to get to E from each of the end points. We transfer these costs from Table 10–3 to Table 10–4 where the costs of each B-C segment and

TABLE 10–3.   STAGE 2 ANALYSIS FOR ROUTING A TRANSMISSION LINE

Beginning of Segment	End of Segment	Cost of Segment	Cost from End of Segment to E	Cost from Beginning of Segment to E	Optimum
C1	D1	20	10	$20 + 10 = 30$	
C1	D2	10	15	$10 + 15 = 25$	√
C2	D1	12	10	$12 + 10 = 22$	√
C2	D2	18	15	$18 + 15 = 33$	

the minimum remaining costs to point E are listed. We check the optimum routings through $B1$ and $B2$ in the last column and move on to Stage 4. The four additions in Stage 3 give eight additions through the first three stages.

TABLE 10–4.   STAGE 3 ANALYSIS FOR ROUTING A TRANSMISSION LINE

Beginning of Segment	End of Segment	Cost of Segment	Minimum Cost From End of Segment to E	Cost from Beginning of Segment to E	Optimum
B1	C1	17	25	$17 + 25 = 42$	
B1	C2	18	22	$18 + 22 = 40$	√
B2	C1	13	25	$13 + 25 = 38$	√
B2	C2	20	22	$20 + 22 = 42$	

Stage 4 analysis.   We are now back to the origin of the transmission line. We have two possible segments in this stage from point A to points $B1$ and $B2$. For each of these end points, the checkmarks in Table 10–4 identify the minimum cost to get to E. These costs are transferred to Table 10–5 where we check the optimum routing from A in the last column. Including the two additions in Stage 4, we used only ten calculations to identify the optimum. This compares quite favorably to the twenty-four additions required using the "brute force" approach.

TABLE 10–5.   STAGE 4 ANALYSIS FOR ROUTING A TRANSMISSION LINE

Beginning of Segment	End of Segment	Cost of Segment	Minimum Cost From End of Segment to E	Cost from Beginning of Segment to E	Optimum
A	B1	16	40	$16 + 40 = 56$	
A	B2	14	38	$14 + 38 = 52$	√

Recapitulation.   We find the optimum routing by selecting the appropriate marked segments from Tables 10–5, 10–4, and 10–3 in that order. From Table 10–5, the optimum segment is A-$B2$. From Table 10–4, with $B2$ as the beginning of the next segment, the optimum is *B2-C1*. Similarly, from Table 10–3 we get *C1-D2*. The cost of *D2-E*, the last segment of the optimum route, is found from Table 10–2. Hence, the optimum route for the transmission line is *A-B2-C1-D2-E*. See Table 10–6 for a summary of the optimum segments and their costs.

TABLE 10–6. OPTIMUM DESIGN OF FOUR-STAGE TRANSMISSION LINE

Stage	Segment	Cost
4	*A-B2*	14
3	*B2-C1*	13
2	*C1-D2*	10
1	*D2-E*	15
	Total	52

Variation of the Transmission Line Problem

The design problem we just completed dealt with selecting the lowest cost route within a defined corridor. A related problem, depicted in Figure 10–3, is to determine the optimum tower spacing along a specified route. Consider a known terrain profile given by $y(x)$. The starting and finishing points at which towers are assumed to be located are given by x_0 and x_k. The candidate locations for intermediate towers, at or near ridgetops, are identified by $x_1, \ldots, x_{k-1}$ in the upper part of Figure 10–3. In its simplest version, all the towers are of the same design, but installation costs may depend on location. Each tower has a maximum load that it can withstand, expressed in terms of the length of cable it can support. In addition, a minimum vertical clearance $z(x) - y(x) \geq h$ must be maintained at every point along the route. As indicated con-

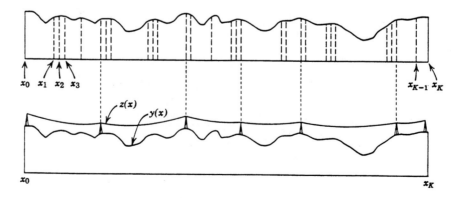

Figure 10–3. Spacing Transmission Towers Along a Known Route *(Reprinted from Pierre, p. 404, with permission from Dover Publications)*

ceptually by the bottom part of Figure 10–3, the least-cost design may involve placing towers at only a small number of the eligible sites. See Pierre[1] for a more detailed discussion of this problem.

Designing a Pollution Control System

Now we apply dynamic programming to another optimum design problem where the efficiencies gained over the "brute force" method are even more impressive than those achieved in the transmission route selection problem. Suppose you are designing a system to remove a pollutant from a fluid moving through a pipe. The system consists of n stages of chemical filters. Your design challenge is to distribute the available amount of the chemical among the n stages to minimize the pollutant remaining in the stream when it leaves the pollution control system.

We will use the index $i = 1, \ldots, n$ to identify the stages. Let P_i represent the amount of the pollutant [measured in parts per million (ppm)] entering stage i and P_{i+1} be the amount of pollutant leaving stage i and entering stage $i + 1$. Our design variable will be C_i, the amount of chemical that is used in stage i.

For a numerical example, assume that we have a five-stage system ($n = 5$) and that one ton of the chemical is available:

$$\sum_{i=1}^{i=5} C_i = 1$$

Further, let's suppose that the amount of chemical that can be used in any stage must be a multiple of 0.1 tons and that a minimum of 0.1 tons must be used in each stage. These utilization requirements pose constraints on the design variable. Also, let the amount of pollutant entering the first stage of the filter system be $P_1 = 100$ ppm.

We'll assume that the effectiveness of the ith filter can be modeled as

$$P_{i+1} = P_i[1 - KC_i(1 - x_i)] \tag{10-1}$$

where

$$x_i = kC_iP_i \tag{10-2}$$

and K and k are constants.

Before working through the numerical example, let's explore the significance of the K and k parameters. First consider the special case of $k = 0$, which leads via Equ. (10–2) to $x_i = 0$. Then we can rearrange Equ. (10–1) as

$$\frac{P_i - P_{i+1}}{P_i} = KC_i$$

This shows that the percentage of reduction in the pollutant at each stage is proportional to the amount of chemical used in that stage. We can then interpret K as the

[1]Pierre, pp. 404–408.

effectiveness of the filter per unit of chemical used. For the general case of $k \neq 0$, the term in parentheses on the right-hand side of Equ. (10–1) has the effect of reducing the value of K. Hence, a nonzero value of k indicates that using too much chemical in a stage can clog the filter and reduce its effectiveness. We'll assume that the filter's effectiveness for this numerical example is characterized by

$$K = 1$$
$$k = 3/200$$

(10–3)

Before illustrating the dynamic programming approach to the optimal design of this system, let us calculate the number of design options that would have to be examined if we used the "brute force" approach to identify the optimum.

List of design options. We start at Stage 1 and consider the number of available design options at each stage.[2] First, the maximum amount of the chemical that we can use in Stage 1 is 0.6 tons since we must reserve at least 0.4 tons for use in Stages 2, 3, 4, and 5. In fact, if we decide to use the maximum allowable 0.6 tons in Stage 1, we have no further alternatives; we must use 0.1 tons in each of Stages 2, 3, 4, and 5. This particular allocation of chemicals among the five stages is listed as Option 1 in Table 10–7.

On the other hand, if we choose to use only 0.5 tons in Stage 1, we have the choice of using either 0.2 tons or 0.1 tons of the chemical in Stage 2. If we use only 0.1 in Stage 2, then we have two more options regarding Stage 3. This also is true at Stage 4. That gives us a total of four designs if we choose 0.5 at Stage 1. These are listed as Options 2 through 6 in Table 10–7.

If we continue this analysis through all subsequent stages for each choice at Stage 1, we identify the 126 design options shown in Table 10–7. To determine the pollutant level remaining in the stream after it exits the last stage of the system (P_6), we have to apply Equ. (10–1) five times for each of the 126 options. If we count each application of Equ. (10–1) as a single mathematical operation, 630 operations are required to find the optimum design. The dynamic programming technique offers the opportunity to find the best of these 126 options without having to examine each one.

Stage 1 analysis. We start by constructing Table 10–8 to describe what is happening at Stage 1 of the system. We choose the amount of pollutant P_i in the water as the variable that describes the condition of the system as it enters each stage. Our objective is to allocate the available chemical to each stage in order to remove the appropriate amount of pollutant at that stage. Hence, a good candidate to serve as our design variable is the amount of chemical used at each stage. However, since the cumulative use of the chemical over all five stages is constrained to be one ton, another

---

[2]We applied dynamic programming to the optimal design for the transmission line route by moving backward through the system starting at point E and ending at point A in Figure 10–2. We could have just as easily moved forward through the system. For the pollution control problem, the form of Equ. (10–1) and the known pollutant level at the entry provide a clear advantage to moving forward through the system.

TABLE 10–7. LIST OF DESIGN OPTIONS FOR FIVE-STAGE POLLUTION CONTROL SYSTEM [AMOUNT OF CHEMICAL USED (TONS)]

Option	1	2	3	4	5		Option	1	2	3	4	5
1	0.6	0.1	0.1	0.1	0.1		64				0.2	0.3
2	0.5	0.2	0.1	0.1	0.1		65				0.1	0.4
3		0.1	0.2	0.1	0.1		66			0.1	0.5	0.1
4			0.1	0.2	0.1		67				0.4	0.2
5				0.1	0.2		68				0.3	0.3
6	0.4	0.3	0.1	0.1	0.1		69				0.2	0.4
7		0.2	0.2	0.1	0.1		70				0.1	0.5
8			0.1	0.2	0.1		71	0.1	0.6	0.1	0.1	0.1
9				0.1	0.2		72		0.5	0.2	0.1	0.1
10		0.1	0.3	0.1	0.1		73			0.1	0.2	0.1
11			0.2	0.2	0.1		74				0.1	0.2
12				0.1	0.2		75		0.4	0.3	0.1	0.1
13			0.1	0.3	0.1		76			0.2	0.2	0.1
14				0.2	0.2		77				0.1	0.2
15				0.1	0.3		78			0.1	0.3	0.1
16	0.3	0.4	0.1	0.1	0.1		79				0.2	0.2
17		0.3	0.2	0.1	0.1		80				0.1	0.3
18			0.1	0.2	0.1		81		0.3	0.4	0.1	0.1
19				0.1	0.2		82			0.3	0.2	0.1
20		0.2	0.3	0.1	0.1		83				0.1	0.2
21			0.2	0.2	0.1		84			0.2	0.3	0.1
22				0.1	0.2		85				0.2	0.2
23			0.1	0.3	0.1		86				0.1	0.3
24				0.2	0.2		87			0.1	0.4	0.1
25				0.1	0.3		88				0.3	0.2
26		0.1	0.4	0.1	0.1		89				0.2	0.3
27			0.3	0.2	0.1		90				0.1	0.4
28				0.1	0.2		91		0.2	0.5	0.1	0.1
29			0.2	0.3	0.1		92			0.4	0.2	0.1
30				0.2	0.2		93				0.1	0.2
31				0.1	0.3		94			0.3	0.3	0.1
32			0.1	0.4	0.1		95				0.2	0.2
33				0.3	0.2		96				0.1	0.3
34				0.2	0.3		97			0.2	0.4	0.1
35				0.1	0.4		98				0.3	0.2
36	0.2	0.5	0.1	0.1	0.1		99				0.2	0.3
37		0.4	0.2	0.1	0.1		100				0.1	0.4
38			0.1	0.2	0.1		101			0.1	0.5	0.1
39				0.1	0.2		102				0.4	0.2
40		0.3	0.3	0.1	0.1		103				0.3	0.3
41			0.2	0.2	0.1		104				0.2	0.4
42				0.1	0.2		105				0.1	0.5
43			0.1	0.3	0.1		106		0.1	0.6	0.1	0.1
44				0.2	0.2		107			0.5	0.2	0.1
45				0.1	0.3		108				0.1	0.2
46		0.2	0.4	0.1	0.1		109			0.4	0.3	0.1
47			0.3	0.2	0.1		110				0.2	0.2
48				0.1	0.2		111				0.1	0.3
49			0.2	0.3	0.1		112			0.3	0.4	0.1
50				0.2	0.2		113				0.3	0.2
51				0.1	0.3		114				0.2	0.3
52			0.1	0.4	0.1		115				0.1	0.4
53				0.3	0.2		116			0.2	0.5	0.1
54				0.2	0.3		117				0.4	0.2
55				0.1	0.4		118				0.3	0.3
56		0.1	0.5	0.1	0.1		119				0.2	0.4
57			0.4	0.2	0.1		120				0.1	0.5
58				0.1	0.2		121			0.1	0.6	0.1
59			0.3	0.3	0.1		122				0.5	0.2
60				0.2	0.2		123				0.4	0.3
61				0.1	0.3		124				0.3	0.4
62			0.2	0.4	0.1		125				0.2	0.5
63				0.3	0.2		126				0.1	0.6

possibility for the design variable is the total amount of chemical consumed through the completion of each stage. This is the approach that we will take. For convenience, we introduce the notation

$$Q_j = \sum_{i=0}^{i=j} C_i$$

to represent the cumulative chemical consumption through stage j. As described earlier, the six possible choices in Stage 1 are $0.1 \leq Q_1 \leq 0.6$ in increments of 0.1 tons. These are listed in the first column in Table 10–8. In the second column, we list the total amount of chemical used in all preceding stages. Of course $Q_0 = 0$ for Stage 1 since there are no preceding stages, but we include this column so that Table 10–8 will have the same format as the tables for the subsequent stages. The third column is the amount of chemical used in the current stage, and is the difference between the first two columns. The pollutant level of the stream entering this stage is listed in the next column. Finally, the pollutant level in the outgoing stream (P_2) is calculated by setting $i = 1$ in Equs. (10–1) and (10–2) and using Equ. (10–3).

TABLE 10–8.   STAGE 1 OF FIVE-STAGE POLLUTION CONTROL SYSTEM

Q_1	Q_0	C_1	P_1	P_2
0.1	0	0.1	100.00	91.50
0.2	0	0.2	100.00	86.00
0.3	0	0.3	100.00	83.50
0.4	0	0.4	100.00	84.00
0.5	0	0.5	100.00	87.50
0.6	0	0.6	100.00	94.00

Note that 0.3 tons of chemicals removes the most pollutant from the stream during Stage 1. This result confirms our earlier interpretation that using too much chemical can decrease the effectiveness of the filter. It is premature to identify $Q_1 = 0.3$ as the optimum design decision for Stage 1 since it could lead to less total pollutant removal after five stages than another choice. Hence, we postpone making choices regarding the best option for Stage 1. Let's keep track of the number of operations as we go through the stages. We have so far used Equ. (10–1) six times.

Stage 2 analysis.   We now move to Stage 2. We set up Table 10–9 using the same format as Table 10–8 and carry forward the relevant information from Table 10–8. Because of the aforementioned restrictions on distribution of chemical use among the stages, our Stage 2 design decisions are limited to $0.2 \leq Q_2 \leq 0.7$.

Let's first examine the option of $Q_2 = 0.2$. In the second column of Table 10–9, we enter the cumulative chemical use through the prior stage. In this case, if the cumulative use through Stage 2 is 0.2, the cumulative use through Stage 1 must have been $Q_1 = 0.1$, and 0.1 tons of chemical must be used in Stage 2. Now we carry forward from Table 10–8 the value of $P_2 = 91.50$ associated with $Q_1 = 0.1$. Finally, we apply

TABLE 10–9.   STAGE 2 OF FIVE-STAGE POLLUTION CONTROL SYSTEM

Q_2	Q_1	C_2	P_2	P_3	Optimum
0.2	0.1	0.1	91.50	83.61	√
0.3	0.1	0.2	91.50	78.22	√
0.3	0.2	0.1	86.00	78.51	
0.4	0.1	0.3	91.50	75.35	
0.4	0.2	0.2	86.00	73.24	√
0.4	0.3	0.1	83.50	76.20	
0.5	0.1	0.4	91.50	74.99	
0.5	0.2	0.3	86.00	70.18	√
0.5	0.3	0.2	83.50	70.98	
0.5	0.4	0.1	84.00	76.66	
0.6	0.1	0.5	91.50	77.15	
0.6	0.2	0.4	86.00	69.35	
0.6	0.3	0.3	83.50	67.86	√
0.6	0.4	0.2	84.00	71.43	
0.6	0.5	0.1	87.50	79.90	
0.7	0.1	0.6	91.50	81.81	
0.7	0.2	0.5	86.00	70.74	
0.7	0.3	0.4	83.50	66.83	√
0.7	0.4	0.3	84.00	68.33	
0.7	0.5	0.2	87.50	74.59	
0.7	0.6	0.1	94.00	85.93	

Equs. (10–1), (10–2), and (10–3) with $i = 2$ to get $P_3 = 83.61$. We conclude that if we design the system so that its cumulative chemical use through the first two stages is 0.2 tons, then we must use 0.1 tons in Stage 1 and 0.1 tons in Stage 2. Of the 100 ppm of pollutant entering the system, 83.61 ppm remain in the flow as it exits Stage 2 ($P_3 = 83.61$).

Of course, we have many choices for our Stage 2 design decision other than $Q_2 = 0.2$. Let's consider the next possibility, $Q_2 = 0.3$. This has two design options associated with it; either $Q_1 = 0.1$ and $C_2 = 0.2$, or $Q_1 = 0.2$ and $C_2 = 0.1$. These are shown in the second and third rows of Table 10–9. We bring forward the corresponding values of P_2 from Table 10–8, and use Equ. (10–1) to calculate the P_3 values shown in the last column of Table 10–9 for both options. We check the preferred option ($Q_1 = 0.1$) since this leaves us with less pollutant (78.22 ppm compared to 78.51 ppm). Thus, although we don't know whether 0.3 is the best choice for Q_2, we have found the preferred design option to get us to $Q_2 = 0.3$. We continue with this approach for the remaining possible values of Q_2 in Stage 2. For each such value, we examine all the possible ways to arrive at that value, calculate the results, and check the best such result in the last column. Stage 2 involves twenty-one arithmetic operations. Including Stage 1, we've conducted twenty-seven operations through Stage 2.

Stage 3 analysis. The Stage 3 analysis is summarized in Table 10–10. The permissible design decisions regarding Q_3 range from 0.3 to 0.8. For each of these values, we list all the Q_2 options but only carry forward from Table 10–9 the best P_3 associated with each Q_2.

For example, if we choose $Q_3 = 0.4$ as our design decision for Stage 3, then the two options compatible with that decision are $Q_2 = 0.2$ or $Q_2 = 0.3$. With $Q_2 = 0.2$, Table 10–9 provides $P_3 = 83.61$. But for $Q_2 = 0.3$, Table 10–9 offers both $P_3 = 78.22$ and $P_3 = 73.24$. We only carry forward the smallest of these, because if the optimum system design calls for $Q_3 = 0.4$, then we want that design to reflect the optimum method to achieve $Q_3 = 0.4$. Again, we use Equs. (10–1), (10–2), and (10–3) to calculate the outgoing pollutant level, P_4. We check those values of P_4 which represent the best outcome for each value of Q_3. These are the only values of P_4 that will carry forward to the next stage. We have used another twenty-one operations for a total of forty-eight operations through the end of Stage 3.

TABLE 10–10.   STAGE 3 OF FIVE-STAGE POLLUTION CONTROL SYSTEM

Q_3	Q_2	C_3	P_3	P_4	Optimum
0.3	0.2	0.1	83.61	76.29	√
0.4	0.2	0.2	83.61	71.08	√
0.4	0.3	0.1	78.22	71.32	
0.5	0.2	0.3	83.61	67.96	
0.5	0.3	0.2	78.22	66.25	√
0.5	0.4	0.1	73.24	66.72	
0.6	0.2	0.4	83.61	66.94	
0.6	0.3	0.3	78.22	63.02	
0.6	0.4	0.2	73.24	61.81	√
0.6	0.5	0.1	70.18	63.91	
0.7	0.2	0.5	83.61	68.02	
0.7	0.3	0.4	78.22	61.62	
0.7	0.4	0.3	73.24	58.51	√
0.7	0.5	0.2	70.18	59.10	
0.7	0.6	0.1	67.86	61.77	
0.8	0.2	0.6	83.61	71.19	
0.8	0.3	0.5	78.22	62.06	
0.8	0.4	0.4	73.24	56.82	
0.8	0.5	0.3	70.18	55.78	√
0.8	0.6	0.2	67.86	57.05	
0.8	0.7	0.1	66.83	60.82	

Stage 4 analysis.   We perform calculations for Stage 4 in a similar manner to those we did for Stage 3. The results are shown in Table 10–11. That brings us up to sixty-nine operations through the end of Stage 4.

TABLE 10–11.   STAGE 4 OF FIVE-STAGE POLLUTION CONTROL SYSTEM

Q_4	Q_3	C_4	P_4	P_5	Optimum
0.4	0.3	0.1	76.29	69.54	√
0.5	0.3	0.2	76.29	64.53	√
0.5	0.4	0.1	71.08	64.73	
0.6	0.3	0.3	76.29	61.26	
0.6	0.4	0.2	71.08	59.89	√
0.6	0.5	0.1	66.25	60.28	
0.7	0.3	0.4	76.29	59.75	
0.7	0.4	0.3	71.08	56.58	
0.7	0.5	0.2	66.25	55.63	√
0.7	0.6	0.1	61.81	56.20	
0.8	0.3	0.5	76.29	59.97	
0.8	0.4	0.4	71.08	54.77	
0.8	0.5	0.3	66.25	52.30	
0.8	0.6	0.2	61.81	51.74	√
0.8	0.7	0.1	58.51	53.17	
0.9	0.3	0.6	76.29	61.95	
0.9	0.4	0.5	71.08	54.48	
0.9	0.5	0.4	66.25	50.28	
0.9	0.6	0.3	61.81	48.42	√
0.9	0.7	0.2	58.51	48.86	
0.9	0.8	0.1	55.78	50.67	

Stage 5 analysis.   The Stage 5 analysis requires examining only six options since we know that we must have $Q_5 = 1$. The results are shown in Table 10–12. We have completed the analysis using seventy-five operations, substantially less than the 630 operations required for the "brute force" approach.

TABLE 10–12.   STAGE 5 OF FIVE-STAGE POLLUTION CONTROL SYSTEM

Q_5	Q_4	C_5	P_5	P_6	Optimum
1	0.4	0.6	69.54	53.93	
1	0.5	0.5	64.53	47.88	
1	0.6	0.4	59.89	44.55	
1	0.7	0.3	55.63	43.12	
1	0.8	0.2	51.74	43.00	√
1	0.9	0.1	48.42	43.93	

Recapitulation.   The optimum design as determined from Table 10–12 is the one that reduces the pollutant level to $P_6 = 43.00$. In order to identify the configuration of each stage of this optimum design we now examine the tables in reverse order. Table 10–12 tells us that we should use 0.2 tons of chemical in Stage 5 and that the optimum design is the one for which $Q_4 = 0.8$. Then, from Table 10–11 with $Q_4 = 0.8$, the optimum chemical use in Stage 4 is $C_4 = 0.2$. This leads us back to Table 10–10 with $Q_3 = 0.6$ to find the optimum chemical use in Stage 3 is $C_3 = 0.2$. Continuing back in the same manner to Tables 10–9 and 10–8, we find $C_2 = 0.2$ and $C_1 = 0.2$. The amount of chemical used in each stage and the outgoing pollutant levels at each stage for the optimum design are summarized in Table 10–13.

TABLE 10–13.   OPTIMUM DESIGN OF FIVE-STAGE
POLLUTION CONTROL SYSTEM

Stage i	Chemical Use (tons) C_i	Outgoing Pollutant Level (ppm) P_{i+1}
1	0.2	86.00
2	0.2	73.24
3	0.2	61.81
4	0.2	51.74
5	0.2	43.00

Generalization

The transmission line and pollution control problems discussed in this section are examples of multi-staged systems suitable for optimal design via dynamic programming. The dynamic programming analysis can proceed forward through the system (as with the pollution control problem) or backward through the system (as with the transmission line problem). The technique can be applied to any optimum design problem for a system with n stages that requires a design decision at each stage i. For the transmission line problem, the design decision involved selecting the physical locations for the intermediate destinations. For the pollution control problem, the design decision involved selecting the cumulative amount of chemical to be consumed through each stage. The design decision at each stage transforms the condition, or state, of the system entering stage i into an outgoing state that serves as the entering state for the next stage $i + 1$ (see Figure 10–4). We used cost as the variable that described the state of the transmission line at each stage. Pollutant level was used as the state variable for the pollution control problem.

Sometimes the most difficult part of dynamic programming is to identify the stages, select the variables that describe the state, and choose the parameters to serve as the design decisions. This points to the importance of mastering the problem for-

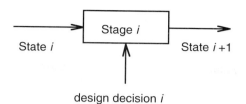

Figure 10–4.   States, Stages, and Design Decisions in Dynamic Programming

mulation concepts covered in Chapter 2. In particular, formulating dynamic programming problems involves establishing an objective (the system state) that is to be optimized, selecting the design variables, and incorporating constraints on the design variables. Let's apply this terminology to the two problems we just solved.

For the transmission line route selection problem, the objective was to minimize the construction costs. The performance characteristic of each line segment was the cost. The design variables were the intermediate locations for the transmission line. The constraints on the design variables were the limited candidate locations provided as part of the problem statement. For the variation of that problem involving the spacing of the transmission towers (see Figure 10–3), the design variable is the intermediate locations of the towers. The distance between towers determines the cable weight supported by each tower, which in turn determines the tower cost. Constraints exist on the maximum weight that can be supported by each tower and on the minimum ground clearance for the cables.

For the pollution control problem, our objective was to minimize the amount of pollutant remaining in the fluid after it leaves the system. The design variable was the amount of chemicals used in each filter. There were constraints on the total amount available, and the amounts that could be used in each filter.

10.3 LINEAR PROGRAMMING

In this and the following several sections we consider techniques for optimizing engineering designs that are defined by n design variables $(x_1, ..., x_n)$ subject to inequality constraints (see Chap. 2 for the role of inequality constraints in defining engineering design problems). The optimum design is defined as that combination of values for the design variables $(x_1, ..., x_n)$ that yield the maximum possible value of an objective function $U(x_1, ..., x_n)$ without violating the constraints. The techniques we will be discussing are limited to design problems for which the objective and constraints can be written in mathematical form.

The most widely studied class of optimization problems involving inequality constraints consists of both an objective function U and a set of constraints that are linear functions of the design variables. This class of problems is called linear programming. The term "linear programming" precedes the widespread use of computers and does not refer in any sense to programming computers. Mathematically, the standard form of the linear programming program is: Maximize the objective function U

$$U(x_1, x_2, \ldots, x_n) = \sum_n k_i x_i$$

where x_i are the n design variables, subject to m constraints on the design variables of the form

$$\sum_i a_{ij} x_i \leqslant r_j \qquad j = 1, 2, \ldots, m$$

In the above expressions, a_{ij} and k_i are constants presumed to have known values in any particular problem. The standard form of linear programming problems also requires that the design variables be non-negative, that is,

$$x_i \geqslant 0$$

Designing Metal Alloys—Geometrical Interpretation

Small amounts of various metals such as chromium and titanium are commonly added to iron and aluminum to provide steel and aluminum alloys with more desirable properties than the parent metal. Suppose we wish to add two additives to the parent metal iron to design a specialty steel with certain enhanced properties. Let x_1 represent the percent composition of the first additive in the alloy and x_2 be the percent composition of the second additive. These will be our two design variables. Let us say that our objective is to maximize the increase in crack resistance that the alloy has over pure iron. We are given that every percent of additive 1 in the alloy increases crack resistance by 2% and every percent of additive 2 increases crack resistance by 1%. If we let U represent the percent increase in crack resistance of the alloy, our design problem is to maximize

$$U = 2x_1 + x_2 \tag{10-4}$$

As you might expect, there are several constraints to the amount of additives we can add. Consider a cost constraint. Each percent increase in additive 1 increases the cost by 3% and each percent increase in additive 2 increases the cost by 1%. In order to remain cost competitive with other materials, we want to limit the overall cost increase to no more than 9%. This constraint can be expressed as

$$3x_1 + x_2 \leqslant 9 \tag{10-5}$$

The second constraint deals with the possible undesirable effects of the alloy additives on the metal's corrosion resistance. Additive 2 decreases the corrosion resistance by 2% for every 1% increase in additive 2. This can be partially offset by the fact that corrosion resistance increases by 1% for every 1% increase in additive 1. The maximum acceptable decrease in corrosion resistance is 4%. We thus have a second inequality constraint

$$2x_2 - x_1 \leqslant 4 \tag{10-6}$$

There is also a constraint on the reduction in melting temperature. Each percent increase in additive 1 decreases the melting temperature by 1% and each percent increase in additive 2 also decreases the melting temperature by 1%. We want to limit the total decrease in melting temperature to no more than 4%. This constraint can be expressed as

$$x_1 + x_2 \leq 4 \tag{10–7}$$

Finally, additive 2 is in great demand right now and our supplier only has enough in stock to provide at most a 5% level in the alloy. Thus

$$x_2 \leq 5 \tag{10–8}$$

The generic linear programming constraints

$$x_1 \geq 0 \tag{10–9}$$

$$x_2 \geq 0 \tag{10–10}$$

also apply.

In order to visually interpret the objective function and the constraints, we can establish a two-dimensional design space in which x_1 and x_2 are the coordinates. Thus, any design defined by values of x_1 and x_2 is represented by a point in the x_1–x_2 plane. If we temporarily ignore the inequality part of each constraint, then each of Equs.

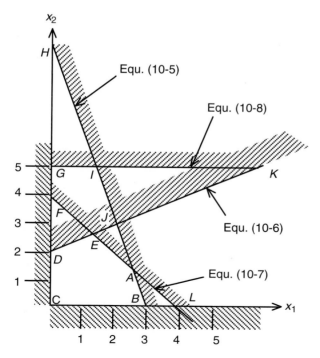

Figure 10–5. Geometrical Representation of Constraints

(10–5)–(10–10) is the equation of a straight line in the x_1-x_2 plane. These lines are shown in Figure 10–5 [Equs. (10–9) and (10–10) coincide respectively with the x_2 and x_1 axes]. Solving any two of Equs. (10–5)–(10–10) simultaneously locates one of the twelve points of intersection $A, \ldots, K$.

The cross-hatched zone to one-side of each line identifies the inaccessible part of the x_1-x_2 plane as determined by the inequality aspect of that constraint.[3] Note that the constraint specified by Equ. (10–8) is redundant since the satisfaction of all the other constraint conditions automatically satisfies Equ. (10–8). Thus we can ignore Equ. (10–8) since it doesn't affect the solution. Similarly, portions of the straight lines representing the other constraints (such as A-J-I-H) are likewise redundant. Eliminating all these redundancies, Figure 10–6 shows the polygonal region A-B-C-D-E of the x_1-x_2 plane within or on which all acceptable design solutions must lie. This region is known as the feasible region. Because of the generic constraints regarding the non-negativity of the design variables [given in this example by Equs. (10–9) and (10–10)],

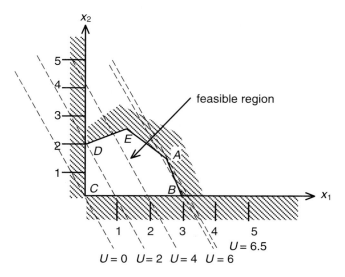

Figure 10–6. Feasible Region and Contour Lines for Objective Function

---

[3]A quick way to locate the inaccessible side of the boundary representing an inequality constraint is to select a convenient point on either side of the boundary and test whether or not the constraint is satisfied. For example, consider the constraint defined by Equ. (10–6). It is clear from Figure 10–5 that the point $x_1 = 1, x_2 = 3$ is above and to the left of the boundary line representing the constraint. It is also clear by substituting these values of x_1 and x_2 into Equ. (10–6) that the constraint is violated by this combination of design values. Hence, the region of the x_1-x_2 plane above and to the left of the line representing Equ. (10–6) is inaccessible.

the feasible region for all two-dimensional linear programming problems lies in the first quadrant of the x_1-x_2 plane.

We continue with our geometrical interpretation of linear programming problems by focusing now on the objective function U. Recall that we are looking for that combination of the design variables x_1 and x_2 which yields the maximum possible value of U. Let's rewrite the objective function [Equ. (10–4)] as

$$x_2 = U - 2x_1 \qquad\qquad (10\text{--}11)$$

This is the equation of a family of straight lines in the x_1-x_2 plane. All lines in this family have a slope of -2; the position of each line is determined by the value of U. Figure 10–6 depicts several members of this family of lines.

Consider the $U = 0$ line. None of the points on this line lie in the interior of the feasible region. Also, there is only one point on this line that lies on the boundary of the feasible region. Thus, the only possible solution for $U = 0$ that is consistent with the constraints is $x_1 = 0$ and $x_2 = 0$.

Let us examine the lines with higher values of U since we want to maximize U. For $U = 2$, there are an infinite number of points on the $U = 2$ line that lie within the feasible region plus two points [(1,0) and (0,2)] that lie on the boundary. Each one of the solutions associated with $U = 2$ is better than the $U = 0$ solution. A similar set of even better solutions are on the $U = 4$ line, and still better solutions lie on the $U = 6$ line.

As we continue moving in the direction of lines of increasing U, a limiting case is reached for which a single solution occurs at the corner point A on the boundary. Increasing U any further cannot lead to any additional solutions that satisfy the constraints. We can locate the corner point A precisely as the intersection of the two boundary lines defined by Equs. (10–5) and (10–7). Solving those two equations simultaneously yields $x_1 = 2.5$ and $x_2 = 1.5$ as the coordinates of A. The value of the U line passing through that point is found from Equ. (10–4) as

$$U = 2(2.5) + 1.5 = 6.5$$

So for this problem, the maximum value is $U = 6.5$.

Significance of Corner Points

The fact that the optimum solution of a linear programming problem occurs at a corner point has great significance for the development of an efficient strategy for locating the optimum. From Figure 10–6, there are clearly an infinite number of points within the feasible region. There are also an infinite number of points on the boundary of the feasible region. In principle (if we did not know the optimum occurs at a corner), the optimum could occur at any one of these interior or boundary points. Thus, a search algorithm would have to systematically calculate U at a large number of points in order to have a reasonable prospect for finding the optimum. However, knowing that the optimum occurs at a corner point means that we only have to evaluate U at each of the corner points. For the problem shown in Figure 10–6, that means we only have to calculate U at the five corner points A, B, C, D, and E. The largest of these five values of U will be the optimum.

Without offering a formal proof, our conclusion that the optimum occurs at a corner point applies to any linear programming problem involving two design variables. In fact, we can prove that the optimum occurs at a corner point in n-dimensional design space when there are n design variables (an n-dimensional point is defined by values for $x_1, x_2, \ldots, x_n$). This property is the basis for the search algorithm we will discuss shortly.

Special Cases

We now examine several exceptions to the rule that the optimum always occurs at a corner point. First, suppose we modify the problem by removing the constraints defined by Equs. (10–5), (10–7), and (10–8). Then Figure 10–5 is transformed into Figure 10–7. Note that the feasible region is unbounded. Clearly it is possible to have the objective function take on arbitrarily large values without violating the constraints. Hence, no optimum exists.

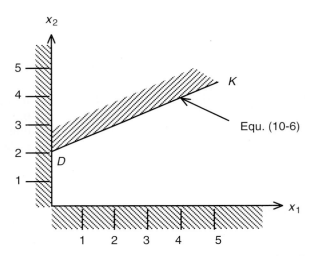

Figure 10–7. Unbounded Feasible Region

We illustrate a second special case by changing the direction of the inequality represented by Equ. (10–8). Then Figure 10–5 is converted into Figure 10–8 and it is clear that no feasible region exists. That is, no solution can satisfy all the constraints.

The final special case that we will consider occurs if we modify the constraint represented by Equ. (10–5) to read

$$4x_1 + 2x_2 \leq 13 \qquad (10\text{–}12)$$

This modifies Figure 10–6 as shown in Figure 10–9 so the contour line associated with $U = 6.5$ is coincident with Equ. (10–12). While $U = 6.5$ is still the maximum possible value of U, that maximum no longer occurs just at the corner point A. There are an infinite number of points along the boundary segment A-B that correspond to the maxi-

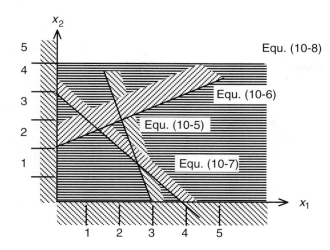

Figure 10–8.   No Feasible Region

mum value of U. Hence we have the special case of a unique maximum associated with a nonunique set of solutions for the design variables.

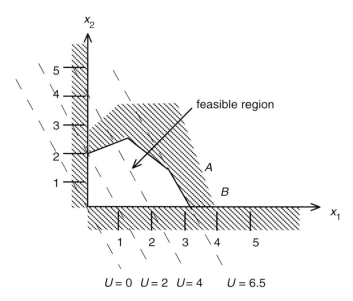

Figure 10–9.   Contour Lines Parallel to Part of Feasible Region Boundary

Production Planning Problem—Algebraic Formulation

The geometrical interpretation serves us well when we are seeking to optimize with respect to only two design variables. In preparation for handling problems with more design variables, we turn now to an algebraic formulation of the linear programming problem. However, we will continue to rely on the geometrical concepts of feasible regions and their corner points to assist in interpreting the significance of the solutions. The geometrical concepts will also help us devise efficient search strategies for obtaining solutions to problems involving many dimensions.

One class of engineering design problems solvable by linear programming is that of optimizing the use of manufacturing resources. Suppose a manufacturing firm has the capability to make three different products in its factory. Fabricating each product requires several manufacturing operations. Table 10–14 shows the time required for each operation per unit of each product manufactured. Also shown is the maximum available time per day for each operation and the unit profit associated with each product. We wish to determine how much of each product we should produce in order to maximize the total profit.

Let x_i represent the amount of product i produced. Then our objective is to maximize

$$U = 3x_1 + 2x_2 + 5x_3 \tag{10-13}$$

where the three design variables x_1, x_2, x_3 are subject to the constraints

$$x_1 + 2x_2 + x_3 \leqslant 430 \tag{10-14}$$

$$3x_1 + 2x_3 \leqslant 460 \tag{10-15}$$

$$x_1 + 4x_2 \leqslant 420 \tag{10-16}$$

The design variables also must satisfy the generic constraints that $x_1, x_2, x_3 \geq 0$. Solution of the problem is facilitated by converting each of the inequality constraints to an equation. This is accomplished without any loss in generality by adding a new variable to each constraint to get

$$x_1 + 2x_2 + x_3 + x_4 = 430 \tag{10-17}$$

$$3x_1 + 2x_3 + x_5 = 460 \tag{10-18}$$

$$x_1 + 4x_2 + x_6 = 420 \tag{10-19}$$

TABLE 10–14.   DATA FOR PRODUCTION PLANNING PROBLEM

Operation	Time per Unit (minutes)			Operation Capacity (minutes)
	Product 1	Product 2	Product 3	
Machining	1	2	1	430
Welding	3	0	2	460
Casting	1	4	0	420
Profit/Unit ($)	3	2	5	

where the new variables x_4, x_5, x_6 are called slack variables.[4] The slack variables must also satisfy the generic constraints; that is, x_4, x_5, $x_6 \geq 0$. From an algebraic perspective, the three slack variables have the same standing as the three design variables. Recalling the geometrical interpretation of the constraint equations from the alloy problem, we have transformed our original three-dimensional feasible region defined by Equs. (10–14)–(10–16) into a six-dimensional feasible region defined by Equs. (10–17)–(10–19). A set of values for x_1, ..., x_6 which satisfies Equs. (10–17)–(10–19) represents a (six-dimensional) corner point on the boundary of our feasible region.

Equs. (10–17)–(10–19) constitute a set of three linear simultaneous equations with six unknowns. Thus, there are an infinite number of possible solutions. We can find any particular solution by assigning values to any three of the variables and then solving Equs. (10–17)–(10–19) for the remaining three variables. We devote the next section to examining a strategy for finding the optimum solution.

10.4. THE SIMPLEX ALGORITHM

The challenge now is to use a systematic and efficient strategy to find the optimum solution from among the infinite set of possible solutions to Equs. (10–17)–(10–19). The most commonly used strategy is called the Simplex method. The Simplex method can be understood in terms of the geometric interpretations in Section 10.3. Specifically, each solution is a vertex of an m-dimensional polyhedron where m is the number of design variables plus slack variables. It also helps to think of these vertices on the m-dimensional polyhedron being connected to each other by edges of the polyhedron. The optimum solution is the solution with the largest value of the objective function.

The Simplex algorithm involves three main phases. First, select an appropriate corner point to begin the analysis and evaluate the objective function at that point. Second, determine which edge of the polygon to move along in order to get to an adjacent corner point. The most efficient strategy is to move along the edge for which the objective function increases most rapidly. Third, decide how far along that edge to move. We will illustrate the application of this search strategy to the production planning problem presented in Section 10.3. Table 10–15 summarizes each step in the process and the application of that step to this particular problem.

The objective function, Equ. (10–13), and the three constraint equations, Equs. (10–17), (10–18), and (10–19), are displayed for convenience at the top of Table 10–15. For convenience in working within the table, we'll renumber the constraint equations as (1), (2), and (3). As indicated in the leftmost column, Table 10–15 divides the solution process into eight steps with each step constituting an action described in the "Action" column. These actions are repeated for as many cycles as needed in order to find the optimum. Let us follow the process one step at time.

---

[4]We can easily verify that Equ. (10–17) is equivalent to Equ. (10–14). If $x_4 = 0$ in Equ. (10–17), then the equality part of Equ. (10–14) is satisfied. If $x_4 > 0$, then the inequality part of Equ. (10–14) is satisfied. Similar verifications exist for the other constraints.

TABLE 10-15. SIMPLEX METHOD EXAMPLE PROBLEM

Objective Function: $U = 3x_1 + 2x_2 + 5x_3$

Subject to m constraints:
$$x_1 + 2x_2 + x_3 + x_4 = 430 \quad (1)$$
$$3x_1 + 2x_3 + x_5 = 460 \quad (2)$$
$$x_1 + 4x_2 + x_6 = 420 \quad (3)$$

Step	Action	Cycle 1	Cycle 2	Cycle 3
1	Select m basis variables and set all other variables to 0. For Cycle 1 choose as basis variables those which appear in only one constraint equation each and whose co-efficients in these equations are unity. For subsequent cycles, obtain variables from Steps 5 and 8 of prior cycle.	basis: x_4, x_5, x_6 $x_1 = x_2 = x_3 = 0$	basis: x_3, x_4, x_6 $x_1 = x_2 = x_5 = 0$	basis: x_2, x_3, x_6 $x_1 = x_4 = x_5 = 0$
2	Solve m constraint equations for values of the m basis variables.	$x_4 = 430$ $x_5 = 460$ $x_6 = 420$	$x_3 = 230$ from (2) $x_4 = 200$ from (1) $x_6 = 420$ from (3)	$x_3 = 230$ from (2) $2x_2 + 230 = 430$, from (1) $x_2 = 100$ $x_6 = 420 - 4(100) = 20$ from (3)
3	Evaluate U for this cycle.	$U = 0$	$U = 1150$	$U = 1350$
4	Begin search for improved solution by expressing U in terms of nonbasis variables, using constraint equations as necessary to eliminate basis variables from U.	$U = 3x_1 + 2x_2 + 5x_3$	using (2) to eliminate x_3 $U = 3x_1 + 2x_2 + 5(230 - 3/2\,x_1 - 1/2\,x_5)$ $U = 1150 - 9/2\,x_1 + 2x_2 - 5/2\,x_5$	eliminate x_3 from (1) and (2), then solve for x_2 in terms of x_1, x_4, x_5 $U = 1150 - 9/2\,x_1 - 5/2\,x_5$ $+ 2(100 + 1/4\,x_1 - 1/2\,x_4 + 1/4\,x_5)$ $U = 1350 - 4x_1 - x_4 - 2x_5$
5	Select as incoming basis variable that variable for which a unit increase has most effect on increasing U.	x_3 is incoming basis variable	x_2 is incoming basis variable	no further increase in U is possible
6	Express constraint equations in terms of current basis variables + incoming basis variable.	$x_3 + x_4 = 430$ (a) $2x_3 + x_5 = 460$ (b) $x_6 = 420$ (c)	$2x_2 + x_3 + x_4 = 430$ (a) $2x_3 = 460$ (b) $4x_2 + x_6 = 420$ (c)	
7	Select as outgoing variable, the current variable which imposes the most severe constraint on the incoming variable.	from (a), x_4 requires $x_3 \le 430$ from (b), x_5 requires $x_3 \le 230$ from (c), x_6 does not limit x_3 x_5 is outgoing basis variable	from (b), x_3 does not limit x_2 from (c), x_6 requires $x_2 \le 105$ using (b) to eliminate x_3 from (a) shows that x_4 requires $x_2 \le (430 - 230)/2$ x_4 is outgoing basis variable	
8	With new basis variables selected, set nonbasis variables to 0 and start next cycle.	$x_1 = x_2 = x_5 = 0$	$x_1 = x_4 = x_5 = 0$	

Cycle 1

As mentioned above, the first phase of the Simplex method is to select a corner point at which the objective function is initially evaluated. Recall our discussion at the end of Section 10.3: A set of values for x_1, ..., x_6 defines a six-dimensional corner point. Also, since we only have three equations involving these six unknowns, we can arbitrarily assign values to any three of the x_i, and solve Equs. (1)–(3) for the remaining three x_i. A key to the success of the Simplex algorithm is the following rule for selecting this initial corner point.

    Step 1.   The three variables x_i to be solved for from Equs. (1)–(3) are called the basis variables. We choose the initial basis variables to be those variables that appear in only one constraint equation each, and whose coefficients in these equations are unity. Thus, x_1 is not a suitable initial basis variable since it appears in all three constraint equations.
    Similarly, x_2 and x_3 are not suitable initial basis variables. However, x_4 is an appropriate initial basis variable since it appears only in Equ. (1) and its coefficient in Equ. (1) is +1. Likewise, x_5 and x_6 can serve as initial basis variables. With x_4, x_5, x_6 selected as the basis variables, we are free to arbitrarily assign values to the nonbasis variables, x_1, x_2, and x_3. The Simplex method assigns values of 0 to each of the nonbasis variables. The result of these actions in Step 1 is summarized in Table 10–15 in the Cycle 1 cell associated with Step 1.

    Step 2.   We now determine the coordinates of the corner point associated with this set of basis variables. Setting $x_1 = x_2 = x_3 = 0$ allows us to solve Equs. (1)–(3) for x_4, x_5, and x_6. The Cycle 1 cell associated with Step 2 shows these results. Note that the values $x_1 = 0$, $x_2 = 0$, $x_3 = 0$, $x_4 = 430$, $x_5 = 460$, and $x_6 = 420$ define a corner point in six-dimensional space.

    Step 3.   We complete Phase 1 of the Simplex method by evaluating the objective function at the corner point we just identified. Substitute these values of x_1, ..., x_6 into the objective function. This yields $U = 0$; this value is displayed in the next cell in the Cycle 1 column.
    Let us summarize what we have accomplished so far. We have found a corner point, defined by the values of x_1, ..., x_6, and have evaluated the objective function at that point. From the physical significance of our design variables and objective function, this solution says our profit will be 0 if we do not produce any of the three products. This is not a surprising conclusion!

    Step 4.   We are now ready to start Phase 2 of our search strategy. The current corner point is connected to nearby corner points by edges of the polyhedron. We know that if we move along any one of these edges, we will eventually arrive at another corner point. As mentioned above, our strategy in this phase of the search is to decide which edge to move along. Since our goal is to find the maximum value of U as efficiently as possible, a good rule to follow is to select the edge along which U increases most rapidly. At the current point, the nonbasis variable x_1, x_2, and x_3 are all equal to 0. Allowing any one of these to vary from 0 can be interpreted as moving along an edge in the direction of that coordinate. The effect of such a motion on U can be determined by expressing U solely in terms of the nonbasis variables. For Cycle 1 in this particular problem, U is already in the desired form. We will see later in Cycles 2 and 3 that we need to use the constraint equations to eliminate all the basis variables from the expression for U. Our expression for U is

$$U = 3x_1 + 2x_2 + 5x_3$$

Step 5. At the current corner point, $x_1 = x_2 = x_3 = 0$. If we allow x_1 to increase from 0 while we keep $x_2 = x_3 = 0$, every unit increase in x_1 will produce an increase of three units in the value of U. But the fastest increase in U will occur in this example if we select x_3 to increase from its current value of 0; therefore, we select x_3 as the incoming basis variable.

Step 6. We can only have three basis variables at a time. Thus, if we bring x_3 into the basis, we have to delete one of the current basis variables x_4, x_5, or x_6. This is the beginning of Phase 3 of the search strategy. Since any increase in x_3 from its current value of 0 produces an increase in U, it makes sense to increase x_3 as much as possible. To see how far we can increase x_3, let us examine the constraint equations by expressing them in terms of the current basis variables x_4, x_5, x_6 and the incoming basis variable x_3. Equations (1), (2), and (3) then take the form

$$x_3 + x_4 = 430 \tag{a}$$

$$2x_3 + x_5 = 460 \tag{b}$$

$$x_6 = 420 \tag{c}$$

Step 7. Recall that all variables are subject to the non-negativity constraint. We examine Equs. (a), (b), and (c) to determine when that constraint is encountered as we allow x_3 to increase. From Equ. (a) we see that x_3 can increase to 430 before x_4 becomes negative. On the other hand, Equ. (b) indicates that if x_3 exceeds 230, the non-negativity constraint on x_5 will be violated. Equ. (c) tells us that x_3 can increase indefinitely without violating the non-negativity constraint on x_6. Since we want x_3 to increase as much as possible, we select x_5 as the variable to be removed from the basis to make room for x_3. The x_5 variable is outgoing because it imposes the most severe constraint on the incoming variable x_3.

Step 8. With x_3, x_4, x_6 as the new set of basis variables, we set the nonbasis variables x_1, x_2, x_5 to 0 in preparation for the next cycle.

Cycle 2

We are now ready to find the next corner point and to evaluate the objective function at that corner point. This is the beginning of Cycle 2 (see middle column in Table 10–15). With $x_1 = x_2 = x_5 = 0$, Equs. (1)–(3) reduce to

$$x_3 + x_4 = 430$$

$$2x_3 = 460$$

$$x_6 = 420$$

Invoking Step 2 of the search strategy, the second of these equations yields $x_3 = 230$. Substituting this into the first equation provides $x_4 = 200$. The third equation of this set immediately gives $x_6 = 420$. These are the coordinates of the next corner point. Applying Step 3, the value of the objective function at this corner point is

$$U = 3(0) + 2(0) + 5(230)$$

$$U = 1150$$

This solution is clearly superior (U is greater) to the Cycle 1 solution. We apply Step 4 to begin our search for an even better solution. We use constraint Equ. (2) to eliminate the basis variable x_3 from the U expression. Then U is rewritten solely in terms of the nonbasis variables x_1, x_2, x_5 as

$$U = 1150 - \frac{9}{2}x_1 + 2x_2 - \frac{5}{2}x_5$$

Clearly, the only possibility for increasing U is to let x_2 increase from its current value of 0 (recall, none of the variables can be negative). With x_2 selected as the incoming basis variable, we invoke Step 6 in order to identify which variable should be removed from the basis to make room for x_2. The three constraint equations take the form

$$2x_2 + x_3 + x_4 = 430 \qquad \text{(a)}$$

$$2x_3 = 460 \qquad \text{(b)}$$

$$4x_2 + x_6 = 420 \qquad \text{(c)}$$

Equ. (b) in this set tells us that x_3 doesn't impose any constraint on the incoming variable x_2. According to the third equation, x_6 requires that $x_2 \leq 105$ in order to prevent x_6 from becoming negative. And according to Equ. (a), with $x_3 = 230$, x_4 limits x_2 to be no greater than 100. Since x_4 imposes the most severe constraint on the incoming variable, we select x_4 as the outgoing variable.

Cycle 3

We then move on to Cycle 3 with x_2, x_3, x_6 as the new basis. This leads, in Steps 2 and 3, to a solution $x_1 = 0$, $x_2 = 100$, $x_3 = 230$, $x_4 = 0$, $x_5 = 0$, $x_6 = 20$, and $U = 1350$. This is a better solution than that obtained in Cycle 2. But as we see in Step 4 of Cycle 3 in Table 10–15, bringing any of the nonbasis variables into the basis can only decrease U. Hence, no further improvements in U are possible. Thus, we have found the optimum production plan. Note that $x_1 = 0$ means that the optimum production plan is not to make any of product 1; we should devote all resources to making 100 units of product 2 and 230 units of product 3.

Thus, we were able to find the optimum solution by examining only three corner points. This is an impressive demonstration of the efficiency of the Simplex algorithm. Of course, changes in the objective function or constraints could cause us to utilize more than three cycles to locate the optimum. Clearly, as the number of design variables or constraints increases, the manual solution method we just demonstrated becomes increasingly cumbersome. Fortunately, the Simplex algorithm is built into the major commercial spreadsheet software packages so it is unlikely that you will ever have to use the manual solution method in a "real" complicated design problem. However, it is highly recommended that you work through the problem just completed and several other exercises. This will help you appreciate the geometric and algebraic significance of linear programming problems and the Simplex solution technique. Table 10–16 provides a template for applying the manual Simplex method.

10.5. VARIATIONS OF THE STANDARD LINEAR PROGRAMMING PROBLEM

There are several important variations of the standard linear programming problem that still allow the Simplex algorithm to be used.

Minimizing the Objective Function

The standard formulation of the linear programming problem involves maximizing the value of the objective function U. Many optimum design problems require that an objective function be minimized (weight, cost, etc.). This variation can easily be accommodated by the Simplex method.

If the objective function U is to be minimized, we can introduce a new objective function U' defined as

$$U' = -U$$

and proceed with the Simplex algorithm to maximize U'.

Negative Design Variables

The standard formulation of the linear programming problem requires that all design variables be non-negative. For design problems involving variables which may take on negative values (stress, velocity, etc.), the following modification can be made.

If x_i is a design variable which may take on negative values, we can express x_i in terms of two new variables x_i' and x_i'':

$$x_i = x_i' - x_i'' \tag{10--20}$$

where

$$x_i', x_i'' \geqslant 0$$

TABLE 10–16. SIMPLEX METHOD TEMPLATE

Objective Function:

Subject to m constraints:

Step	Action	Cycle 1	Cycle 2	Cycle 3
		basis:	basis:	basis:
1	Select m basis variables and set all other variables to 0. For Cycle 1 choose as basis variables those which appear in only one constraint equation each and whose coefficients in these equations are unity. For subsequent cycles, obtain variables from Steps 5 and 8 of prior cycle.			
2	Solve m constraint equations for values of the m basis variables.			
3	Evaluate U for this cycle.			
4	Begin search for improved solution by expressing U in terms of nonbasis variables, using constraint equations as necessary to eliminate basis variables from U.			
5	Select as incoming basis variable that variable for which a unit increase has most effect on increasing U.	is incoming basis variable		
6	Express constraint equations in terms of current basis variables + incoming basis variable.			
7	Select as outgoing variable, the current variable which imposes the most severe constraint on the incoming variable.	is outgoing basis variable		
8	With new basis variables selected, set nonbasis variables to 0 and start next cycle.			

Equ. (10–20) can be used to replace x_i wherever x_i appears in the objective function and constraint equations. Also, Equ. (10–20) must be treated as an additional constraint equation. With these adjustments, the Simplex method can then be used directly to solve for the optimum in terms of the original non-negative design variables and the new non-negative variables x_i' and x_i''. Once the values of x_i' and x_i'' that are associated with the optimum are determined, Equ. (10–20) can be used to recover the value of the design variable x_i. Obviously, if $x_i'' > x_i'$, then x_i will be negative.

Reversal of Constraint Direction

The constraints in the standard formulation of the linear programming problem are all of the form "equal to or less than." Many design problems require constraints of the form "equal to or greater than." The adaptation of the Simplex method to accommodate this variation is demonstrated by reworking the example problem represented by Equs. (10–4)–(10–8), but reversing the direction of the constraint represented by Equs. (10–5). The modified optimization problem is to maximize

$$U = 2x_1 + x_2 \tag{10–21}$$

subject to

$$3x_1 + x_2 \geqslant 9 \tag{10–22}$$

$$-x_1 + 2x_2 \leqslant 4 \tag{10–23}$$

$$x_2 \leqslant 5 \tag{10–24}$$

$$x_1 + x_2 \leqslant 4 \tag{10–25}$$

The constraint boundaries are depicted in Figure 10–10. We see geometrically that reversing the direction of the constraint modifies which constraints and portions of constraints are active. Now the active constraints are Equs. (10–22) and (10–25).

The new feasible region A-B-L is shown in Figure 10–11. Also shown on Figure 10–11 are several contour lines of the objective function U. In the original formulation of the problem shown in Figure 10–5, the optimum occurred at point A with $U = 6.5$. It is clear from Figure 10–11 that now the optimum occurs at point L with $U = 8.0$. While it is easy to geometrically interpret this modification because the problem only involves two dimensions, it is worth explaining the required algebraic modifications in order to appreciate the implications for problems involving more than two design variables.

The first step in preparing to use the Simplex method is to convert the active inequality constraints [Equs. (10–22) and (10–25)] to equations. This is accomplished using a technique similar to what we used in Section 10.3. We add a slack variable x_3 to Equ. (10–25). However, because of the reversal in the direction of the constraint, we have to subtract what we call a "surplus" variable x_4 from Equ. (10–22). The constraint equations thus take the form

$$3x_1 + x_2 - x_4 = 9$$
$$x_1 + x_2 + x_3 = 4 \tag{10–26}$$

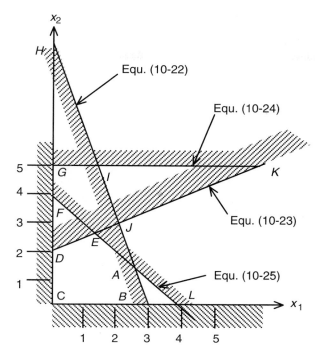

Figure 10–10. Geometrical Representation of Modified Constraints

If we tried now to begin Step 1 in the Simplex method, we would find that we do not have a suitable set of initial basis variables [while x_4 appears only in Equ. (10–26), its coefficient in that equation is -1 instead of $+1$]. To provide a suitable initial basis, we add an auxiliary variable x_5 to Equ. (10–26). Once x_5 helps to launch the Simplex method, we want to drive it out of the basis as soon as possible so that it doesn't affect

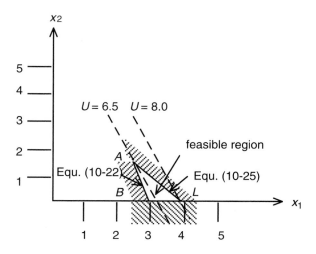

Figure 10–11. Feasible Region and Contour Lines for Modified Problem

TABLE 10-17. BIG M METHOD EXAMPLE PROBLEM

Objective Function: $U = 2x_1 + x_2 - 100x_5$

Subject to m constraints: $3x_1 + x_2 - x_4 + x_5 = 9$ (1)
$x_1 + x_2 + x_3 = 4$ (2)

Step	Action	Cycle 1	Cycle 2	Cycle 3
1	Select m basis variables and set all other variables to 0. For Cycle 1 choose as basis variables those which appear in only one constraint equation each and whose coefficients in these equations are unity. For subsequent cycles, obtain variables from Steps 5 and 8 of prior cycle.	basis: x_3, x_5 $x_1 = x_2 = x_4 = 0$	basis: x_1, x_3 $x_2 = x_4 = x_5 = 0$	basis: x_1, x_4 $x_2 = x_3 = x_5 = 0$
2	Solve m constraint equations for values of the m basis variables.	$x_3 = 4$ $x_5 = 9$	$x_1 = 3$ $x_3 = 1$	$x_1 = 4$ $3(4) - x_4 = 9$ $x_4 = 3$
3	Evaluate U for this cycle.	$U = -900$	$U = 6$	$U = 8$
4	Begin search for improved solution by expressing U in terms of nonbasis variables, using constraint equations as necessary to eliminate basis variables from U.	using (1) $U = 2x_1 + x_2$ $- 100(9 - 3x_1 - x_2 + x_4)$ $U = 302x_1 + 101x_2 - 100x_4 - 900$	using (1) $U = 2(3 - 1/3\,x_2 + 1/3\,x_4$ $- 1/3\,x_5) + x_2 - 100x_5$ $U = 6 + 1/3\,x_2 + 2/3\,x_4 - 302/3\,x_5$	using (2) $U = 2(4 - x_2 - x_3)$ $+ x_2 - 100x_5$ $U = 8 - x_2 - 2x_3 - 100x_5$
5	Select as incoming basis variable that variable for which a unit increase has most effect on increasing U.	x_1 is incoming basis variable	x_4 is incoming basis variable	no further increase in U is possible
6	Express constraint equations in terms of current basis variables + incoming basis variable.	$3x_1 + x_5 = 9$ (a) $x_1 + x_3 = 4$ (b)	$3x_1 - x_4 = 9$ (a) $x_1 + x_3 = 4$ (b)	
7	Select as outgoing variable, the current variable which imposes the most severe constraint on the incoming variable.	from (a), x_5 requires $x_1 \leq 3$ from (b), x_3 requires $x_1 \leq 4$ x_5 is outgoing basis variable	from (b), x_1 does not limit x_4 from (a), in order for $x_4 \geq 0$, $x_1 \geq 3$. x_1 cannot leave the basis x_3 is outgoing basis variable	
8	With new basis variables selected, set nonbasis variables to 0 and start next cycle.	$x_2 = x_4 = x_5 = 0$	$x_2 = x_3 = x_5 = 0$	

the solution. This can be done by multiplying x_5 by a large number M and subtracting it from the objective function. For this example, let's choose $M = 100$. The reformulated problem then consists of maximizing

$$U = 2x_1 + x_2 - 100x_5 \tag{10-27}$$

subject to

$$3x_1 + x_2 - x_4 + x_5 = 9 \tag{10-28}$$

$$x_1 + x_2 + x_3 = 4 \tag{10-29}$$

The solution is summarized in Table 10–17.

Equality Constraints

Linear programming problems sometimes have constraints in the form of equalities. Consider the problem defined by Equs. (10–21)–(10–25), and suppose we replace the inequality constraint given by Equ. (10–25) by the equality constraint

$$x_1 + x_2 = 4$$

No slack variable or surplus variable is needed to prepare this constraint for a Simplex solution. However, an auxiliary variable has to be added to this constraint in order to provide a suitable initial basis. The big M solution method can then be used to drive out the auxiliary variable.

10.6. NONLINEAR PROGRAMMING

The property primarily responsible for the usefulness of the linear programming concept is that the optimum solution occurs at the corner points of the feasible region. This allows us to utilize a very efficient search procedure to find the optimum design for problems in which both the objective function and the constraints are linear in the design variables. In this section we explore the complications that occur when either the objective function or the constraints are nonlinear in the design variables.

Nonlinear Constraint

Suppose you are director of public works for a small but rapidly growing city. One of your responsibilities is to develop and oversee all street paving projects. You have developed an approach to deploying work crews and paving equipment in an optimum fashion. The first opportunity to test your approach is a project to pave First Avenue and Second Street (see Figure 10–12).

First Avenue has three lanes of traffic and Second Street has two lanes. You realize that this would be a good opportunity to have the experienced work crew do the First Avenue paving while the crew of trainees pave Second Street. This works out well because the experienced crew can pave three lanes in the same time that the

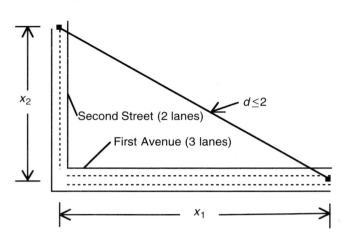

Figure 10–12. Layout of Street Paving Project

trainees can pave two. You expect substantial savings by sending both crews to the work site at the same time and having them both start paving from the intersection. The only constraint is that the supervisor of the experienced work crew must remain in constant radio contact with the supervisor of the trainees. The radio they use for communications has a maximum range d of two miles. What road length should each crew pave to maximize the total amount of street surface paved on this shift?

If we let x_1 represent the length of First Avenue paved and x_2 be the length of Second Street paved, we can formulate the problem as maximizing the linear objective function

$$U = 3x_1 + 2x_2 \tag{10-30}$$

subject to the nonlinear constraint

$$x_1^2 + x_2^2 \leq 4 \tag{10-31}$$

Obviously the traditional non-negativity constraints on the design variables x_1 and x_2 also hold.

Equ. (10–31) is the equation of a circle of radius 2 with center at the origin. Figure 10–13 shows the feasible region and two contour lines, $U = 4$ and $U = 8$.

With the benefit of the geometrical interpretation, the optimum clearly occurs when one of the contour lines $4 < U < 8$ is tangent to the circle defined by Equ. (10–31). As demonstrated in the Appendix to this section, we can locate that point of tangency using standard techniques of differential calculus.

Suppose that the problem had more than two design variables so that a geometrical interpretation such as provided in Figure 10–13 was not available. Alternatively, suppose the equations of the boundary curves were very complicated so the tangency points could not be located using traditional calculus techniques (we will examine a calculus-based technique later in Sec. 10.8). For these reasons we decide to examine the feasibility of applying a Simplex-type search strategy to find the optimum.

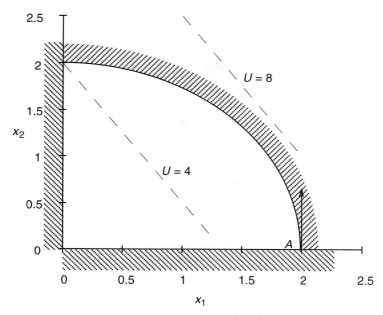

Figure 10–13.   Feasible Region and Contour Lines with a Nonlinear Constraint—Single Optimum

Generalizing from this example we conclude that the optimum of a linear objective function with nonlinear constraints occurs either at a corner point (as it does when all the constraints are linear), or at some intermediate point on one of the curvilinear edges of the feasible region. We now turn our attention to devising an efficient strategy for locating that optimum point on the boundary.

The initial corner point identified via a Simplex-type strategy applied to the optimum street paving problem might be the origin ($x_1 = 0$, $x_2 = 0$). Then the most efficient strategy would take us along the $x_2 = 0$ edge until we encountered corner point A ($x_1 = 2$, $x_2 = 0$). We could not advance any further along the $x_2 = 0$ edge without violating the constraint. However, a good search technique would tell us that U will increase if we move away from point A in the direction tangent to the circular arc (represented by the arrow in Figure 10–13). So far, so good. But now a Simplex-type method fails us, because its success relies on continuing to move in the specified direction until we reach the next corner point. However, in this case, we will never encounter the next corner point ($x_1 = 0$, $x_2 = 2$) if we continue to move in the direction of the arrow.

It is not difficult to envision developing a modified search strategy that accounts for the fact that the direction of the boundary changes continuously. Such a strategy, depicted in Figure 10–14, might include this rule:

Move away from the corner point A in the direction of the arrow by a specified small increment Δ, then take corrective action k_1 to get back on the boundary. Once back on the

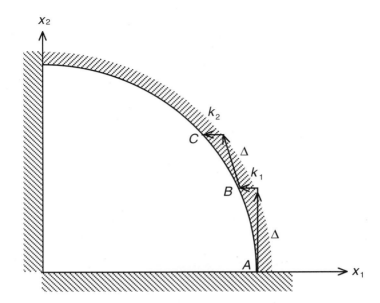

Figure 10–14. Modified Search Strategy Along a Curved Boundary of the Feasible Region

boundary at B, evaluate both U and the slope of the boundary at B to determine whether it is possible to increase U by moving away from B in the direction of the tangent. If so, again move in the prescribed direction by the amount Δ. Then stop and take corrective action k_2 to get back on the boundary at C. Continue the process until it is no longer possible to increase U by continuing to move tangent to the edge.

It is plausible to argue that (for this particular problem at least) such a search strategy will usually lead us to the optimum. In the interest of locating the optimum quickly, we are tempted to use a large value of Δ. However, we could get into trouble if Δ is too large. For this example, if we chose $\Delta > 2$, a strategy for k_1 based on the correction occurring parallel to the x_1 axis will be unsuccessful since the correction would not bring us back to the edge. The smaller we choose Δ, the smaller each correction k_i will be, and we are less likely to "overshoot" the optimum. However, the smaller Δ is, the more cycles, or iterations, needed to find the optimum. Slow convergence to an optimum can use up expensive computer time on large problems. Refinements in the search strategy, such as changing the size of Δ or the direction of k_i as we progress, can improve the efficiency.

Local and Global Optimum

Let's see now whether this modified Simplex-type search strategy will be successful with a somewhat different version of the street paving problem. In this case, as depicted in Figure 10–15, First Avenue is 4.8 miles long and Second Street is 3.5 miles

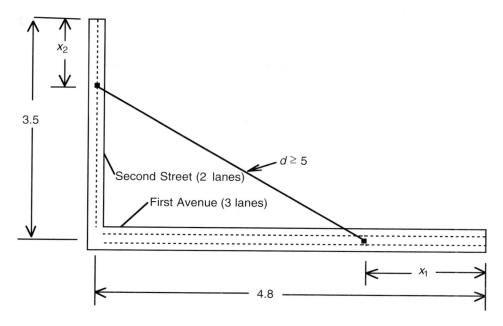

Figure 10–15. Layout of Modified Street Paving Project

long. Our modified plan is to deploy the two crews at the end of the streets and have them both work toward the intersection. We have replaced the radio in favor of a new communications satellite to keep the crews in touch with each other. However, the satellite's fine resolution system is temporarily disabled. We must use the coarse resolution system, which cannot transmit messages between objects that are closer than 5 miles to each other. What length of each road should we pave to maximize the total amount of street surface paved on this shift?

The objective function is the same as in the original paving problem:

$$U = 3x_1 + 2x_2 \tag{10–32}$$

but now the nonlinear constraint is of the form

$$(4.8 - x_1)^2 + (3.5 - x_2)^2 \geq 25 \tag{10–33}$$

We show the feasible region and several of the contour lines for the objective function in Figure 10–16. The modified Simplex-type search strategy takes us, as before, to the corner point A $(x_1 = 1.23, x_2 = 0)$ where $U = 3.7$. Now U decreases if we move away from point A along the tangent to the curvilinear edge (see arrow in Figure 10–16). This leads us to conclude that the optimum occurs at A. In a sense A is an optimum since the value of U at any feasible point in the neighborhood of A is less than the value of U at A. But of course, the true optimum occurs at point B $(x_1 = 0, x_2 = 2.1)$ where $U = 4.2$. We will use the term "local optimum" to describe situations such as that at A. We will call the optimum at B the "global optimum" for this problem. If there is only one local optimum (such as in Figure 10–13), then the local optimum is

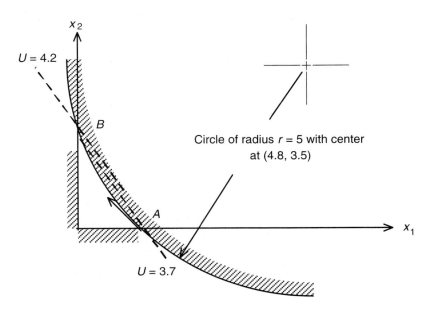

Figure 10–16. Feasible Region and Contour Lines with a Nonlinear Constraint—
Multiple Optima

also the global optimum. If there is more than one local optimum, the global optimum
is the local optimum that has the highest value of U.

The two versions of the street paving problem give us confidence that the optimum for this class of problems (linear objective function and nonlinear constraints)
occurs either at a corner point or at some intermediate point on a curvilinear edge of
the feasible region. However, a Simplex-type search strategy may fail to locate the
global optimum. Before drawing general conclusions regarding general nonlinear
programming problems, let's explore another special class of problems involving nonlinearities.

Nonlinear Objective Function

Suppose we want to design a cantilever beam to support the set of loads shown in Figure 10–17. As happens many times in design, we are not always sure what the loads
are going to be. In this case, we have two distributed loads of intensity 2
newtons/meter in the downward direction. One of them extends a distance x_1 from the
wall and the other extends a distance x_2. At the ends of the distributed loads are up-
ward acting concentrated loads of 4 newtons and 6 newtons respectively. In addition,
there is a 9 newton/meter clockwise couple acting on the beam. The constraints on the
loading are as follows:

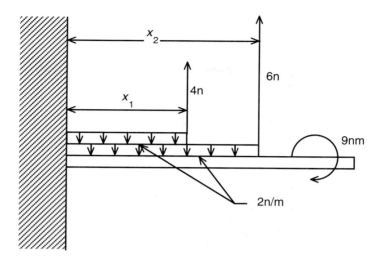

Figure 10–17.   Loads on a Cantilever Beam

$$x_2 \leqslant 4 \tag{10-34}$$

$$x_2 - x_1 \geqslant 0 \tag{10-35}$$

$$x_1, x_2 \geqslant 0 \tag{10-36}$$

We want to design the bracket that attaches the beam to the wall so that it can withstand the maximum bending moment at the wall caused by this loading.

From Figure 10–17, we can express the bending moment U at the wall as

$$U = -(2x_1)\left(\frac{x_1}{2}\right) - (2x_2)\left(\frac{x_2}{2}\right) + 4x_1 + 6x_2 - 9$$

or

$$U = -x_1^2 - x_2^2 + 4x_1 + 6x_2 - 9 \tag{10-37}$$

Thus, our optimization problem consists of maximizing Equ. (10–37) subject to the constraints given by Equs. (10–34)–(10–36). We can gain insight into the location of the maximum by examining the geometrical form of the problem in $x_1 - x_2$ space. See Figure 10–18 for the feasible region and several of the contour curves.

Since U is a nonlinear function of the design variables, the curves of constant U are not straight lines, as they were in the previous several problems. In this case, the contour curves are concentric circles[5] centered at $x_1 = 2$, $x_2 = 3$. The maximum possible

---

[5]The form of the contour curves can be more easily identified if we rewrite Equ. (10–37) as

$$U = (-x_1^2 + 4x_1 - 4) + 4 + (-x_2^2 + 6x_2 - 9) + 9 - 9$$

which simplifies to

$$U = 4 - (x_1 - 2)^2 - (x_2 - 3)^2$$

This clearly represents a family of circles of radius $(4 - U)$ with centers at $x_1 = 2$ and $x_2 = 3$.

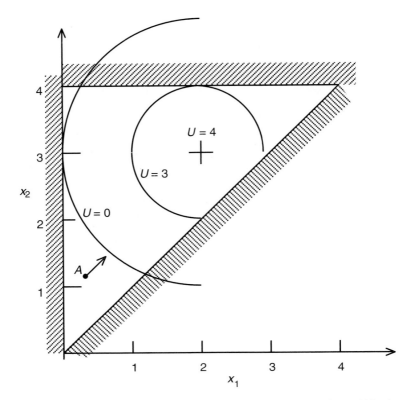

Figure 10–18. Feasible Region and Contour Curves with a Nonlinear Objective Function—Single Optimum at an Interior Point

value of U is $U = 4$ and is associated with the circle of 0 radius. We have a situation not previously encountered in our discussion of feasible regions—the optimum does not occur on the boundary of the feasible region. Clearly, any Simplex-type search strategy that relies on searching along the boundary will not be able to find such an optimum. Developing an efficient search strategy that can find optima at interior points as well as at boundary points is a very difficult task. We will now introduce some key ingredients of a search strategy without becoming bogged down in the details of the calculations.

An efficient search strategy for this problem is as follows: (a) Select a starting point either in the interior or on the boundary and evaluate U at that point; (b) Move a predetermined distance Δ in the direction for which U increases most rapidly; (c) Evaluate U at this new point and repeat the process until U cannot be increased any further. An arbitrary starting point for this problem, such as A in Figure 10–18, would yield a search direction normal to the contour curve through the starting point. This search direction remains unchanged in subsequent iterations and the speed with which we locate the optimum depends on the size of Δ. A small Δ requires many iterations, but a large Δ runs the risk of overshooting and failing to converge. For example, if our starting point A was on the contour curve $U = 3$ and we chose $\Delta = 2$, the search will never converge.

We will need a more sophisticated search strategy if we modify the problem by replacing Equ. (10–34) by

$$x_2 \leqslant 2.5$$

We show the modified feasible region in Figure 10–19. The search strategy described above leads us to point B on the boundary, and we have to change directions in order to reach the optimum at C. Hence we amend part (c) of the search strategy as: (c) Evaluate U at this new point and repeat the process until U cannot be increased any further or until a boundary is reached. If a boundary is reached first, move along the boundary in the direction of increasing U and repeat the process.

Investigating a slightly more complicated nonlinear objective function will reveal additional difficulties in locating the global optimum. Let's find the optimum design for the system whose objective function is

$$U = 6x_1^2 - 3x_2^2 - x_1^3 + x_1x_2 - 9x_1 - 9x_2 \tag{10–38}$$

subject to the constraints

$$x_1 \leqslant 5$$

$$x_2 \leqslant 4$$

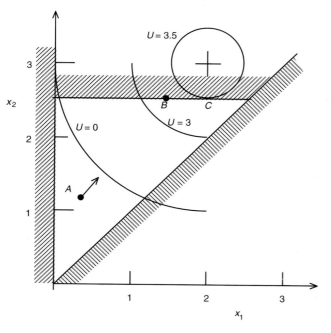

Figure 10–19. Feasible Region and Contour Curves with a Nonlinear Objective Function—Single Optimum at a Non-Corner Boundary Point

The traditional non-negativity constraints are also in effect. Figure 10–20 shows the feasible region and the contour curves for $U = 0, 4, 8, 12$. Notice that two optima are present: O_1 and O_2. If we start our search for an optimum at point A, we will most likely converge (assuming an appropriate choice for Δ) to the local optimum at O_1: we will find $4 < U < 8$ in the vicinity of $x_1 = 0$, $x_2 = 1.5$. On the other hand, if we start our search at B, we are likely to converge to the global optimum at O_2: we will find $U > 12$ in the vicinity of $x_1 = 3.5$, $x_2 = 2.0$. So, for nonlinear objective functions that have more than one optimum, the optimum that is reached via the search strategy may depend on the starting point.

Thus, while commercial spreadsheet software packages contain nonlinear programming solution algorithms, there is no assurance that they will converge to the global optimum. For problems involving three or more design variables, a geometrical depiction such as Figure 10–20 is not available to help identify the number of local optima and their general locations. In the absence of these geometrical clues, the ability of a search strategy to converge to a global optimum may depend on the starting point and choice of the search strategy.

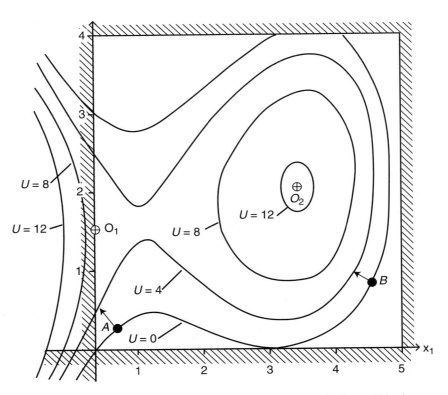

Figure 10–20. Feasible Region and Contour Curves for a Nonlinear Objective Function—Multiple Optima

Optimum design of a spring. As an example of computational difficulties that may be encountered, consider the problem of minimizing the weight of a coil spring under tension, as shown in Figure 10–21.[6] The three design variables are the number of coils n, the wire diameter d, and the mean coil diameter D.

Figure 10–21. A Coil Spring Under Tension

The objective function to be minimized is

$$U = (n + 2)Dd^2$$

The constraints involve limits on the deflection, shear stress, surge frequency, and outside diameter in addition to restrictions on the design variables. After inserting the numerical values for the material properties, the seven constraints take the form

$$1 - \frac{D^3n}{71{,}875d^4} \leqslant 0$$

$$\frac{4D^2 - dD}{12{,}566(Dd^3 - d^4)} + \frac{1}{5{,}108d^2} - 1 \leqslant 0$$

$$1 - \frac{140.45d}{D^2n} \leqslant 0$$

$$\frac{D + d}{1.5} - 1 \leqslant 0$$

$$d \geqslant 0.05$$

$$D \geqslant 0.005$$

$$n \geqslant 1$$

Eight different algorithms were used to search for the optimum design. Two of them failed to converge, two others converged to a local minimum near the starting point, two others found other local minima, and two arrived at the global minimum. We now turn our attention to tests that may help us to identify such situations.

Convexity in Two Dimensions

As shown earlier in this section, the ability of a search strategy to locate a global optimum may depend on the shape of the feasible region (compare Figure 10–13 with Fig-

[6]Belegundu and Arora, pp. 1601–1623.

ure 10–16) and the form of the objective function (compare Figure 10–18 with Figure 10–20). Let's first address the issue of the shape of the feasible region.

Convex feasible region.   The feasible regions from two linear programming problems discussed earlier (see Figs. 10–6 and 10–11) plus those from Figure 10–13 and Figure 10–16 are reproduced in Figure 10–22. By inspection of Figs. 10–22(a)–(c), any two points P_1 and P_2 within or on the boundary of the feasible region can be connected by a straight line such that no point on the line lies outside the feasible region. Such regions are called convex regions.

It is also clear from Figure 10–22(d) that we can select two points P_1 and P_2 within or on the boundary of the feasible region such that a portion of the straight line connecting those points lies outside the feasible region. We call such a region a concave region.

The concepts of convex region and concave region are crucial elements of the tests for global optimum. Since we cannot visualize the shape of the feasible region when there are more than three design variables, we need a mathematical basis for

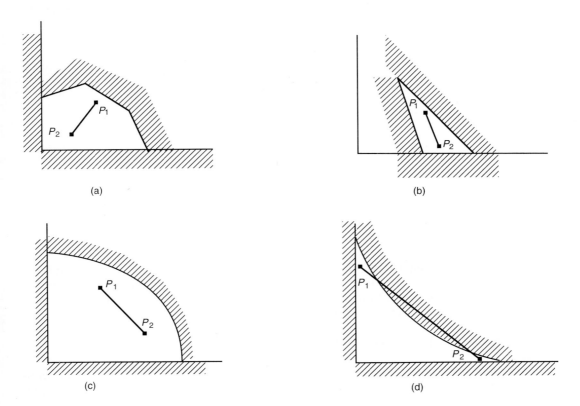

Figure 10–22.   Examples of Convex and Concave Feasible Regions

determining when a feasible region is convex or concave. To do this we have to examine the shape of the boundary curves that define the region.

Convex and concave boundary curves. Consider any two-dimensional boundary curve expressed as

$$x_2 = f(x_1)$$

Three such boundary curves are shown in Figure 10–23. Consider any two points P_1 and P_2 lying on such a curve. If the straight line connecting P_1 and P_2 lies entirely above or on the curve, we call the function a convex function. Such is the case for the function f_1 depicted in Figure 10–23(a). On the other hand, if the straight line connecting P_1 and P_2 lies entirely below or on the curve, we call the function a concave function. Such is the case for the function f_2 depicted in Figure 10–23(b). The function f_3 shown in Figure 10–23(c) is neither concave nor convex. From these definitions, we conclude that a straight line is both convex and concave.

In preparation for dealing with problems involving more than two design variables, we want to convert these definitions of concave and convex functions into mathematical tests. To do this, let's look at mathematical representations for the three parts of Figure 10–23.

Tests for convexity and concavity. The function f_1 in Figure 10–23(a) might have an equation of the form

$$x_2 = 3x_1^2 - 2x_1 + 4$$

The slope of the curve is given by the first derivative

$$\frac{dx_2}{dx_1} = 6x_1 - 2$$

Note that the slope is negative for small values of x_1, but continuously becomes more positive as x_1 increases. The rate of change of the slope is given by

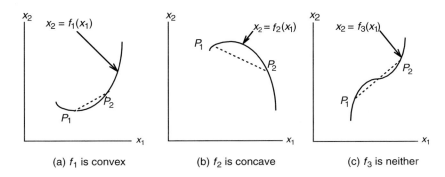

(a) f_1 is convex   (b) f_2 is concave   (c) f_3 is neither

Figure 10–23. Convex and Concave Functions

$$\frac{d^2x_2}{dx_1^2} = 6$$

Thus, the second derivative of the function f_1 depicted in Figure 10–23(a) is never negative. This turns out to be true for all convex functions; which leads us to the following test for convexity:

A function $f(x)$ is convex if $\dfrac{d^2f}{dx^2} \geqslant 0$

The function f_2 in Figure 10–23(b) might have an equation of the form

$$x_2 = -3x_1^2 + 2x_1 + 4$$

The slope of the curve is given by the first derivative

$$\frac{dx_2}{dx_1} = -6x_1 + 2$$

Note that the slope is positive for small values of x_1, but continuously becomes more negative as x_1 increases. The rate of change of the slope is given by

$$\frac{d^2x_2}{dx_1^2} = -6$$

Thus, the second derivative of the function f_2 depicted in Figure 10–23(b) is never positive. This turns out to be true for all concave functions; which leads us to the following test for concavity:

A function $f(x)$ is concave if $\dfrac{d^2f}{dx^2} \leqslant 0$

The function f_3 in Figure 10–23(c) might have an equation of the form

$$x_2 = 4x_1^3 - 3x_1^2 + 2x_1 + 4$$

The slope of the curve defined by this equation is given by the first derivative

$$\frac{dx_2}{dx_1} = 12x_1^2 - 6x_1 + 2$$

Note that the slope is positive for small values of x_1, becomes negative for intermediate values of x_1, then becomes positive for larger values of x_1. The rate of change of the slope is given by

$$\frac{d^2x_2}{dx_1^2} = 24x_1 - 6$$

We see that for $d^2x_2/dx_1^2 \leq 0$ for $x_1 \leq 1/4$ and $d^2x_2/dx_1^2 \geq 0$ for $x_1 \geq 1/4$. Hence, the function is neither convex nor concave. From another perspective we recognize that the boundaries of feasible regions are formed by the intersection of boundary curves with each other. Hence the convexity and concavity properties are primarily of interest within certain ranges of the design variables. From this perspective we can characterize f_3 as concave for $x_1 \le 1/4$ and convex for $x_1 \ge 1/4$.

Test for convex and concave feasible regions.   Now that we have a test for convexity and concavity of boundary curves, we can develop a convexity test for the feasible region formed by a set of boundary curves. The three examples of feasible regions shown in Figure 10–24 will assist us in this effort.

Figure 10–24 demonstrates that it is not just the convexity and concavity of the boundary curves, but the manner in which they combine to form the feasible region, that determines the convexity or concavity of the feasible region. Figures 10–23(a)–(c) suggest the following test:

> A feasible region is convex if it is defined by less-than-or-equal-to constraint functions that are concave functions and greater-than-or-equal-to constraint functions that are convex functions.

The ability to determine that a feasible region is convex is crucial to identifying whether a local optimum is also a global optimum. In particular, if the feasible region is concave, there is no known test that can determine whether any local optimum is also a global optimum. As depicted in Figure 10–16, a search strategy that identifies a local optimum at A of a concave feasible region may not be successful in locating the global optimum at B. So, for purposes of testing for a global optimum, we will now limit our subsequent discussion to optimal design problems in which the feasible region is convex. As seen by comparing Figures 10–18 and 10–20, finding the global op-

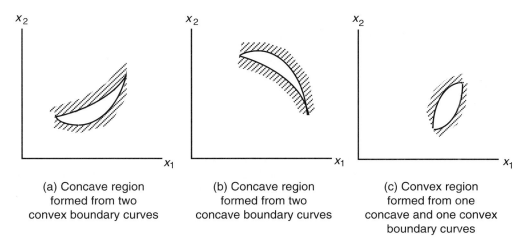

(a) Concave region
formed from two
convex boundary curves

(b) Concave region
formed from two
concave boundary curves

(c) Convex region
formed from one
concave and one convex
boundary curves

Figure 10–24.   Convex and Concave Regions Formed from Convex and Concave Boundary Curves

timum depends not only on the shape of the region (both of these feasible regions are convex), but also on the form of the objective function. We turn our attention now to the issue of convexity and concavity properties of an objective function.

Convex and concave objective functions. Our recent discussion focused on convex and concave boundary curves in a two-dimensional $x_1 - x_2$ design space. We can build upon these concepts to examine convexity and concavity properties of the surface defined by U in the three-dimensional $U - x_1 - x_2$ space. Let's use the objective function defined by Equ. (10–38), which we reproduce here:

$$U = 6x_1^2 - 3x_2^2 - x_1^3 + x_1x_2 - 9x_1 - 9x_2 \qquad (10\text{–}38)$$

In our discussion of boundary curves, we saw that we could test for convexity and concavity of a function by examining the sign of the second derivatives of the function. In this case, our U surface is a function of two independent variables x_1 and x_2. However, if we construct a plane defined by $x_2 = C$, this plane will intersect the U surface to define a two-dimensional $U - x_1$ curve. We can draw some conclusions about the shape of this curve by examining the first and second derivatives:

$$\frac{\partial U}{\partial x_1} = 12x_1 - 3x_1^2 + x_2 - 9$$

$$\frac{\partial^2 U}{\partial x_1^2} = 12 - 6x_1 \qquad (10\text{–}39)$$

Applying the test developed in the previous subsection, we see that this curve is convex for $x_1 \leq 2$ and concave for $x_1 \geq 2$.

Similarly, if we construct a plane defined by $x_1 = C$, this plane will intersect the U surface to define a two-dimensional $U - x_2$ curve. We can draw some conclusions about the shape of this curve by examining the first and second derivatives:

$$\frac{\partial U}{\partial x_2} = -6x_2 + x_1 - 9$$

$$\frac{\partial^2 U}{\partial x_2^2} = -6 \qquad (10\text{–}40)$$

The test developed in the previous subsection reveals that this curve is concave for all values of x_2.

This information is useful, but it is not sufficient for characterizing the convexity or concavity of the U surface. In particular, we also have to examine the second mixed partial derivative $\partial^2 U / \partial x_1 \partial x_2$. Without going into the details, we present the convexity/concavity test as follows:

A function $U(x_1, x_2)$ is convex if $\partial^2 U / \partial x_1^2$ and $\partial^2 U / \partial x_2^2$ are both non-negative and $\partial^2 U / \partial x_1^2\, \partial^2 U / \partial x_2^2 > (\partial^2 U / \partial x_1 x_2)^2$. A function $U(x_1, x_2)$ is concave if $\partial^2 U / \partial x_1^2$ and $\partial^2 U / \partial x_2^2$ are both non-positive and $\partial^2 U / \partial x_1^2\, \partial^2 U / \partial x_2^2 > (\partial^2 U / \partial x_1 x_2)^2$.

We apply this test to Equ. (10–38). From Equ. (10–40) we see that $\partial^2 U/\partial x_2^2$ is negative. From Equ. (10–39) we see that $\partial^2 U/\partial x_1^2$ is negative as long as $x_1 > 2$. Since our feasible region extends from $0 < x_1 < 5$, U is neither convex nor concave over the feasible region.

If, however, we modify the x_1 constraint to read $2.5 < x_1 < 5$, then $\partial^2 U/\partial x_1^2$ and $\partial^2 U/\partial x_2^2$ are both negative over the feasible region. The second mixed partial derivative is

$$\frac{\partial^2 U}{\partial x_1 \partial x_2} = 1$$

Since

$$\frac{\partial^2 U}{\partial x_1^2}\frac{\partial^2 U}{\partial x_2^2} = (12 - 6x_1)(-6) > 1$$

if

$$x_1 > \frac{13}{6}$$

U is concave if the feasible region is defined by $2.5 < x_1 < 5$ rather than by $0 < x_1 < 5$ as shown in Figure 10–20.

Test for Global Optimum with Two Design Variables

We have now developed convexity and concavity tests for the feasible region and for the objective function. We now offer the following test for a global optimum:

> If the feasible region is convex, a local minimum of a convex objective function is also a global minimum. If the feasible region is concave, a local maximum of a concave objective function is also a global maximum.

It is important to recognize several implications and limitations of this test:

1. If the feasible region is not convex, we can draw no conclusions about a global minimum regardless of the form of the objective function.
2. If the feasible region is convex, but the objective function is neither convex nor concave, we can draw no conclusions about a global optimum.
3. We cannot draw any conclusions regarding a global maximum of a convex objective function or a global minimum of a concave objective function.

With regard to the optimization problem depicted in Figure 10–20, the feasible region is convex, but the objective function is neither convex nor concave. Thus, a search technique might successfully lead us to a local optimum, but we have no analytical way to test whether we have reached a global optimum. We have to rely on successive searches with different starting points to perhaps lead us to more than one local opti-

mum. A direct comparison of the local optima to each other will then allow us to select the best of the identified local optima.

Global Optimum of *n* Design Variables

The concepts of convexity and concavity of regions and surfaces apply in n-dimen-
sional space. However, without the benefit of a geometrical interpretation, we have to
rely solely on an algebraic concept. For any function $F(x_1, x_2, ..., x_n)$, we can define the
Hessian matrix $\mathbf{H}$ as

$$\mathbf{H} = \begin{bmatrix} \dfrac{\partial^2 F}{\partial x_1^2} & \dfrac{\partial^2 F}{\partial x_1 \partial x_2} & \cdots & \dfrac{\partial^2 F}{\partial x_1 \partial x_n} \\ \dfrac{\partial^2 F}{\partial x_2 \partial x_1} & \dfrac{\partial^2 F}{\partial x_2^2} & \cdots & \dfrac{\partial^2 F}{\partial x_2 \partial x_n} \\ \vdots & \vdots & \cdots & \vdots \\ \dfrac{\partial^2 F}{\partial x_n \partial x_1} & \dfrac{\partial^2 F}{\partial x_n \partial x_2} & \cdots & \dfrac{\partial^2 F}{\partial x_n^2} \end{bmatrix}$$

It can be shown that F is a convex function if the eigenvalues of its Hessian matrix $\mathbf{H}$
are non-negative. Similarly, it can be shown that F is a concave function if the eigenvalues of its Hessian matrix $\mathbf{H}$ are non-positive. Thus, we can use the Hessian matrix
to characterize the convexity and concavity of the n-dimensional feasible region and
the objective function $U(x_1, x_2, ..., x_n)$. Then we can use the above test for the existence of a global optimum.

Appendix 10A: Using Calculus to Find the Optimum

As indicated in Figure 10–13, the optimum design occurs when a contour line is tangent to the curved part of the boundary. From Equ. (10–30) we can write the equation
of the contour lines as

$$x_2 = \frac{U}{2} - \frac{3x_1}{2} \qquad (10\text{–}41)$$

We find the slope of the contour lines by taking the derivative of Equ. (10–41) with
respect to x_1. This yields

$$\frac{dx_2}{dx_1} = -\frac{3}{2} \qquad (10\text{–}42)$$

Similarly we obtain the slope of the circular part of the boundary by differentiating
Equ. (10–31) with respect to x_1 to get

$$2x_1 \frac{dx_1}{dx_1} + 2x_2 \frac{dx_2}{dx_1} = 0 \qquad (10\text{–}43)$$

We can satisfy the requirement that the contour line and the boundary curve have the same slope by substituting from Equ. (10–42) into Equ. (10–43). This yields

$$2x_1 + 2x_2 \left(\frac{-3}{2}\right) = 0$$

which can be solved to give

$$x_1 = \frac{3}{2} x_2 \qquad\qquad (10\text{–}44)$$

Now, to ensure that the contour line and the boundary curve have a point in common, we substitute Equ. (10–44) into Equ. (10–31). This leads to

$$\left(\frac{3}{2} x_2\right)^2 + x_2^2 = 4$$

which simplifies to

$$\left(\frac{13}{4}\right) x_2^2 = 4$$

Solving this for x_2 gives

$$x_2 = 4 \frac{\sqrt{13}}{13} = 1.11 \qquad\qquad (10\text{–}45)$$

Substituting this back into Equ. (10–44) and solving for x_1 provides

$$x_1 = 6 \frac{\sqrt{13}}{13} = 1.66$$

Thus the optimum design is to pave a length $x_1 = 1.66$ of First Avenue and pave a length $x_2 = 1.11$ of Second Street. We find the total amount of surface paved by substituting these values for x_1 and x_2 into Equ. (10–30) to obtain

$$U = 3 \left(6 \frac{\sqrt{13}}{13}\right) + 2 \left(4 \frac{\sqrt{13}}{13}\right)$$

This reduces to

$$U = 2\sqrt{13} = 7.21$$

10.7. SEPARABLE PROGRAMMING

We saw in the last section the complications that arise when optimum design problems involve nonlinear terms in either the objective function or a constraint equation. Not only can we not always be sure that a local optimum is also a global optimum, a search algorithm may fail to converge, or convergence may be very slow. In this section we discuss a class of nonlinear programming problems which can be converted to a form suitable for application of linear programming techniques. In particular, we focus on those problems in which the nonlinear terms involve only one design variable. For example, suppose the objective function was of the form

$$U = 3x_1{}^2 + \cos x_2 - \ln (x_3)^2$$

This equation involves each of the design variables x_1, x_2, and x_3 in a nonlinear manner. But the nonlinearities are "separated" in the sense that each nonlinear term involves only one design variable. An example of a nonlinear objective function that is not separable is

$$U = 6x_1 \ln x_2 + x_3$$

In order for a nonlinear programming problem to be separable, all nonlinear terms in both the objective function and the constraints must be in separable form.

In general, a separable programming problem consists of maximizing

$$U = \sum_{i=1}^{M} f_i(x_i) \qquad\qquad (10\text{–}46)$$

subject to N constraints of the form

$$\sum_{j=1}^{M} g_{ij}(x_j) \leq b_i \quad i = 1, \ldots, N \qquad\qquad (10\text{–}47)$$

The approach to solving this problem is to approximate each of the nonlinear $f_i(x_i)$ and $g_{ij}(x_j)$ as piecewise linear functions. This will allow us to convert the original nonlinear programming problem into a related linear programming problem. The solution to the related linear programming problem will be an approximate solution to the original nonlinear problem. We now take a slight diversion to develop a particular piecewise linear representation of nonlinear functions.

Piecewise Linear Representation of Nonlinear Functions

Consider the nonlinear function $f(x)$ shown in Figure 10–25. The interval over which we want to construct a piecewise linear approximation is defined by a lower limit x^L and an upper limit x^U. We subdivide that interval into $(I - 1)$ subintervals, with the lower and upper limits of the ith subinterval designated by $x^{(i)}$ and $x^{(i+1)}$ respectively. Note that these subintervals do not have to be of equal length. In fact, in places where $f(x)$ is changing rapidly, we can establish small subintervals, while in places where $f(x)$ is changing slowly we can afford to use large subintervals.

Consider the task of representing $f(x)$ by a straight line approximation in the interval between $x^{(i)}$ and $x^{(i+1)}$. Let $f^{(i)}$ and $f^{(i+1)}$ be the values of $f(x)$ at $x^{(i)}$ and $x^{(i+1)}$.

We begin by introducing two new variables $\alpha^{(i)}$ and $\alpha^{(i+1)}$ that are related to x by

$$x = \alpha^{(i)}x^{(i)} + \alpha^{(i+1)}x^{(i+1)} \qquad\qquad (10\text{–}48)$$

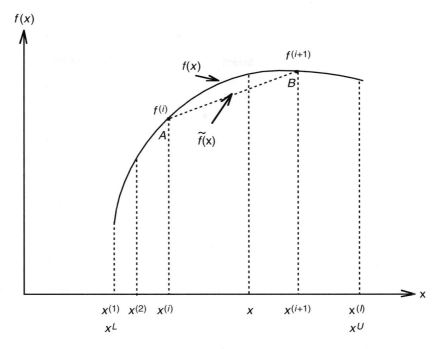

Figure 10–25.   Piecewise Linear Approximation to a Nonlinear Function

and to each other by

$$\alpha^{(i)} + \alpha^{(i+1)} = 1 \tag{10-49}$$

We impose the additional requirement that

$$0 \leq \alpha^{(i+1)} \leq 1 \tag{10-50}$$

As shown in the Appendix to this section, the equation of the straight line connecting points A and B can then be written in terms of these new variables as

$$\tilde{f}(x) = f^{(i)}\alpha^{(i)} + f^{(i+1)}\alpha^{(i+1)} \tag{10-51}$$

Since the range $x^L < x < x^U$ within which we want to write the piecewise linear approximation contains $(I-1)$ intervals, we can write the entire $(I-1)$ sets of Equs. (10–48), (10–49) and (10–51) as

$$x = \sum_i \alpha^{(i)} x^{(i)} \tag{10-52}$$

$$\tilde{f}(x) = \sum_i f^{(i)}\alpha^{(i)} \tag{10-53}$$

and

$$\sum_i \alpha^{(i)} = 1 \qquad (10\text{–}54)$$

The non-negativity requirement on the new variables can be written as

$$\alpha^{(i)} \geqslant 0 \qquad (10\text{–}55)$$

Since Equs. (10–52)–(10–54) only apply within a given interval $x^{(i)} < x < x^{(i+1)}$, we must require that all the $\alpha^{(i)}$ are 0 except for the two consecutive $\alpha^{(i)}$ that are associated with the particular interval being examined.

For the original optimization problem expressed by Equs. (10–46)–(10–47), we can replace each nonlinear term $f_i(x_i)$ in the objective function by its piecewise linear approximation $\tilde{f}_i(x_i)$ given by Equ. (10–53), where we've replaced the design variable x_i by the new variable $\alpha^{(i)}$. The number of new variables introduced by Equ. (10–53) depends on the number of subintervals established. The introduction of the $\alpha^{(i)}$ is accompanied by the additional constraint Equs. (10–54) and (10–55). If any of the original constraint equations contain separable nonlinear terms $g_{ij}(x_j)$, each of those terms is replaced by a piecewise linear approximation $\tilde{g}_{ij}(x_j)$ in the same manner.

An Approximate Solution to a Nonlinear Optimization Problem

Suppose that we want to maximize

$$U = 30x_1 + x_2^3 \qquad (10\text{–}56)$$

subject to

$$\frac{x_1^2}{2} - x_1 + x_2 - 5 \leqslant 0 \qquad (10\text{–}57)$$

This problem involves a nonlinear function of x_2 in the objective function and another nonlinear function of x_1 in the constraint. Both of these nonlinearities are separable. Figure 10–26 shows the feasible region and several contour curves. By inspection, the optimum occurs in the vicinity of $x_1 = 1.4$ and $x_2 = 5.5$.

We can obtain an exact solution by equating the boundary and contour curve equations (to determine the points of intersection), and also equating their first derivatives (to determine the point of tangency). Using this technique, we find the exact solution for the optimum to be $U_{\max} = 201$. This occurs at $x_1 = 1.34$ and $x_2 = 5.44$.

To get an approximate solution to this problem using the technique of separable programming, let

$$f(x_2) = x_2^3 \qquad (10\text{–}58)$$

and

$$g(x_1) = \frac{x_1^2}{2} \qquad (10\text{–}59)$$

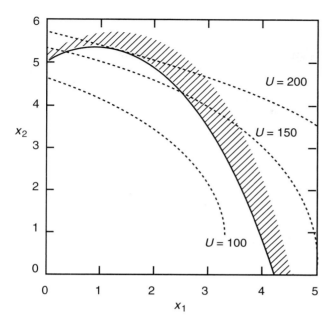

Figure 10–26.   Contour Lines and Feasible Region for Example Problem

We want to replace these nonlinear terms by piecewise linear expressions of the form

$$\tilde{f}(x_2) = \sum_i f^{(i)} \alpha^{(i)} \qquad\qquad (10\text{–}60)$$

$$\tilde{g}(x_1) = \sum_i g^{(i)} \beta^{(i)} \qquad\qquad (10\text{–}61)$$

where $\alpha^{(i)}$ are the new variables replacing the design variable x_2 and $\beta^{(i)}$ are the new variables replacing the design variable x_1. Since there is only one nonlinear term in the objective function, and the only constraint equation has only one nonlinear term, we drop the subscripts on f_i and g_{ij}.

The first step in the approximation operation is to select an interval of analysis and a set of subintervals into which we shall divide that interval. In this example, non-linearities occur in both of the design variables x_1 and x_2. This means that we have to develop appropriate intervals of analysis and piecewise approximations for both variables. With the benefit of Figure 10–26, let us construct our approximation over the intervals $0 \le x_1 \le 4$ and $2 \le x_2 \le 6$. We choose to subdivide each of these intervals into four equal subintervals so that

$$x_1^{(1)} = 0, \quad x_1^{(2)} = 1, \quad x_1^{(3)} = 2, \quad x_1^{(4)} = 3, \quad x_1^{(5)} = 4,$$
$$x_2^{(1)} = 2, \quad x_2^{(2)} = 3, \quad x_2^{(3)} = 4, \quad x_2^{(4)} = 5, \quad x_2^{(5)} = 6$$

These values are listed in Table 10–18. We can now use Equ. (10–48) to express x_1 and x_2 in terms of these coefficients and the new variables $\alpha^{(i)}$ and $\beta^{(i)}$ as

TABLE 10–18.　DETERMINATION OF COEFFICIENTS
IN PIECEWISE APPROXIMATION

i	$x_1^{(i)}$	$g^{(i)}$	$x_2^{(i)}$	$f^{(i)}$
1	0	0.00	2	8
2	1	0.50	3	27
3	2	2.00	4	64
4	3	4.50	5	125
5	4	8.00	6	216

$$x_1 = \beta^{(2)} + 2\beta^{(3)} + 3\beta^{(4)} + 4\beta^{(5)}$$

$$x_2 = 2\alpha^{(1)} + 3\alpha^{(2)} + 4\alpha^{(3)} + 5\alpha^{(4)} + 6\alpha^{(5)}$$

Next we use Equs. (10–58) and (10–59) to calculate the coefficients $f^{(i)}$ and $g^{(i)}$ at the subinterval boundaries $x_1^{(1)}, \ldots, x_1^{(5)}$ and $x_2^{(1)}, \ldots, x_2^{(5)}$. These calculations are shown in Table 10–18.

Inserting the values from the appropriate columns in Table 10–18 into Equs. (10–60) and (10–61) leads to

$$\tilde{f}(x_2) = 8\alpha^{(1)} + 27\alpha^{(2)} + 64\alpha^{(3)} + 125\alpha^{(4)} + 216\alpha^{(5)}$$

$$\tilde{g}(x_1) = 0.5\beta^{(2)} + 2\beta^{(3)} + 4.5\beta^{(4)} + 8\beta^{(5)}$$

We have replaced the original nonlinear terms $f(x_2) = x_2^3$ and $g(x_1) = x_1^2/2$ by the above expressions that are linear in the new variables $\alpha^{(1)}$ through $\alpha^{(5)}$ and $\beta^{(1)}$ through $\beta^{(5)}$. The original problem is therefore converted to that of maximizing

$$\tilde{U} = 30[\beta^{(2)} + 2\beta^{(3)} + 3\beta^{(4)} + 4\beta^{(5)}] + 8\alpha^{(1)} + 27\alpha^{(2)} + 64\alpha^{(3)} + 125\alpha^{(4)} + 216\alpha^{(5)}$$

subject to

$$0.5\beta^{(2)} + 2\beta^{(3)} + 4.5\beta^{(4)} + 8\beta^{(5)} - [\beta^{(2)} + 2\beta^{(3)} + 3\beta^{(4)} + 4\beta^{(5)}]$$
$$+ [2\alpha^{(1)} + 3\alpha^{(2)} + 4\alpha^{(3)} + 5\alpha^{(4)} + 6\alpha^{(5)}] - 5 + x_3 = 0 \tag{10–62}$$

$$\alpha^{(1)} + \alpha^{(2)} + \alpha^{(3)} + \alpha^{(4)} + \alpha^{(5)} - 1 = 0 \tag{10–63}$$

$$\beta^{(1)} + \beta^{(2)} + \beta^{(3)} + \beta^{(4)} + \beta^{(5)} - 1 = 0 \tag{10–64}$$

This is a linear programming problem where the x_3 term in Equ. (10–59) is a slack variable introduced to convert the original inequality constraint to an equation. Equations (10–63) and (10–64) are applications of Equ. (10–51) to this particular problem.

We can now apply the Simplex method provided that any basis involving the $\alpha^{(i)}$ and $\beta^{(i)}$ must involve two consecutive $\alpha^{(i)}$ and $\beta^{(i)}$ with all other $\alpha^{(i)}$ and $\beta^{(i)}$ set equal to 0. The price paid for converting the original nonlinear problem to a linear problem is that the two variables x_1 and x_2 are replaced by the ten new variables $\alpha^{(1)}$ through $\alpha^{(5)}$ and $\beta^{(1)}$ through $\beta^{(5)}$. However these additional dimensions can be readily handled by the Simplex algorithm. The solution to the linear programming problem is

$$\tilde{U}_{max} = 200$$

at

$$\alpha^{(1)} = 0.0$$
$$\alpha^{(2)} = 0.0$$
$$\alpha^{(3)} = 0.0$$
$$\alpha^{(4)} = 0.5$$
$$\alpha^{(5)} = 0.5$$
$$\beta^{(1)} = 0.0$$
$$\beta^{(2)} = 1.0$$
$$\beta^{(3)} = 0.0$$
$$\beta^{(4)} = 0.0$$
$$\beta^{(5)} = 0.0$$

Using Equ. (10–48) and Table 10–18, the values of the design variables at which the optimum occurs are

$$x_1 = \alpha^{(2)} x_1^{(2)} = 1.0$$
$$x_2 = \beta^{(4)} x_2^{(4)} + \beta^{(5)} x_2^{(5)} = 5.5$$

Thus, separable programming yields an approximate solution $\tilde{U}_{max} = 200$ for this example problem that is very close to the exact value of $U_{max} = 201$. And the approximate values of the design variables $x_1 = 1.0$ and $x_2 = 5.5$ at which the optimum occurs are reasonably close to their exact values of $x_1 = 1.34$ and $x_2 = 5.44$.

Conversion to Separable Form

As discussed at the beginning of this section, we can use the separable programming technique only when the nonlinear terms contain the design variables in separable form. However, many nonlinearities that are not in separable form can be converted to separable form. For example, consider the nonlinear term

$$f(x_1, x_2) = x_1 \ln x_2 \qquad (10\text{–}65)$$

used earlier as an example of a nonlinearity that is not separable. We can write this as

$$f(x_1, x_2) = f'(x_1) f''(x_2) \qquad (10\text{–}66)$$

where

$$f'(x_1) = x_1$$

and

$$f''(x_2) = \ln x_2$$

Clearly, both f' and f'' are separable. For any $f'(x_1)$ and $f''(x_2)$, we can introduce two new variables, y and z, defined as

$$f'(x_1) = y + z \tag{10–67}$$

$$f''(x_2) = y - z \tag{10–68}$$

For this particular example, these relationships take the specific form

$$x_1 = y + z \tag{10–69}$$

$$\ln x_2 = y - z \tag{10–70}$$

Substituting Equs. (10–67) and (10–68) into Equ. (10–66), we can write the original nonlinear term as

$$f(x_1, x_2) = (y + z)(y - z)$$

which simplifies to

$$f(x_1, x_2) = y^2 - z^2 \tag{10–71}$$

Now, $f(x_1, x_2)$, when written in terms of y and z, is separable. Hence, we have converted the nonseparable nonlinearity involving the design variables x_1 and x_2 [given by Equ. (10–65)] into a separable form involving two new variables y and z. Accompanying Equ. (10–71) are two new separable constraints, Equs. (10–69) and (10–70).

In general, if we can express a nonseparable term as the product of two separable terms, we can convert the nonseparable term to the separable form given by Equ. (10–71) involving two new variables y and z. Additional separable constraints of the form given by Equs. (10–67) and (10–68) have to be added to the existing constraint conditions.

Appendix 10B: Derivation of Equation (10–51)

With reference to Figure 10–25, the equation of the straight line connecting points A and B is

$$\tilde{f}(x) = f^{(i)} + \{[f^{(i)} - f^{(i+1)}]/[x^{(i)} - x^{(i+1)}]\}[x - x^{(i)}] \tag{10–72}$$

Note that x is the only variable on the right side of this equation. All the other terms on the right side represent known values at the specified end points on the straight line. We would like to replace x, $x^{(i)}$, and $x^{(i+1)}$ in this equation by expressions involving $\alpha^{(i)}$ and $\alpha^{(i+1)}$, the two new variables defined by Equs. (10–48), (10–49), and (10–50).

We can interpret $\alpha^{(i)}$ and $\alpha^{(i+1)}$ geometrically by using Equ. (10–49) to eliminate $\alpha^{(i)}$ from Equ. (10–48). This yields

$$x = x^{(i)} + \alpha^{(i+1)}[x^{(i+1)} - x^{(i)}] \tag{10–73}$$

Thus, the value of $\alpha^{(i+1)}$ locates x between the end points of the interval. When $\alpha^{(i+1)} = 0$, Equ. (10–73) reduces to $x = x^{(i)}$ and we are at the left end of the interval. When $\alpha^{(i+1)} = 1$, Equ. (10–73) reduces to $x = x^{(i+1)}$ and we are at the right end of the interval. In order to ensure that we don't go outside the interval, we must impose the limits on $\alpha^{(i+1)}$ given in Equ. (10–48). Then from Equ. (10–47) we see that the same restriction must apply to $\alpha^{(i)}$. So both $\alpha^{(i)}$ and $\alpha^{(i+1)}$ satisfy the non-negativity requirement for linear programming design variables.

Now substitute x from Equ. (10–47) into Equ. (10–69) to get

$$\tilde{f}(x) = f^{(i)} + \{[f^{(i)} - f^{(i+1)}]/[x^{(i)} - x^{(i+1)}]\}\{\alpha^{(i+1)}x^{(i+1)} - [1 - \alpha^{(i)}]x^{(i)}\}$$

Then use Equ. (10–47) to replace $\alpha^{(i)}$ in the above expression by $\alpha^{(i+1)}$. With some rearranging we arrive at

$$\tilde{f}(x) = [1 - \alpha^{(i+1)}]f^{(i)} + \alpha^{(i+1)}f^{(i+1)}$$

which, after application of Equ. (10–47) one more time, leads to the desired result, Equ. (10–49).

10.8. LAGRANGE MULTIPLIERS

In this section, we examine a calculus based approach to optimum design involving constraints. We saw in Section 10.6 that one of the problems encountered in nonlinear programming is the uncertainty over whether the search algorithm converges to the global optimum. The method described here offers the potential for being able to analytically identify the global optimum.

One Equality Constraint

We wish to select the dimensions of an open top rectangular storage container to minimize its fabrication cost. The container must be able to store V ft.^{3} of supplies, and is to be fabricated from a steel bottom costing S \$/ft.^{2} and wooden sides costing W \$/ft.^{2}. As shown in Figure 10–27, we'll let x and y be the dimensions of the base and z be the height of the container.

The objective function to be minimized is the fabrication cost C, where

$$C = Sxy + 2Wz(x + y) \tag{10–74}$$

The requirement that the container must have a storage capacity of V can be expressed as the constraint

$$xyz = V \tag{10–75}$$

One approach to dealing with the constraint on the volume is to solve Equ. (10–75) for one of the variables, say z in terms of x and y, then substitute back to eliminate z from Equ. (10–74). Then C, a function of the two remaining variables x and y, can be minimized by

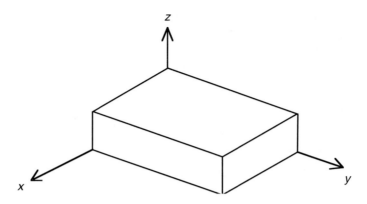

Figure 10–27.   Configuration of Storage Container

$$\frac{\partial C}{\partial x} = 0, \frac{\partial C}{\partial y} = 0$$

In principle this substitution approach will always work. However, in practice, it may not be possible to solve the constraint equation explicitly for one of the design variables. For example, suppose the constraint equation was of the form

$$\cos(xy) + z^2x = e^{yz}$$

There is no way this transcendental equation can be solved symbolically to express any one of the design variables in terms of the other two; hence, the substitution approach cannot be used. Even in circumstances where substitution can be used, doing so may complicate the objective function so that taking the partial derivatives becomes very tedious.

The Lagrange multiplier technique for optimizing an objective function subject to a constraint avoids substituting the constraint into the objective function. To illustrate its application, let's first rewrite the constraint, Equ. (10–75), as

$$xyz - V = 0 \tag{10–76}$$

Now we construct a Lagrangian function L by multiplying Equ. (10–76) by a new variable called a Lagrange multiplier λ and add the resulting expression to the objective function C. We get

$$L = C + \lambda(xyz - V)$$

or, using Equ. (10–74)

$$L = Sxy + 2Wz(x + y) + \lambda(xyz - V) \tag{10–77}$$

Note that the expression inside the parentheses in the last term on the right side of Equ. (10–77) is 0. So all we have really done in constructing L is to add 0 to our original objective function C. It follows then that optimizing L is the same as optimizing C. We can now optimize L, remembering that L is a function of four variables (x, y, z, λ). Since we have added the additional variable λ to our original problem, you might be tempted to argue that we've made the problem more complicated. But to offset this additional complication, we have, in effect, transformed the original *constrained* optimization problem to an *unconstrained* optimization problem. We proceed with the solution as follows:

$$\frac{\partial L}{\partial x} = 0 = Sy + 2Wz + \lambda yz \qquad (10\text{–}78)$$

$$\frac{\partial L}{\partial y} = 0 = Sx + 2Wz + \lambda xz \qquad (10\text{–}79)$$

$$\frac{\partial L}{\partial z} = 0 = 2W(x + y) + \lambda xy \qquad (10\text{–}80)$$

$$\frac{\partial L}{\partial \lambda} = 0 = xyz - V \qquad (10\text{–}81)$$

Note that Equ. (10–81) is just a repeat of Equ. (10–76). (In general, carrying out the $\partial L/\partial \lambda$ operation always leads to the constraint equation.) Equs. (10–78)–(10–81) are a set of four nonlinear equations involving the four unknowns x, y, z, and λ. The solution to these give the values of x, y, z, and λ for which the objective function [Equ. (10–74)] is optimized. The optimal value of the objective function is obtained by substituting these values of x, y, z into Equ. (10–74). This is easier said than done because the equations that have to be solved for x, y, z, and λ are nonlinear. That means there may not be a closed form solution. Also, there may be multiple solutions, each one corresponding to a local optimum. If a numerical technique must be used to search for the solutions, it may not be possible to converge to the global optimum.

   We pursue a closed form solution noting that Equs. (10–78)–(10–80) are linear in the Lagrange multiplier λ. Hence, let's first solve for λ in terms of x, y, and z as

$$\lambda = -\frac{2W(x + y)}{xy}$$

To eliminate x and y from the denominator, we'll multiply both top and bottom by z and use Equ. (10–81) to get

$$\lambda = -\frac{2Wz(x + y)}{V} \qquad (10\text{–}82)$$

If we substitute Equ. (10–82) back into Equs. (10–78) and (10–79) we get

$$Sy + 2Wz - \frac{2Wyz^2(x + y)}{V} = 0 \qquad (10\text{–}83)$$

$$Sx + 2Wz - \frac{2Wxz^2(x + y)}{V} = 0 \qquad (10\text{–}84)$$

Subtracting Equ. (10–83) from Equ. (10–84) gives

$$S(x - y) - \frac{2Wz^2(x + y)(x - y)}{V} = 0$$

or

$$(x - y)\left[S - \frac{2Wz^2(x + y)}{V}\right] = 0$$

This has two solutions:

$$x = y \tag{10–85}$$

and

$$S - \frac{2Wz^2(x + y)}{V} = 0 \tag{10–86}$$

Let's deal first with the solution given by Equ. (10–85). This tells us that the optimum design for the container is one in which the bottom plate is square. To find z, we substitute Equ. (10–85) into Equ. (10–81) to eliminate y and obtain

$$z = \frac{V}{x^2} \tag{10–87}$$

Now inserting Equs. (10–85) and (10–87) into Equ. (10–83) and simplifying yields

$$x = \left(\frac{2VW}{S}\right)^{1/3} \tag{10–88}$$

By virtue of Equ. (10–85), we also get

$$y = \left(\frac{2VW}{S}\right)^{1/3} \tag{10–89}$$

These can now be used to solve for z from Equ. (10–81) as

$$z = \left(\frac{S}{2W}\right)^{2/3} V^{1/3} \tag{10–90}$$

Equs. (10–88)–(10–90) give the dimensions of the least cost container in terms of the design parameters V, S, and W. The cost of the optimal container is found by substituting Equs. (10–88)–(10–90) into Equ. (10–74) to arrive at

$$C = 3S^{1/3}(2VW)^{2/3} \tag{10–91}$$

We will consider two special cases. First, if the cost per square foot of the container sides is the same as the bottom plate, $S = W$, then Equs. (10–88)–(10–90) reduce to

$$x = (2V)^{1/3}$$
$$y = (2V)^{1/3}$$
$$z = \frac{(2V)^{1/3}}{2}$$

Thus, the container height should be half the bottom dimensions, and Equ. (10–91) reduces to

$$C = 3S(2V)^{2/3}$$

A second special case occurs if the cost per square foot of the container sides is twice that of the bottom plate, that is, $S = 2W$. Then Equs. (10–88)–(10–90) reduce to

$$x = V^{1/3}$$

$$y = V^{1/3}$$

$$z = V^{1/3}$$

For this special condition, the optimal container is a cube and Equ. (10–91) reduces to

$$C = 3SV^{2/3}$$

Before leaving this example, we return to the solution associated with Equ. (10–86). It can be shown that Equ. (10–86) can be reduced to $x = y$. Hence, Equ. (10–86) doesn't provide any new information.

There are three major advantages of the Lagrange multiplier technique over nonlinear programming to find the optimal design of the storage container. First, we were able to find a general expression for the optimum for any value of V, S, and W [Equs. (10–88)–(10–91)]. In order to use nonlinear programming to find the optimal design, we would have to specify numerical values for these three parameters. Second, our optimum design was found analytically. We did not have to worry about convergence problems that could occur with a nonlinear programming approach. Third, we could easily identify the global optimum, something that could prove difficult to do using nonlinear programming.

Multiple Constraints and Inequalities

Consider a somewhat more complicated version of the optimal container design problem just discussed. To prevent the problem from becoming too unwieldy, we'll take advantage of the result that the bottom plate should be a square. Also, instead of minimizing the cost subject to the volume constraint, let's formulate the problem as that of maximizing the volume subject to a cost constraint.[7] In addition, we'll assume that we have a limited amount of the special epoxy glue needed to glue the edges of the side panels and bottom plate to each other. Further, let's treat the cost and glue constraints as inequalities; a design is acceptable as long as we do not exceed the stipulated cost C, and as long as we don't require any more than the available amount of glue G.

---

[7]This is an example of the situation discussed in Section 2.3 where the role of the objective and constraints are interchanged. We do this here in the interest of illustrating a particular application of the Lagrange multiplier method. In the real world, the client should be consulted regarding which objective function and constraint conditions represent the best formulation of the problem.

The problem is to maximize

$$U = x^2 z \tag{10-92}$$

subject to

$$Sx^2 + 4Wxz \leqslant C \tag{10-93}$$

$$4x + 4z \leqslant G \tag{10-94}$$

We introduce slack variables s_1 and s_2 to deal with the inequality constraints and square them before adding them to Equs. (10–93) and (10–94) to get[8]

$$Sx^2 + 4Wxz + s_1^2 = C \tag{10-95}$$

$$4x + 4z + s_2^2 = G \tag{10-96}$$

Since there are two constraint equations, we introduce two Lagrange multipliers, λ_1 and λ_2, and write the Lagrangian function as

$$L = x^2 z + \lambda_1 (Sx^2 + 4Wxz + s_1^2 - C) + \lambda_2 (4x + 4z + s_2^2 - G) \tag{10-97}$$

This is a function of six variables x, z, λ_1, λ_2, s_1, and s_2. Its maximum is found from

$$\frac{\partial L}{\partial x} = 0 = 2xz + \lambda_1 (2Sx + 4Wz) + 4\lambda_2 \tag{10-98}$$

$$\frac{\partial L}{\partial z} = 0 = x^2 + 4Wx\lambda_1 + 4\lambda_2 \tag{10-99}$$

$$\frac{\partial L}{\partial s_1} = 0 = 2s_1\lambda_1 \tag{10-100}$$

$$\frac{\partial L}{\partial s_2} = 0 = 2s_2\lambda_2 \tag{10-101}$$

plus Equs. (10–95) and (10–96).

The following four possible solution sets arise from Equs. (10–100) and (10–101):

$$s_1 = 0, \lambda_1 \neq 0, s_2 \neq 0, \lambda_2 = 0 \tag{10-102}$$

$$s_1 \neq 0, \lambda_1 = 0, s_2 = 0, \lambda_2 \neq 0 \tag{10-103}$$

$$s_1 = 0, \lambda_1 \neq 0, s_2 = 0, \lambda_2 = 0 \tag{10-104}$$

$$s_1 \neq 0, \lambda_1 = 0, s_2 \neq 0, \lambda_2 = 0 \tag{10-105}$$

---

[8]We square the slack variables in order to avoid having to introduce the additional constraints that they must be non-negative. We couldn't square the slack variables when we first introduced them in Section 10.4, because in that context they had to appear linearly in the constraint equations.

Each one of these has to be explored in turn. Since there are only two design variables (x, z), we can resort to geometric reasoning to enhance our understanding of the various solutions. As we will see, in order for each solution set to occur, certain relationships between the design parameters must exist.

Case 1: $s_1 = 0$, $\lambda_1 \neq 0$, $s_2 \neq 0$, $\lambda_2 = 0$. Let's look first at the solutions associated with Equ. (10–102). With $s_1 = 0$ and $s_2 \neq 0$, that means that we are looking for optimal solutions that lie on the boundary defined by Equ. (10–93) but do not lie on the boundary defined by Equ. (10–94). These represent designs for which the maximum cost is incurred but not all the available glue is utilized. As an example of this situation, see Figure 10–28 where the feasible region is the region to the left and below the boundaries associated with $C/W = 8$, $S/W = 2$ in Equ. (10–93) and $G = 15$ in Equ. (10–94). The contour curves associated with $U = 1$ and $U = 2$ are also shown. Part of the $U = 1$ curve passes through the feasible region. However, all points on the $U = 2$ curve are outside the feasible region. It is clear that the optimum design has a value $1 < U < 2$ and occurs when a contour curve is tangent to the boundary defined by Equ. (10–93).

Let's solve for this optimum algebraically so that we can gain some insight into the circumstances under which this type of optimum design can occur. As before, our solution strategy takes advantage of the fact that the Lagrange multipliers appear linearly in the governing equations. With $\lambda_2 = 0$, Equ. (10–99) can be solved directly for λ_1 as

$$\lambda_1 = -\frac{x}{4W}$$

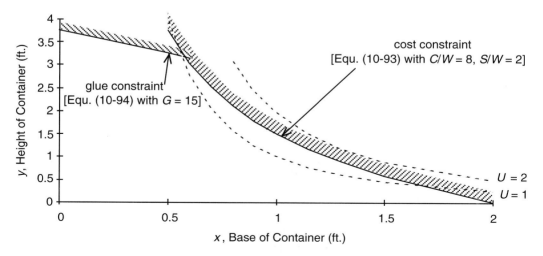

Figure 10–28.   Geometric Depiction of Optimum Storage Container Design ($s_1 = 0$, $\lambda_2 = 0$)

Substituting this into Equ. (10–98) leads to (with $\lambda_2 = 0$)

$$2xz - \frac{x}{4W}(2Sx + 4zW) = 0$$

or, after simplifying,

$$z = \frac{Sx}{2W} \tag{10–106}$$

With this, Equ. (10–95) can be solved (with $s_1 = 0$) for x to get

$$x = \left(\frac{C}{3S}\right)^{1/2} \tag{10–107}$$

Invoking Equ. (10–94), we see that in order for this solution to occur, the amount of available glue G must satisfy the following relationship:

$$G > 4\left(\frac{C}{3S}\right)^{1/2}\left(1 + \frac{S}{2W}\right) \tag{10–108}$$

Note that the strict nature of the above inequality is needed to ensure that the optimum design does not occur on the boundary defined by Equ. (10–94). For the particular numerical example depicted in Figure 10–28, $S/W = 2$ and $C/W = 8$; so Equ. (10–108) reduces to

$$G > 9.24 \text{ ft.}$$

Our choice of $G = 15$ as the boundary depicted in Figure 10–28 clearly satisfies this requirement. Substituting the numerical values of the design parameters into Equ. (10–107) yields

$$x = 1.15 \text{ ft.}$$

Then Equ. (10–106) gives

$$z = 1.15 \text{ ft.}$$

Using these with Equ. (10–98) yields

$$U = 1.52 \text{ ft.}^3$$

as the volume of the optimal container.

Case 2: $s_1 \neq 0$, $\lambda_1 = 0$, $s_2 = 0$, $\lambda_2 \neq 0$. We now turn to the second set of solutions; those associated with Equ. (10–103). With $s_1 \neq 0$ and $s_2 \neq 0$, that means that we are looking for optimal solutions that lie on the boundary defined by Equ. (10–94) but do not lie on the boundary defined by Equ. (10–93). This represents situations for which all the available glue is used but the cost is less than the maximum allowable cost. An example of this situation is shown in Figure 10–29, where the feasible region is the region to the left and below the boundaries associated with $C/W = 24$ and $S/W = 3$

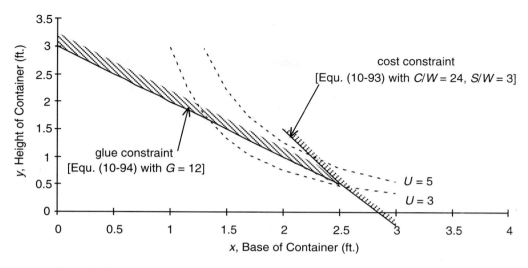

Figure 10–29.   Geometric Depiction of Optimum Storage Container Design ($\lambda_1 = 0$, $s_2 = 0$)

in Equ. (10–93) and $G = 12$ in Equ. (10–94).[9] The contour curves associated with $U = 3$ and $U = 5$ are also shown. It is clear from Figure 10–29 that the optimum occurs at a value $3 < U < 5$ when a contour curve is tangent to the boundary defined by Equ. (10–94).

Lets solve for this optimum algebraically. With $\lambda_1 = 0$, Equ. (10–99) can be solved directly for λ_2 as

$$\lambda_2 = -\frac{x^2}{4}$$

Substituting this into Equ. (10–98) leads to (with $\lambda_1 = 0$)

$$2xz + 4\left(-\frac{x^2}{4}\right) = 0$$

or, after simplifying,

$$z = \frac{x}{2} \tag{10–109}$$

Substituting this into Equ. (10–96) (with $s_2 = 0$) provides

[9]Note that the scales of Figure 10–28 and Figure 10–29 are the not the same; we've adjusted them to enhance the readability of each one, at the expense of easy comparisons between them.

$$x = \frac{G}{6} \qquad (10\text{--}110)$$

The condition under which this type of optimal design can be realized is obtained by inserting Equs. (10–109) and (10–110) into Equ. (10–93). After some rearranging we obtain

$$\frac{C}{W} > \left(\frac{G}{6}\right)^2 \left(\frac{S}{W} + 2\right) \qquad (10\text{--}111)$$

For the specific case illustrated in Figure 10–29 ($G = 12$, $S/W = 3$) we see that this type of solution exists only if

$$\frac{C}{W} > 20$$

The value of $C/W = 24$ used to construct the boundary in Figure 10–29 clearly satisfies this condition. With $G = 12$ we arrive at the optimum container design of $x = 2$ ft, $z = 1$ ft. The volume of the optimum container is $U = 4$ ft^3.

Case 3: $s_1 = 0$, $\lambda_1 \neq 0$, $s_2 = 0$, $\lambda_2 \neq 0$.   We now turn to the third set of solutions, those associated with Equ. (10–104). With $s_1 = 0$ and $s_2 = 0$, that means that we are looking for optimal designs that lie on the boundary defined by Equ. (10–94) and also lie on the boundary defined by Equ. (10–93). Thus we are seeking a solution in which a contour curve passes through the point of intersection of the two boundary segments and which otherwise is outside the feasible region. This cannot be thought of as the limiting case of both previous cases as the tangency point approaches the corner point, since this set of solutions may include contour lines that are not tangent to either of the boundary lines.

Rather than examining these solutions first from a geometric perspective, as was done in the prior two cases, we'll search for these optimal designs algebraically. With $s_1 = 0$ and $s_2 = 0$, we can solve Equs. (10–95) and (10–96) directly for x and z. Solving Equ. (10–96) for z (with $s_2 = 0$) as

$$z = \frac{G}{4} - x \qquad (10\text{--}112)$$

and substituting this into Equ. (10–95) we obtain

$$\left(\frac{S}{W} - 4\right) x^2 + Gx - \frac{C}{W} = 0 \qquad (10\text{--}113)$$

We now seek the conditions under which there is at least one real, positive root to this equation. From the properties of quadratic equations, we know that the solutions to this equation are real only if

$$G \geq 2 \sqrt{\left[\left(\frac{C}{W} \right) \left(4 - \frac{S}{W} \right) \right]} \tag{10-114}$$

In turn, this requires that

$$\frac{S}{W} \leq 4 \tag{10-115}$$

(otherwise G will be imaginary).

If Equs. (10–114) and (10–115) are satisfied, then the only way that x can be positive is to use the positive root from the quadratic formula applied to Equ. (10–113):

$$x = \frac{-G \pm \sqrt{\left[G^2 + 4 \left(\frac{C}{W} \right) \left(4 - \frac{S}{W} \right) \right]}}{2 \left(4 - \frac{S}{W} \right)}$$

Combining this with Equ. (10–112) and the requirement that z must also be positive leads to

$$G^2 > \frac{16 \left(\frac{C}{W} \right) \left(4 - \frac{S}{W} \right)}{\left[\left(6 - \frac{S}{W} \right)^2 - 4 \right]} \tag{10-116}$$

To focus on a specific numerical example, let's assign values to the design parameters as follows:

$$\frac{S}{W} = 2, \quad \frac{C}{W} = 4 \tag{10-117}$$

Then Equ. (10–116) requires that

$$G^2 > \frac{64}{6}$$

or

$$G > 3.27$$

The subsequent algebra will be simplified if we satisfy this condition by setting $G = 6$. Together with Equ. (10–117), this transforms Equ. (10–113) to

$$x^2 - 3x + 2 = 0$$

This factors as

$$(x - 2)(x - 1) = 0$$

The $x = 2$ root leads to a negative value of z. But $x = 1$ provides

$$z = \frac{1}{2}$$

and

$$U = \frac{1}{2}$$

Thus, for the particular values of the design parameters $S/W = 2$, $C/W = 4$, $G = 6$, an optimum design for which all the available glue is used and for which the material cost is the maximum allowable cost is $x = y = 1$ ft. and $z = 0.5$ ft. The volume of such a container is 0.5 ft³. The graphical solution is displayed in Figure 10–30.

Case 4: $s_1 \neq 0$, $\lambda_1 = 0$, $s_2 \neq 0$, $\lambda_2 = 0$.   We turn, finally, to the solution set provided by Equ. (10–104). With $\lambda_1 = 0$ and $\lambda_2 = 0$, Equ. (10–99) gives $x = 0$. Then Equ. (10–94) tells us that

$$0 \leq z \leq \frac{G}{4}$$

From Equ. (10–92) the value of the objective function is $U = 0$. Clearly, this solution corresponds to a minimum rather than a maximum so we don't examine it any further.

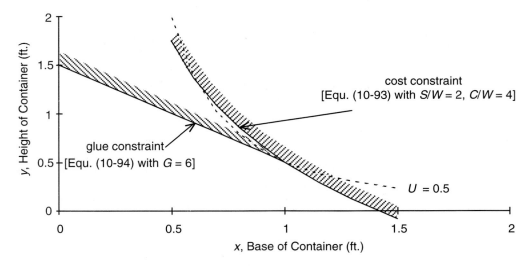

Figure 10–30.   Geometric Depiction of Optimum Storage Container Design ($s_1 = 0$, $s_2 = 0$)

General Considerations

The Lagrange multiplier concepts revealed in the treatment of the optimum container design problem can be generalized. The Lagrange multiplier method can be used to optimize any objective function U of p design variables

$$U = U(x_1, \ldots, x_p)$$

subject to any number of equality or inequality constraints. Suppose there are l equality constraints of the form

$$f_i(x_1, \ldots, x_p) = 0 \quad i \leq l$$

Suppose further that there are m inequality constraints of the form

$$g_j(x_1, \ldots, x_p) \leq 0 \quad l \leq j \leq l + m$$

Each of these inequality constraints can be converted to an equality constraint by adding a slack variable s_j:

$$g_j(x_1, \ldots, x_p) + s_j^2 = 0 \quad l \leq j \leq l + m$$

Suppose there are n inequality constraints of the form

$$h_k(x_1, \ldots, x_p) \geq 0 \quad l + m \leq k \leq l + m + n$$

Each of them is first converted to an equality by subtracting a surplus variable s_k:

$$h_k(x_1, \ldots, x_p) - s_k^2 = 0 \quad l + m \leq k \leq l + m + n$$

We then form a Lagrangian function L by multiplying each constraint by a Lagrange multiplier and adding the resulting expressions to the objective function:

$$L = U(x_1, \ldots, x_p) + \lambda_i \sum_{i=1}^{i=l} f_i(x_1, \ldots, x_p) + \lambda_j \sum_{j=l}^{j=l+m} [g_j(x_1, \ldots, x_p) + s_j^2]$$

$$+ \lambda_k \sum_{k=l+m}^{k=l+m+n} [h_k(x_1, \ldots, x_p) - s_k^2]$$

The equations for determining the optimum design are found by setting the partial derivatives of L equal to zero:

$$\frac{\partial L}{\partial x_r} = 0 \quad r = 1, \ldots, p \tag{10-118}$$

$$\frac{\partial L}{\partial \lambda_i} = 0 \quad i = 0, \ldots, l \tag{10-119}$$

$$\frac{\partial L}{\partial \lambda_j} = 0 \quad j = l, \ldots, l + m \tag{10-120}$$

$$\frac{\partial L}{\partial \lambda_k} = 0 \quad k = l + m, \ldots, l + m + n \qquad (10\text{–}121)$$

$$\frac{\partial L}{\partial s_j} = 0 \quad j = l, \ldots, l + m \qquad (10\text{–}122)$$

$$\frac{\partial L}{\partial s_k} = 0 \quad k = l + m, \ldots, l + m + n \qquad (10\text{–}123)$$

This gives us a set of $p+l+m+n$ simultaneous equations to solve for the values of the p design variables that identify the optimum design, the associated $l+m+n$ Lagrange multipliers, the m slack variables, and the n surplus variables.

Equs. (10–122) and (10–123) will always be of the form

$$s_j \lambda_j = 0$$

$$s_k \lambda_k = 0$$

and can be used to classify solutions according to which of the constraints are the active constraints.

Equs. (10–118) will always be linear in the Lagrange multipliers so they can be solved for the Lagrange multipliers in terms of the design variables. If Equs. (10–119), (10–120), and (10–121) are nonlinear in the design variables, it may not be possible to solve them in closed form. In that case, numerical techniques may be needed to find the solutions and the Lagrange multiplier approach loses its advantage over nonlinear programming as an optimal design tool. In particular, any numerical technique requires an initial estimate of a solution, and convergence to a global optimum is not guaranteed (see the discussion in Sec. 10–6).

10.9. CLOSURE

In this chapter we examined techniques that can efficiently identify the best design from among a large number, or even an infinite set, of options. Each technique is applicable to a specific class of optimum design problems.

We began with the dynamic programming method for finding the best design for a multi-stage system in which either the number of stages and the design options available at each stage is large, but finite. We illustrated both forward and backward sweeps through the stages as well as handling of constraints.

We then turned our attention to selecting the best design when there are several continuous design variables. We first examined the formulation and characteristics of linear programming problems, wherein the design variables are linear, and we explored the classical Simplex solution technique in considerable detail. Implications of nonlinearities were explored qualitatively, and we indicated the types of situations where slow convergence or multiple optima could occur. We then discussed separable programming as an approach to using linear programming algorithms for obtaining approximate solutions to problems involving certain types of nonlinearities. Finally, we explored a calculus based optimization procedure, LaGrange multipliers, that can be applied in principle to handle very general constrained nonlinear problems.

As we have mentioned in earlier chapters in connection with other aspects of design, the most important stage of using optimal design tools is to make sure that the problem is properly formulated. Does the objective function truly represent the characteristic of the design that we wish to optimize? Are all the constraints included, and are they in the proper form? Is the problem well-posed in the sense that the forms of the feasible region and the objective function allow for a finite number of optimal designs with finite values for the design variables? Of course, when the number of design variables exceeds two, it is not possible to rely on geometrical representations to answer these questions. In those circumstances, a physical feel for the system being designed can provide useful insights.

10.10. REFERENCES

ARORA, J. S. 1989. *Introduction to Optimum Design.* New York: McGraw Hill Book Co.

BELEGUNDU, A.D. and J.S. ARORA. 1985. A Study of Mathematical Programming Methods for Structural Optimization. Part II: Numerical Results. *International Journal for Numerical Methods in Engineering.* Vol. 21.

BRADLEY, S. P., A. C. MAX, and T. L. MAGNANTI. 1977. *Applied Mathematical Programming.* Reading, MA: Addison-Wesley Publishing Co.

GOTTFRIED, B., and J. WEISMAN. 1973. *Introduction to Optimization Theory.* Englewood Cliffs, New Jersey: Prentice Hall, Inc.

JEWELL, THOMAS K. 1986. *A Systems Approach to Civil Engineering Planning and Design.* New York: Harper & Row, Publishers.

PIERRE, DONALD A. 1986. *Optimization Theory with Applications.* New York: Dover Publications, Inc.

REKLAITIS, G. V., A. RAVINDRAN, and K. M. RAGSDELL. 1983. *Engineering Optimization: Methods and Applications.* New York: John Wiley and Sons.

SIDDALL, J. N. 1982. *Optimal Engineering Design: Principles and Applications.* New York: Marcel Dekker, Inc.

STOECKER, W. F. 1989. *Design of Thermal Systems, 3rd Edition.* New York: McGraw Hill Book Co.

WALTON, JOSEPH W. 1991. *Engineering Design: From Art to Practice.* St. Paul, MN: West Publishing Company.

10.11. EXERCISES

1. The total pressure drop from point 1 to point 5 in the multi-branch heating duct system shown is to be 500 Pa. The table presents the costs for various duct sizes in each of the sections as a function of the pressure drop in the section. Use dynamic programming to select the pressure drop in each section that results in the minimum total cost of the system. *Hint:* use the *cumulative* pressure drop at the end of each section as the state variables.

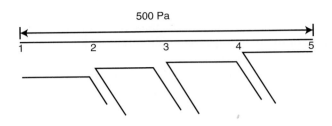

Section	Pressure drop (Pa)	Cost ($)
1–2	100	222
	150	205
	200	193
2–3	100	180
	150	166
	200	157
3–4	100	135
	150	125
	200	117
4–5	100	93
	150	86
	200	81

◆ *Source:* Adapted from Stoecker, p. 233 with permission from McGraw-Hill Companies.

2. A commercial airliner is to take off from airport A and land at airport E, a distance of 1000 km. At each of the equally spaced intermediate check points B, C, and D, the plane can be at an altitude of 2.5 km, 5.0 km, 7.5 km, or 10.0 km. Use dynamic programming to design the flight plan (in terms of the airplane's altitude at B, C, and D) that minimizes fuel consumption over the entire trip.

altitude (km)	A	B	C	D	E
10.0		●	●	●	
7.5		●	●	●	
5.0		●	●	●	
2.5		●	●	●	
0.0	●	●	●	●	●

The fuel consumption (kg) between any two consecutive checkpoints depends on the altitude change between those points according to the following table.

	to altitude (km)				
	0.0	2.5	5.0	7.5	10.0
0.0		1500	2000	2300	2500
from 2.5	200	500	1200	1600	1900
altitude 5.0	60	180	400	900	1200
(km) 7.5	0	50	160	250	600
10.0	0	0	40	80	100

♦ *Source:* Adapted from Stoecker, p. 234 with permission from McGraw-Hill Companies.

3. A rocket starts from rest with 8,000 kg of fuel. The 80-s burning time is divided into four 20-s intervals. The table presents the increase in velocity in each 20-s interval as a function of the mass of fuel in the rocket at the start of the interval and the mass of fuel burned during the interval. All 8,000 kg is to be expended in 80-s, and at least 1000 kg is to be burned each time interval. Use dynamic programming to determine the fuel-burning plan that results in the highest velocity of the rocket in 80-s. Hint: Use the cumulative fuel consumption since launch as the design variable.

	Increase in velocity in 20-s interval, m/s				
Fuel mass at start of interval, kg	Fuel burned in 20s, kg				
	1,000	2,000	3,000	4,000	5,000
1,000	180				
2,000	165	341			
3,000	152	312	486		
4,000	143	288	445	617	
5,000	136	269	412	567	737
6,000	131	253	385	525	679
7,000	127	241	362	491	630
8,000	124	231	343	462	589

♦ *Source:* Adapted from Stoecker, p. 237 with permission from McGraw-Hill Companies.

4. You are designing the wall of a new high-rise office building to consist of two layers, each of a different material (hypothetically, one of the layers could be of zero thickness, i.e., only one of the two available materials is used). The wall has to support a structural load L of at least 12,000 lb. and provide a thermal resistance R of at least 30 ft²-hr-°F/BTU. The parameters of the two candidate materials per inch of layer thickness t are summarized below.

Material	R/t (ft.²-hr-°F/BTU-in.)	L/t (10^3lb./in.)	Cost/t ($1000/in.)
1	15	3	5
2	10	6	4

Graphically determine the design for the least-cost wall. What is the thickness of each layer? What is the cost of the wall? What are the values of R and L for the wall?

♦ *Source:* Adapted from Stoecker, p. 293 with permission from McGraw-Hill Companies.

5.   Use a graphical technique to find the maximum value of $U = 6x_1 + 10x_2 + 7$ subject to the constraint $2x_1 + x_2 \geqslant 6$ and $4x_1 + 5x_2 \leqslant 24$.

6.   The daily fresh water requirement for a town is at least 4 million gallons. The town draws its water supply from two sources: a nearby river and an underground acquifer. The cost of pumping the water from the acquifer is \$100/million gal. The pumping cost from the river is \$50/million gal. The river's water has a pollutant concentration of 200 mg/gal. The water from the acquifer has a pollutant concentration of 50mg/gal. The water from each sources is mixed together before delivery to the town. The town's Public Health Department imposes the requirement that the pollutant concentration level of water supplied to the town be no greater than 100mg/gal. Use a graphical method to find the amount of water to be drawn from each source in order to minimize the total pumping cost.

♦ *Source:* Jewell, p. 161.

7.   Use the manual Simplex method to find the maximum value of $U = 2x_1 + x_2$ subject to the constraints $3x_1 + x_2 \leqslant 9$ and $x_1 + x_2 \leqslant 4$.

8.   Use the manual Simplex method to find the maximum value of $U = x_1 + 4x_2$ subject to the constraints $2x_1 + x_2 \leqslant 10$ and $x_1 + 2x_2 \leqslant 5$.

9.   Use the manual Simplex method to find the minimum value of $U = 6x_1 + 10x_2 + 7$ subject to the constraints $2x_1 + x_2 \geqslant 6$ and $4x_1 + 5x_2 \leqslant 24$.

10.   A boiler for a power plant is capable of burning coal, oil, and natural gas simultaneously. In order to achieve the required 2.4 Mw output from the 75 percent efficient boiler, the input heat rate of the fuel must be 3.2 Mw. The local air-pollution regulation requires that the sulfur content of the boiler fuel not exceed 2 percent. Based on the fuel characteristics shown below, formulate the problem so it is suitable for application of the simplex method to determine the least-cost fuel mix. Make sure all variables and units of measure are clearly defined. Only go as far as selecting an initial basis—do not actually use the Simplex algorithm.

Fuel	Sulfur Content (%)	Cost (¢/kg)	Heating Value (Mj/kg)
Coal	3.0	2.4	35.0
Oil	0.4	3.6	42.0
Natural Gas	0.2	4.2	55.0

♦ *Source:* Adapted from Stoecker, p. 294 with permission from McGraw-Hill Companies.

11.   Use the manual Simplex method to maximize $U = 5x_1 + 7x_2$ subject to the constraints $x_2 \leqslant 5$ and $x_1 - x_2 \leqslant 2$ where $x_1 \geqslant 0$ but x_2 is permitted to be negative.

12.   We wish to maximize $U = 6 - (x_1 - 1)^2 - (x_2 - 2)^2$ subject to the constraints $0 \leqslant x_1 \leqslant 3$, $0 \leqslant x_2 \leqslant 4$, and $x_2 \geqslant 3 - x_1$. Plot the boundaries of the feasible region and the contour

curves of the objective function. Use those plots as a guide for obtaining the exact solution to the optimization problem.

13. Sketch the feasible region and contour curves for finding the maximum value of $U = y - x/10$, subject the the following constraints: $x \geqslant 0$, $y \geqslant 0$, $x^2 + y^2 - 6x - 2y - 6 \leqslant 0$, $y \geqslant x$, $2x + 5y - 10 \geqslant 0$. Based on the clues revealed in the sketch, use calculus to locate the maximum. Verify the result using a commercial software optimization package.

14. We wish to maximize $U = x_1^4 + x_2$ subject to the requirements that x_1 and x_2 are non-negative and $2x_1^2 + 3x_2 - 9 \leqslant 0$. Plot the boundaries of the feasible region and the contour curves for the objective function. Use those plots as a guide for obtaining the exact solution to the optimization problem. Also, use the appropriate mathematical tests to verify that this optimum is indeed a global optimum.

15. A cantilever beam of length L and rectangulare cross-section is subject to a concentrated load P at its free end. The cost of the beam consists of a material cost K_1 \$/in.^{3} and the cost of coating all exposed surfaces with a fire retardant K_2 \$/in.^{2}. The maximum stress in the beam must remain below the known yield strength S of the material. We want to select the base b and height h of the cross-section to minimize the total cost. In preparation for using a nonlinear programming technique for finding the optimum design, what insights can the mathematical test for a global optimum provide?

16. We wish to maximize $U = x_1^4 + x_2$ subject to the requirements that x_1 and x_2 are nonnegative and $x_1^2 + x_2 - 4 \leqslant 0$. Obtain an approximate solution by using the technique of separable programming. Let the interval of analysis be $0 \leqslant x_1 \leqslant 3$ and use four equidistant grid points.

17. A company is redesigning its shell and tube heat exchanger of radius R and length L to maximize the surface area of the tubes. An end view of the unit is shown. The smallest available conducting tube has a radius of 0.5 cm, the largest available tube has a 10 cm radius, and all tubes must be of the same size. Further, the total cross-sectional area of all the tubes cannot exceed 2000 cm^2 to ensure adequate space inside the outer shell. Select the design variables and specify an objective function and constraint conditions for an optimum design.

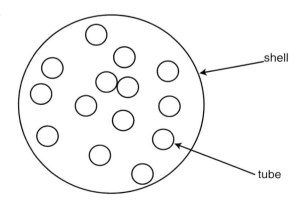

◆ *Source:* Arora, p. 60. Reprinted with permission of the McGraw-Hill Companies.

18. A wooden I-beam is to be assembled from a web and two flanges cut from a log of diameter d. If the three pieces of the beam are to have the same thickness t, we want to determine t such that the assembled beam has the highest possible moment of inertia. Formulate this problem as a nonlinear programming problem and use Excel Solver to find the optimum design when $d = 20$.

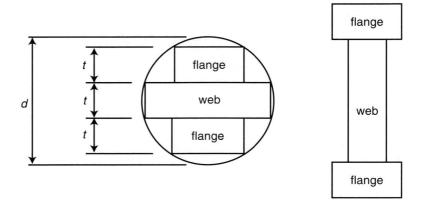

◆ *Source:* Walton, p. 529. © 1991. Reprinted with permission from PWS Publishing Co., a division of International Thomson Publishing, Inc.

19. We wish to select the radius R, length L, and thickness t of a steel cylinder which has hemispherical end closures (also of thickness t). This pressure vessel must be able to contain a specified volume V of steam at a known pressure p. The cost of the pressure vessel consists of: (a) a material cost proportional to the volume of steel used in its construction; (b) a forming cost which is proportional to R for the cylinder and proportional to R^3 for the end closures; and (c) an assembly cost proportional to the length of the welds that join the two end closures to the cylindrical body. The vessel is to be designed so that the maximum stress $[\sigma_{max} = 1.29\, pR/2t]$ is equal to half the known yield strength (σ_y) of the steel. Use the method of Lagrange multipliers to obtain five nonlinear algebraic equations whose solution will yield the design which minimizes the cost of the pressure vessel. You do not have to solve the equations.

20. A rectangular ventilation duct of width w and height h is to be run through the triangular legs of a roof truss as shown. Use Lagrange multipliers to select w and h in terms of the base b and altitude a of the truss triangles so the duct cross-sectional area is maximized.

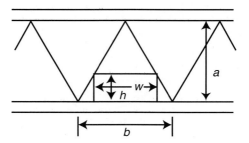

◆ *Source:* Stoecker, p. 177. Reprinted with permission of the McGraw-Hill Companies.

Design Projects

IMPROMPTU DESIGN COMPETITIONS

This section presents several suggestions for in-class projects that are intended to foster teamwork, provide opportunities to apply concepts of creative thinking, provide experience in dealing with uncertainty, demonstrate the nonuniqueness of design "solutions," and show the relationship between design plans and actual performance of systems constructed from those plans. The projects are intended to be completed in one 45–90 minute session, depending on the number of teams participating.

Project 1: Mining Oxygen Tanks

Design and construct a device to transport oxygen supplies to miners stranded at the bottom of a vertical mine shaft. Toxic gas is venting out of the top of the mine and prevents rescue workers from approaching the top of the mine. To be functional after delivery, the oxygen tanks must not hit the mine shaft walls on the way down and must arrive with low impact. The oxygen tanks will be represented by marbles on a table which will serve as ground level. The bottom of the mine will be represented by a target on the floor. The size of the target and the toxic zone around the mine opening will not be known until approximately half way through the design/construction process. We expect the target to be 4–6″ in diameter while the width of the toxic zone, w, should be 8–10″.

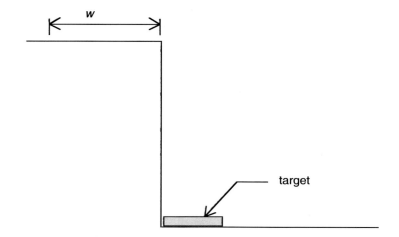

You will be allowed to start with one worker to operate the device. Additional workers may be hired for 10¢ each. Marbles may be purchased for 1¢ each. A worker entering the toxic zone will be terminated. Available materials and tools will be as follows:

1 paper bag	scissors
24″ masking tape	4 paper clips
24″ string	6 toothpicks
6 straws	$1.00
2 balloons	

Success will be determined by the following formula:

$$S = 0.5 \cdot \left(\frac{marbles}{30}\right) + 0.25 \cdot cents + 0.25 \cdot \left(1 - \frac{t}{120}\right)$$

where *marbles* are the number of marbles on the target, *cents* is the amount of money left over in cents, and *t* is the time expired. To receive credit for *t* and *cents*, at least one marble must be in the target.

Project 2: Nuclear Waste Storage Tank

Design and construct a device to cover a leaking nuclear waste storage tank before it explodes. The tank is surrounded by a radioactive zone of width *x* (see sketch). It is estimated that *x* will be between 4″ and 6″ but data is still being collected and the exact value will not be known until after construction of the device begins. The storage tank is represented by a 2″ × 2″ piece of engineering paper containing 100 cells taped to the center of a 2′ × 5′ table. The cover is represented by a piece of cloth of the same size.

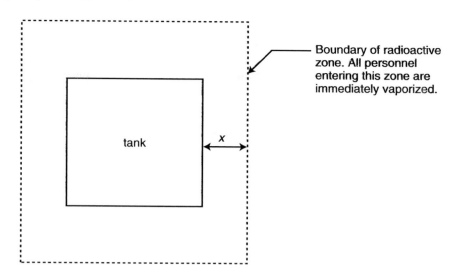

Boundary of radioactive zone. All personnel entering this zone are immediately vaporized.

tank

x

The device must be entirely self-powered and self-controlled. The following materials and tools will be provided:

1 paper bag	5 straight pins
24″ masking tape	4 rubber bands
24″ string	6 straws
2 paper clips	scissors

No other materials or tools may be used.

Success of the design will be determined by: 1) the surface area of the tank covered (to be determined by the number of complete blocks, *B*, on the engineering paper remaining uncovered after the cloth has come to rest); and 2) how quickly the device is installed and the cover deployed (the tank explodes at $t = 180$ s). Success will be calculated as

$$S = 0.5 \cdot \left(1 - \frac{B}{100}\right) + 0.5 \cdot \left(1 - \frac{t}{180}\right)$$

To receive credit for time, at least one block must be covered.

Project 3: String Roll Up

A 20′ piece of string is at rest in a straight line with the string centered on a table and with the axis of the string parallel to the long sides of the table top. The string is marked every 2″ along its length and all markers are in contact with the table. Design a system for minimizing contact between the string and table surface. You will have two attempts; minor repairs but no redesign is allowed between attempts. Extent of contact will be determined by counting markers on the string.

Both the string and the system for moving the string may be initially held in place by human contact, but both the string and the system for moving it must be free from human contact once the string starts to move. No part of the string or the system for moving the string can be affixed to the table. At the end of its motion, the entire string must be at rest somewhere in the space bounded by the table top and vertical extensions of the edges of the table top.

The system for moving the string, and all tools for fabricating the system and attaching it to the string can consist only of the following items:

1 paper bag	2 plastic straws
2 paper cups	4 rubber bands
28″ of masking tape	4 paper clips style A
2 paper clips style B	

Project 4: Liquid Hazardous Waste

Design and construct a vehicle and launching system to transport a hazardous waste liquid cargo from an origin point A to a waste repository at point B. A and B are both at ground level and x ft. apart, separated by an immovable and impenetrable barrier y ft. high located half way between them (see sketch).

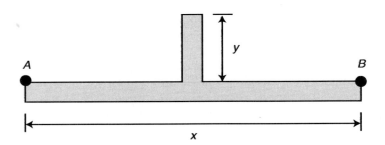

The cargo and its container are simulated by a small paper cup filled approximately half way with an amount of water w_i. Because of the political controversy surrounding the location of the hazardous waste repository and the aesthetic impact of the barrier, the values of x and y will not be finalized until approximately half way through the design/construction process. The range of values under consideration are 0.1 ft. $< x <$ 1.5 ft. and 0.5 ft. $< y <$ 2.0 ft.

The vehicle must start from rest at point A and must be launched without any energy input provided by human motion. Once launched, the cargo, cargo container, vehicle, and launching system cannot make contact with humans. The vehicle and launching system, and all tools for fabricating them, can consist only of the following objects:

1 paper bag   8 rubber bands

24″ of string   4 paper clips style A

2 plastic cups   1 paper clip style B

5 plastic straws   24″ of masking tape

5 toothpicks   6 cotton balls

1 plastic "ziploc" bag (suspected of being defective)

Success of each mission will be measured in terms of two equally weighted criteria: (a) percent of cargo delivered, (determined from the weight of the remaining water in the cargo container (w_f) after the cargo has come to rest); and (b) closeness, d, of the center of mass of the cargo container to point B. The success rating of each mission will be calculated as

$$ S = 0.5 \cdot \left(\frac{w_f}{w_i} \right) + 0.5 \cdot \left(1 - \frac{d}{x} \right) $$

You will be entitled to two attempts, with the final score being the average of the score on each attempt. Minor repairs and replenishment of the cargo, but no redesign, are permitted between attempts. Failure to clear the barrier will result in a score of $S = 0$.

MINI-DESIGN PROJECTS

Each project in this section is intended to focus on one or two aspects of design that are covered in this book. They are not intended to be comprehensive or "capstone" projects, but opportunities to reinforce specific material treated earlier in this book and apply that material in a project setting. Each project is intended to be assigned to groups of 3–5 students, with a duration of 2–4 weeks. Documentation of project results can be in the form of a technical report, short memo with supplementary analytical material, oral presentaion, or technical poster presentation, at the discretion of the instructor.

Project 5: Automatic Egg Cooker

You are employed by an innovative restaurant supply company. The president of the company recently went out for breakfast and had to send his eggs back three times before they were prepared to his satisfaction. The difficulty seems to be that the restaurant was not big enough for two cooks, but during peak times, the load was too much for one cook. The cook complained that eggs require supervision while they are cooking, but the cook does not have the time to monitor a large number of small orders during peak times. The president of the company wants to investigate the possibility of designing, building, and marketing an automatic egg cooker. Your boss asks that you develop three preliminary designs. The device should easily fit on a counter top, and should be inexpensive to manufacture and maintain.

Project 6: Drilling for Oil

You are chief engineer for an oil production company which holds a lease on some promising oil property. Geological data indicate that there is a 40% chance of an average size reservoir being found which can be opened up for production within two years of a decision to begin drilling production wells. The capital investment required to drill production wells is $50,000,000. Over the expected 20 year production lifetime of the reservoir, the net annual profits from such an operation are $20,000,000. The geological data also suggests a 10% chance of finding a "gusher" on the property, which will produce a net annual profit of $100,000,000.

Both of these estimates are based on the assumption that oil prices will increase only moderately over the lifetime of the investment. Under a 10% chance that prices will skyrocket, the expected payoffs described above will triple; but there is also a 5% chance that the payoffs will be cut in half by sharply falling oil prices.

One alternative to immediate drilling of production wells is to first invest $5,000,000 in seismic analysis that will further characterize the geological properties of the site as excellent, medium, or poor. Data from production wells that were drilled at other sites with similar geological characteristics are summarized in the table below. The seismic analysis project will delay the onset of production drilling by three years.

Type of Well	Percentage of Production Wells of a Given Type Found in Different Geological Zones		
	Excellent	Medium	Poor
Gusher	60	30	10
Average	25	50	25
Dry	10	20	70

You could also sell the lease immediately at a price of $10,000,000 or wait two years to see what happens to the price of oil before proceeding either with the sale, seismic analysis, or production drill options. Each design team shall use its own estimate that there is an x% chance that oil prices will rise substantially, y% chance that they will remain relatively stable, and a z% chance that they will fall during the next two years. If they rise, the selling price of the lease will increase to $15,000,000 and the estimates for future price trends over the 20 year lifetime of the project will be modified to 20% skyrocket, and 2% falling. If oil prices fall during those two years, the value of the lease is reduced to $2,000,000 and the estimates of lifetime price trends revised to 5% skyrocket and 30% fall. If you conduct the seismic analysis, the value of the lease doubles for an "excellent" geological formation and is cut in half for a "poor" geological formation.

Using a maximum expected utility decision rule and an appropriate discount rate, identify the desired strategy for dealing with this property. Each design team has a choice of selecting either a significantly risk averse, slightly risk averse, slightly risk prone, or significantly risk prone utility function to reflect its firm's approach to this issue.

Project 7: Minimum Weight Beam

Design a minimum-weight simply supported beam of length 12″ to support a concentrated load of 50 lb. at midspan. The beam is to be fabricated from a 1.0″ diameter Plexiglas rod with an allowable stress of S_a = 12,000 psi. There are several options available for reducing the weight, but they are limited by machining costs. The outer diameter of the rod may be machined down to 13/16″ diameter and/or 5/8″ diameter over any portion of the length. However, at least a 1/2″ long segment of the original 1.0″ diameter must remain at midspan to accommodate the applied load. Also, a hole may be drilled through the center of the rod to convert it to a hollow tube. Any such hole must go the entire length of the beam and have a diameter of at least 3/16″. If the hole is drilled, the minimum permissible wall thickness of the tube is 1/8″. As a result of these weight reduction measures, the beam may consist of up to three different cross-sectional dimensions along its length.

Project 8: Automobile Suspension System

The sketch shows a simplified spring-mass-damper model of the suspension system of an automobile. The goal of this project is to design the suspension system so that the system's response to road bumps is optimized. Our objective is to minimize the maximum vertical acceleration of the mass (automobile chassis) when the vehicle encounters road bumps. To prevent bottoming out of the spring, we must limit the maximum compression of the spring to one-half of the static spring length L_{static} = 0.2 m. Also, to provide appropriate clearances between the tires and the frame, the static length can be no less than 80% of the free length L_{free}.

The response to a road bump is to be simulated by having the system, traveling at a constant horizontal speed v_h, encounter a ramp inclined at an angle θ. Further as-

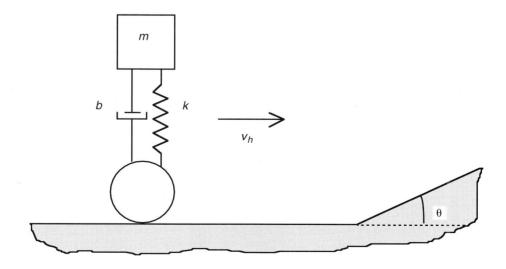

sume that the suspension system being designed supports 1/4 of the mass of the car, or $m = 250$ kg.

Your design challenge is to choose the values for b and k which achieve the design objective while satisfying the constraints. Your design approach should utilize at least one of the optimization techniques covered in class. Brute force, or trial and error, approaches to selecting b and k are not acceptable.

For the first phase of the design effort, assume the car is moving with a velocity of $V_h = 10$ m/s and $\theta = 20°$. After you have achieved an optimum design under those circumstances, consider other values of v_h and θ so you can select a set of b and k values that will attenuate the input over a wide range of bump profiles and car velocities. You may design a two-mode suspension system—a sport mode and a luxury mode. Therefore, you may have two k values and two b values.

Project 9: Electric Delivery Vehicle

Part A. As part of a government-funded experiment, you have been asked to help design a battery-powered electric delivery van. The vehicle will travel daily between two locations specified by your instructor. Your job is to specify a fixed route between the two points and design the electric propulsion system for the vehicle by specifying the minimum battery weight and motor power rating needed to reliably navigate the route while carrying a payload. The propulsion system will be installed on a cargo van that has a Gross Vehicle Weight Rating of 6000 lb. and weighs 2000 lb. stripped. The payload varies on each trip, following a normal distribution with mean 2000 lb. and standard deviation 500 lb. The vehicle must be able to climb the steepest hill on the route, and carry enough energy to travel the total round-trip distance, while carrying the 99.9th percentile payload. Be sure that the van can keep up with prevailing traffic speeds along the route you choose, and that it can climb the steepest grade on the route at no less than 75% of the local speed limit.

Several preliminary tests on a similar vehicle have been conducted according to the SAE Electric Vehicle Acceleration, Gradeability, and Deceleration Test Procedure. Those tests determined that the maximum tractive force at very low speeds is 600 lb. Also, at an average speed of 30 mph on level ground, the vehicle can accelerate at 2.16 mph/sec; at 20 mph, the vehicle can accelerate at 2.25 mph/sec.

Assume that e, the energy efficiency (mi/kW of electricity delivered by the battery) is a function of total vehicle weight and driving environment according to the following equations:

$$\text{for steady speed} - e = 2.74 \times 10^{-5} \, w + 0.0375$$
$$\text{for stop and go} - e = 6.16 \times 10^{-5} \, w + 0.0843$$
$$\text{composite} - e = 5.14 \times 10^{-5} \, w + 0.0701$$
$$\text{where } w \text{ is the vehicle weight in lbs.}$$

Consider the following data:

Road load power demand $= [0.5 \, r \, C_d \, A_f \, V^2 + m \, g \sin \theta + m \, g \, C_{rr} + m \, a] \, V$
where r = density of air (kg/m³)
    C_d = drag coefficient

A_f = frontal area (m²)

V = velocity (m/s)

m = total vehicle mass as tested

g = gravitational constant

θ = road grade angle measured from horizontal (degrees)

C_{rr} = rolling resistance coefficient

a = acceleration rate (m/s²)

The batteries will actually need to deliver about 30% more power than this, to account for drivetrain friction, battery efficiency, and motor efficiency.

You will use lead acid batteries that have an energy density of 36 kg/kWh of energy, and a specific power of 0.11 kW/kg. The specially designed electric motor weighs 3 lb./kW of rated power. The weight of structural enhancements needed to carry the additional weight of the batteries can be estimated as 0.3 times the portion of fully loaded vehicle weight that exceeds the manufacturer's gross weight specification.

Part B. Answer the following questions:

(a) Based on your vehicle's final curb weight, what is its Gross Vehicle Weight Rating Class that the National Highway Traffic Safety Administration would require be embedded in the Vehicle Identification Number (VIN)?

(b) What is the equivalent petroleum-based fuel economy as calculated according to Department of Energy (DOE) research and development standards?

(c) The DOE technique ignores the efficiency of the power plant that burns oil to generate the electricity (40%), the efficiency of transmitting current along utility lines (86%), and the efficiency of charging (85%) and discharging (85%) the battery. Adjust your DOE answer for these factors. If a gasoline powered vehicle got the resulting fuel economy, would it be subject to the Federal Internal Revenue Service Gas Guzzler tax? If so, how much would be assessed?

Project 10: Natural Gas Purchase

Natural gas, as transmitted through pipelines, typically consists of 90% methane (CH_4), 8% ethane (C_2H_6), and 2% propane (C_3H_8) by volume. You are chief engineer for a company which plans to buy natural gas at a flow rate x_o from the pipeline at a price p_o. Your firm will process the gas at a rate x_1 in an extractor which extracts the methane and sells pure methane back to the pipeline at a rate x_2 and price p_2. The extractor has a capital cost C_e, a capacity to process a flow rate of natural gas x_e, and an operating cost o_e. Natural gas at a rate of x_5 must be diverted from the purchased stream to operate the extractor.

The ethane/propane mixture coming out of the extractor can then either be flared at a rate x_7 and/or sent to another unit called a separator at a flow rate x_8. In the separator, the ethane/propane mixture can be separated into two streams which your company plans to sell to neighboring firms. One stream consists of pure propane at a rate x_4 which you can sell at a price p_4; the other stream is a mixture of ethane and residual propane at a rate x_3 which you can sell at a price p_3. The separator has a capital cost C_s, can process up to a flow rate x_s of ethane/propane mixture with an operating cost o_s. Natural gas at a rate of x_6 must be diverted from the purchased stream to operate the separator.

Your plant operates three shifts round the clock, seven days a week, except for an annual three-week shut down for preventive maintenance. The economic and performance characteristics of the equipment alternatives are shown below along with the purchase/selling prices of the various gas streams. For what equipment alternatives and flow rates will your company maximize its profits? Assume all equipment has a ten-year life; ignore the time value of money and the effect of taxes.

	EQUIPMENT OPTIONS			
	Capacity x_e or x_s (m^3/s)	Capital Cost C_e or C_s (10^6\$)	Operating Cost o_e or o_s (¢/s)	Natural Gas Consumed x_5 or x_6 (m^3/s)
Extractor 1	125	15.0	$4.00 + 0.050\, x_1$	$0.120\, x_1$
Extractor 2	200	20.0	$5.00 + 0.030\, x_1$	$0.080\, x_1$
Separator 1	8.00	5.0	$3.00 + 0.020\, x_4$	$0.150\, x_8$
Separator 2	14.0	8.0	$4.00 + 0.010\, x_4$	$0.120\, x_8$

PURCHASE/SELLING PRICES

Natural gas: $p_o = 5$ ¢/m^3

Methane: $p_2 = 6.5$ ¢/m^3

Ethane/residual propane mixture: $p_3 = 4$ ¢/m^3

Propane: $p_4 = 10$ ¢/m^3

Project 11: Conversion to Compressed Natural Gas

You are Chief Engineer for the city of Belkirk and have just received the following memo from the Manager of Fleet Operations, Rocky Waters.

TO: Chief Engineer
FROM: Rocky Waters, Fleet Operations Manager
SUBJECT: Conversion to Compressed Natural Gas (CNG)

In an effort to reduce overall fleet operations cost, we are interested in possibly converting some or all of the vehicles in our fleet to run on CNG. We currently purchase fuel at the market rate minus state and local taxes. A breakdown of our fleet and their typical operating conditions is shown below.

Thank you for your efforts.

Vehicle Type	Current Monthly Utilization (miles per vehicle)	Average Fuel Economy (mpg)	Number of Vehicles
Police cars	3000	12	60
Sedans	2000	30	200
Pickup trucks (under 2 tons)	1500	15	100
Trucks greater than 2 tons (diesel)	1500	4	30
Vans	4000	10	40
Miscellaneous heavy equipment (diesel)	500	0.5	20

Use an appropriate engineering economics decision rule to determine the best course of action. Use appropriate assumptions based on actual and projected costs. Your analysis should be based on probabilistic expectations for future events.

Project 12: Product Redesign

Part A. Describe three distinctively different commercially available products whose performance you believe can be improved. Your description of each product should be complete and may include sketches and photographs. Your only limitation in selecting the products is that the improvements in each product can reasonably be expected to be done by a mechanical engineering design firm; and that the redesigned product can reasonably be expected to cost only slightly more than the current version.

Express your dissatisfaction with each product's current design in the form of a needs statement. Indicate your assessment of the opportunity to improve each product in the form of a problem statement that includes a goal, one or more objectives, and a set of constraints. The needs statements and problem statements may be developed with the aid of interviews with users and manufacturers. Organize your material with the objective of convincing the manufacturer of each product that they should hire your firm to redesign the product.

Part B. Select any one of the three products described in Part A and assume that your mechanical engineering design firm has been hired by the product's manufacturer to redesign the selected product. Your current contract calls for you to develop three alternative preliminary designs to improve that product. The emphasis here is on conceiving fundamentally different approaches to satisfying one of the problem statements developed in Part A. Develop each alternative to the point that the manufacturer can select one of the three alternatives and hire your firm to develop a detailed design of that alternative.

Project 13: Refinery Pipeline

Your consulting firm has been hired to design a processing plant to recover marketable gases from the gaseous waste stream at a nearby petroleum refinery. The new processing plant is located one mile from the existing refinery, on the other side of a four lane highway (see map). Your specific assignment is to design a steel pipe to transport 3000 ft^3/min. of the toxic, corrosive gas at 15 mi./hr, 300 °F, and 500 psi from the refinery to the processing plant. The plan is to traverse the highway by suspending the pipe from an existing railroad bridge over the highway.

The only existing structure near the path of the proposed pipe is an outdoor drive-in movie theater located along the highway 250 ft. from the centerline of the pipe crossing. The theater operates only on weekends from April to October.

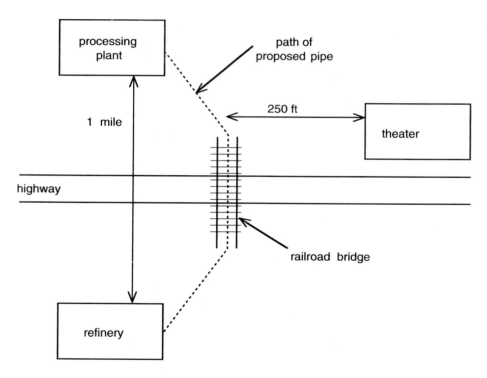

1. Select a diameter D and wall thickness t for the pipe so that it satisfies the federal minimum safety standards for transporting gas in or affecting interstate commerce. The materials engineer in your firm has recommended that you use ASTM A691 Class 20 Grade 5CR pipe. Treat the pipe diameter and wall thickness as deterministic design variables.

2. Now treat D and t as normally distributed random variables. Select values for D and t so that the probability of violating the federal requirements is < 0.0005. Treat all parameters other than D and t as deterministic.

Justify all design decisions by complete and unambiguous reference to the applicable standards and specifications. For purposes of this assignment, you may use updated versions of cited material specs.

Project 14: The Engineer's Role in Halting AIDS

Part A.   Read the attached excerpts from an article, "The Engineer's Role in Halting AIDS." Assume that your mechanical engineering design firm has been hired by a biomedical equipment manufacturer to design a "self-destruct" needle. Express your dissatisfaction with the current situation in the form of a needs statement. Indicate your assessment of the opportunity to improve on the current situation in the form of a problem statement that includes a goal, an objective, and a set of constraints.

Part B.   Your contract calls for you to develop five alternative preliminary designs for a "self-destruct" needle. The emphasis here is on conceiving fundamentally different approaches to satisfying the problem statements developed in Part A. Develop each alternative to the point that the manufacturer can select one of the five alternatives and hire your firm to develop a detailed design of that alternative.

The Engineer's Role in Halting AIDS

David R. Zimmerman

When a terrible cholera outbreak erupted in London in 1854, Dr. John Snow suspected that sewage was contaminating the district's drinking water and spreading the disease. When Snow found that cases were clustered around a water pump on Broad Street, he persuaded skeptical authorities to remove the pump handle, forcing residents to draw water elsewhere. The epidemic stopped as suddenly as it had started.

Today we are in a similar situation with AIDS. We need to isolate the narrowest and most vulnerable link of the disease's chain of transmission— and interrupt it. Nationwide, hypodermic needles used by drug addicts will soon be the main route for the spread of AIDS. In New York City and northern New Jersey, they already are. Before large percentages of addicts are infected in communities across the country, we should promote the development of needles that can be used only once, and induce or coerce addicts to use them.

This would be a simple solution to a complex problem, yet it has not been vigorously pursued.

Until very recently, the federal government has refused to grasp the nettle and deal with AIDS' spread among addicts as a technical problem, on the grounds that to do so would be to condone drug abuse. Instead, Washington urges addicts to, "Just say no!" Most can't, and there are too few treatment slots for addicts who wish to quit.

A few enlightened communities have bucked this political and moral agenda. Public agencies in San Francisco and some New Jersey communities distribute vials of household bleach to allow addicts to disinfect their "works." And a program in Portland, OR, lets addicts exchange dirty needles for clean ones. Data reported at the Fourth International Conference on AIDS, held in Stockholm in June, suggested that needle-exchange programs, in particular, slow the spread of the virus.

These epidemic control methods have picked the right target, the shared needle. But they leave too much to chance or choice among notoriously self-destructive people. An addict who really

◆ *Source:* Excerpted from *Technology Review,* Vol. 97, No. 7, Oct. 1988, pp. 22–23. Reprinted with permission from MIT's *Technology Review Magazine,*© 1997.

needs a fix will use any needle, clean or dirty. At this critical—and frequent—moment, addicts won't protect themselves, their lovers, or society at large from AIDS.

Needles that self-destruct after a single use would be a better solution. Yet in the report by the presidential commission on the epidemic, which contains hundreds of proposals for stepping up the war against AIDS, there is no demand that the government help develop such needles. Indeed, the concept is not even mentioned.

The private sector has been only a little less reticent about self-destruct needles. Representatives of two major medical instrument makers . . . say their biomedical engineers believe such needles can be made. But, they maintain, their companies are stumped by problems such as how to test-market and merchandise a product intended for illicit use. . . . also worries about how a self-destruct syringe could compete in price with a simple disposable syringe, which costs less than a dime wholesale. . . .

How could addicts be induced to use self-distruct needles? The most politically feasible solution could grow out of the safer needle designs being developed to protect health workers from accidental transmission of AIDS. These changes will be mandated under federal occupational safety laws. If the federal government stipulates that the safer needles should also self-destruct,

AIDS transmissions among addicts will drop as the more dangerous needles fall out of the market. Other simple answers also come to mind: self-destruct needles could be government subsidized and sold cheaply or given away, without a prescription.

The laws making it a criminal offense to carry a syringe without medical reason—common to many jurisdictions—could be amended. Possession of a self-destruct needle could be made legal, while the penalty for carrying reusable equipment could be greatly toughened.

Our failure to seize a technological opportunity to combat AIDS transmission by needles raises an old question with new urgency: how can critical advances that are not obviously profitable be developed? We need laws to promote such work. The model could be the Orphan Drug Act, which offers incentives to develop drugs with markets too small to be profitable.

The more immediate issue is how the war against AIDS will be fought. Will the agenda be set by rational calculation? If so, the prospects are good, because the spread of the AIDS virus is governed by fathomable natural rules. Or will it be set by politics and morality? If that is the case, control efforts will fail—for humans respond to moral law inconsistently, viruses not at all. Technology's advocates can no longer afford to be silent.

Project 15: Pressure Vessel

Design a pressure vessel that must hold 70 lbs. of superheated steam at 400 psi. The vessel is to be cylindrical with hemispherical heads. You should restrict your choice of material to SA-414, 515, or 516 carbon or low alloy steel plate.

Also, design the insulation system for the vessel, which is heated internally at a maximum input of 3,000 kW. For calculating the amount of insulation necessary consider only heat loss by conduction through the insulating layer.

Consider a space cost associated with your design, calculated from the maximum projected area of the vessel plus 6″ of clearance on each side and the ends of the vessel. The estimated value of this space is the average site cost for industrial buildings of $5.70/sq. ft. (means construction cost data).

Optimize your design with respect to cost using the method of Lagrangian multipliers. Use your best estimates of real material, forming, and welding costs. Use Section VIII, Division I, of the *ASME Boiler and Pressure Vessel Code* to determine allowable strengths and shell thicknesses.

Project 16: Horizontal Spacer Bar

The steel tower shown is one of a series of towers that support the overhead power cables for electric busses. The tension in the wires that support the power cables results in a force of 300 lb. applied to the top of the tower at a 10° angle to the horizontal. A guy wire for this tower runs from the top of the tower to an anchor in the ground at D. To prevent the guy-wire from obstructing the sidewalk, a horizontal spacer bar is attached to the tower.

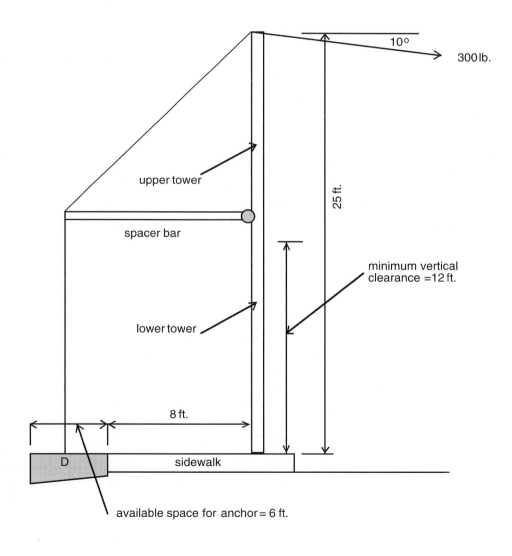

The upper tower segment, the lower tower segment, and the horizontal spacer bar each have a solid circular cross-section, but not necessarily all of the same diameter. The bar, tower segments, and guy-wire are made from the same material ($E = 30 \cdot 10^6$ psi, $\sigma_y = 50 \cdot 10^6$ psi).

Locate the anchor and the height of the spacer bar, and determine the diameters of the spacer bar, tower segments, and guy wire so as to minimize the total cost of the structure. Find this minimum-cost design under each of the following three cost assumptions:

1. The wire costs twice as much per unit volume as the bar and tower segments.
2. The wire has the same per unit volume cost as the bar and tower segments.
3. The wire costs half as much per unit volume as the bar and tower segments.

Use a safety factor of 2.0 in your design. Verify your design approach by building and testing a wooden scale model. The dimensions of the scale model are the same as those shown in the diagram except that the units are inches instead of feet.

$$\text{Score} = \frac{\text{Actual Failure Load (lb.)}}{\text{Weight of Tower + Bar + Wire (lb.)}} - 10 \cdot |\text{Predicted Failure Load} - \text{Actual Failure Load}|$$

Project 17: Secondary Oil Recovery

You are managing an oil field that is in its decline. You are considering "enhanced oil recovery" (EOR): the injection of carbon dioxide into nonproducing oil wells. The CO_2 lowers the viscosity of the oil, increases pressure, and swells its volume, all of which help mobilize it toward a producing well so you can recover it. Your job is to design an enhanced oil recovery installation for your thirty-well field.

An EOR project requires continuous injection of CO_2 along with continuous oil production. Typically, half of the wells are designated for injection while the remaining wells are used for oil extraction. When oil is recovered or "produced" from these wells, large amounts of "produced gas" come out with the oil. It is mostly carbon dioxide (which you injected) plus marketable gases such as butane, propane, and methane or natural gas. Typically, 60% of the carbon dioxide injected per barrel of oil recovered will resurface this way. To keep CO_2 flowing into the reservoir, fresh CO_2 may be purchased and injected, or produced gas may be reinjected (after scrubbing out the marketable gases for market, if desired, and removing any methane, which will prevent the CO_2 from mixing with the oil).

You can get the CO_2 to the site in any of three ways: 1) truck in liquid CO_2, 2) pipe in compressed CO_2 from an industrial plant via an existing pipeline that passes within 5 miles of your property along hilly terrain, or 3) pipe it in gas form from your CO_2 well 30 miles away over flat terrain. The gas from the CO_2 well contains 8 vol% hydrogen sulfide (H_2S) which must be removed first, and the CO_2 must be compressed before injection.

The purity of CO_2 you inject has a direct effect on the amount of oil it will help you recover. In addition, CO_2 coming from the plant pipeline varies in purity, either 95% pure or 98% pure, depending on whether the flow is coming from a power plant or an ammonia plant respectively. The ammonia plant is the source 60% of the time.

Your scientists tell you that the oil field still contains 85% of the original amount of oil in place (OOIP) when it was discovered (that is, 15% of the OOIP was removed already), and that CO_2 injection can yield another 15% of OOIP before it ceases to be effective (production will decline in a linear fashion until zero production at the end of seven years).

Based on the Department of Energy's oil price forecasts for the next seven years, design a plan that would yield the most profit over the seven year span of the project. Assume a 20% chance that the "low" DOE forecast is correct, 40% for the "medium," and 40% for the "high." Be sure to account for inflation by adjusting prices from the attached reference to current dollars.

Consider the following data:

MSCF = thousand standard cubic feet
MMSCF = million standard cubic feet
D = day
X = current price of oil, per barrel

There is a 70% chance that the CO_2 will react with the oil to produce the following results:

Purity of injected CO_2	CO_2 injected per barrel oil yielded	Produced Gas Composition		
		Propane	Butane	Methane
	(MSCF per bbl)	(vol %)	(vol %)	(vol %)
99.9%	4	2%	3%	3%
99%	6	2%	3%	2%
98%	7	2%	2%	1%
97%	9	1%	1%	0%
96%	12	1%	1%	0%
95%	15	0%	1%	0%
94% or below	20	0%	0%	0%

There is a 30% chance that unexpected reservoir conditions will cause the following results instead:

Purity of injected CO_2	CO_2 injected per barrel oil yielded	Produced Gas Composition		
		Propane	Butane	Methane
	(MSCF per bbl)	(vol %)	(vol %)	(vol %)
99.9%	4	1%	2%	0%
99%	6	1%	2%	0%
98%	7	1%	1%	1%
97%	9	0%	1%	2%
96%	12	0%	1%	3%
95%	15	0%	0%	4%
94% or below	20	0%	0%	5%

	Purity	Cost per MSCF	Capital cost
Truck	99.9%	$2.00	0
Plant pipeline	95 or 98%	$0.50 + 0.03*X	See reference
Pipeline from well	92%	0	See reference

Selling prices

Propane	$X/4 per MSCF
Methane	$X/7 per MSCF
Butane	$X/3 per MSCF

Costs

CO_2 recycling plant	See reference
Hydrocarbon or Methane separation	See reference
H_2S separation	See reference (this figure is per MSCF CO_2 per vol% H_2S)
Pressurization	See reference
Daily operating costs (overhead)	$5,000 per day
Injection equipment, per injection well	See reference
Annual operation of enhanced recovery per year per well	See reference

Reservoir data

OOIP	75,000,000 barrels	Reservoir pressure	1400 psi
Oil previously removed	15% of OOIP	Reservoir temperature	120° F
Maximum EOR yield	15% of OOIP	Depth	5000 ft
Vacant pore volume at start of project	15% of OOIP	Vacant pore volume at end of project	30% of OOIP

Assumptions. When scrubbing produced gas and reinjecting, the CO_2 you reinject returns to the original purity (99.9% if you got it from the truck source).

If you scrub, you scrub and sell all the marketable gases and you must repressurize.

If you don't reinject, you don't scrub.

Watch for situations where the "don't scrub" path is invalid because it would result in methane being injected, which is against the rules. Eliminate these options.

When you reinject without scrubbing, you will have to find the long-term average purity of the CO_2 you are injecting. This is 40% fresh CO_2 plus 60% unscrubbed, produced gas. Since the purity of the produced gas is a function of the purity of the injected gas (which in turn is a function of the produced gas, etc.), you need to perform a short iterative calculation to determine the equilibrium purity. Start with 40% fresh + 60% fresh. Then look up purity of produced gas based on that net purity going in. Then revise the purity of the produced gas, perform the calculation again, and look up purity again, and repeat.[1]

In the above iteration process, if you find that methane is being injected in the first couple steps, don't stop immediately. Go a few more steps and methane may no longer be a factor.

REFERENCE

KLINS, M. *Carbon Dioxide Flooding: Basic Mechanisms and Project Design.* Boston: International Human Resources Development Corp. pp. 227–233.

Project 18: Refrigerated Shipping Container

Develop the thermal design for a $20' \times 10' \times 8'$ refrigerated shipping container using dry ice as the refrigerant. Thermal design of the container involves selecting the insulation type and thickness, and the size of the dry ice compartment to maximize the shipping company's net income over the ten year expected lifetime of the container, based on a 15% discount rate.

Your client has provided the following table as her best estimate of the shipping opportunities for each of three products.

Product	Number of Annual Shipping Opportunities
Non-frozen Butter	75
Frozen Meat	50
Ice Cream	25

The shipping company receives 4¢/ft.^{3} for each shipment of butter, 6¢/ft.^{3} for each shipment of frozen meat, and 8¢/ft.^{3} for each shipment of ice cream. Each shipment involves one product (no mixing of products in a given shipment).

The choice for insulation materials has been narrowed down to three sheets of standard thickness: 4″ thick R-20 sheets at $5/ft.^{2}, 2″ thick R-12 sheets at $3/ft.^{2}, or 1″ R-3 at $2/ft.^{2}; all made of closed-shell polystyrene. You may use multiple layers of these sheets in any combination. Each layer after the first requires an extra $1/ft.^{2} for adhesive. Another insulation option is foam insulation which can be applied at any desired thickness. The foam has a thermal conductivity of 0.02 BTU/hr-ft. and costs $500 for set-up plus 10¢/in³.

Neglect all other fabrication and operational costs, and any income tax effects. Assume that the dry ice compartment is a cube and each shipment requires a new charge of dry ice. The current cost of dry ice is 50¢/kg and is expected to decrease at an annual rate of 1.5% during the next ten years.

Your design must be compatible with all relevant provisions of the multilateral ATP treaty for international shipment of perishable foodstuffs. Also, the container must be packed such that 20% of the available interior storage space must be vacant to permit adequate circulation of cold air.

Project 19: Space Shuttle Cargo Bay Truss

Design a two-bar truss to support equipment in the cargo bay of the space shuttle. The truss is to be attached to existing tie-down points 30″ apart on the cargo bay wall. The apex of the truss is to be located at a distance e (to be selected as part of the design) along a line 20″ from the wall. The truss is required to support a load P acting at the apex at an angle θ to the wall as shown below. P and θ are independent normally distributed random variables ($\bar{P} = 10,000$ lb., $s_P = 1,000$ lb., $\bar{\theta} = 30°$, $s_\theta = 5°$). All design parameters are deterministic. Your design should be adequate to reduce the probability of truss failure to less than 1 in 1000. The legs of the truss are to be made from solid rods of circular cross-section. Data for the available materials and sizes are shown in the table on the next page.

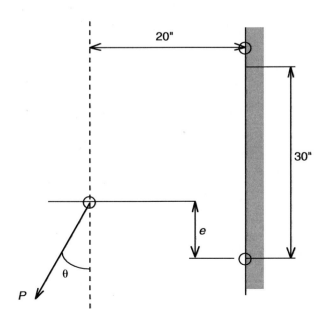

Material	Yield Strength (10^3 psi)	Young's Modulus (10^6 psi)	Density (lb./in.^{3})	Rod Diameter (in.)	Cost Length ($/in.)
				0.50	0.12
Aluminum	40	10	.101	0.75	0.35
				1.00	0.55
				0.25	0.10
Steel	62	30	.284	0.50	0.30
				0.75	0.60

Your design will be evaluated in terms of total cost, calculated as

Total Cost = (Cost of Materials for Truss Legs) + Weight Penalty @ $50/lb.

Project 20: Oil Pipeline Route

A pipeline company proposes to build an oil pipeline to pump oil from an oil tanker unloading facility west of City A (node A on the sketch) to the oil refineries located 65 miles away near Port E (node E on the sketch). The cost of the piping will be $10D$ per foot of length where D is the pipe diameter in feet. The annual pumping cost is estimated to be $150h$ where h is the head loss measured in feet.

The pipe and pumps are expected to last 30 years and the discount rate is currently estimated to be between 10 and 20%. If the company wants to pump 12,000 gallons per minute, what route and size pipe should the company use to minimize cost?

The pipeline route must be located within a designated five-mile-wide corridor and pass successively through three pumping stations, one each at node B, C, and D,

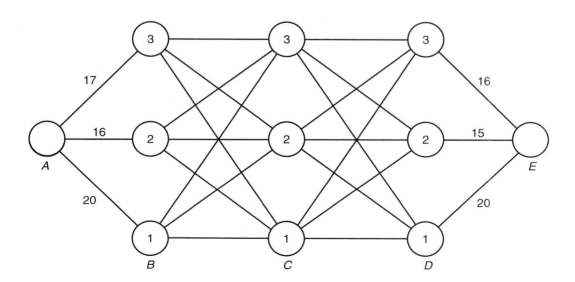

as shown. The construction costs from A to B and from D to E are shown in the sketch and the construction costs between nodes B and C and between nodes C and D are given in the accompanying table.

CONSTRUCTION COSTS
BETWEEN B, C, AND D ($/FT)

		To		
		1	2	3
	1	25	15	21
From	2	15	35	17
	3	21	17	30

Index